Implementation of Digital Technologies on Beverage Fermentation

Implementation of Digital Technologies on Beverage Fermentation

Editors

Claudia Gonzalez Viejo
Sigfredo Fuentes

MDPI • Basel • Beijing • Wuhan • Barcelona • Belgrade • Manchester • Tokyo • Cluj • Tianjin

Editors
Claudia Gonzalez Viejo
The University of Melbourne
Australia

Sigfredo Fuentes
The University of Melbourne
Australia

Editorial Office
MDPI
St. Alban-Anlage 66
4052 Basel, Switzerland

This is a reprint of articles from the Special Issue published online in the open access journal *Fermentation* (ISSN 2311-5637) (available at: https://www.mdpi.com/journal/fermentation/special_issues/digital_fermentation).

For citation purposes, cite each article independently as indicated on the article page online and as indicated below:

LastName, A.A.; LastName, B.B.; LastName, C.C. Article Title. *Journal Name* **Year**, *Volume Number*, Page Range.

ISBN 978-3-0365-3655-2 (Hbk)
ISBN 978-3-0365-3656-9 (PDF)

Contents

About the Editors

Claudia Gonzalez Viejo is a Postdoctoral Fellow in Digital Agriculture, Food and Wine Sciences. Dr Gonzalez Viejo's main research interests lie on the development of emerging technologies based on artificial intelligence such as robotics, sensors, computer vision, biometrics and machine learning modelling and their application in the field of agricultural, food and beverage sciences and engineering.

Sigfredo Fuentes is an Associate Professor in Digital Agriculture, Food and Wine Sciences. Professor Fuentes'scientific interests range from climate change impacts on agriculture, development of new computational tools for plant physiology, food, and wine science, new and emerging sensor technology, proximal, short and long-range remote sensing using robots and UAVs, machine learning, and artificial intelligence.

 *fermentation*

MDPI

Editorial

Editorial: Special Issue "Implementation of Digital Technologies on Beverage Fermentation"

Claudia Gonzalez Viejo * and Sigfredo Fuentes

Digital Agriculture, Food and Wine Research Group, Faculty of Veterinary and Agricultural Sciences, School of Agriculture and Food, The University of Melbourne, Parkville, VIC 3010, Australia; sfuentes@unimelb.edu.au
* Correspondence: cgonzalez2@unimelb.edu.au

In the food and beverage industries, implementing novel methods using digital technologies such as artificial intelligence (AI), sensors, robotics, computer vision, machine learning (ML), and sensory analysis using augmented reality (AR) has become critical to maintaining and increasing the products' quality traits and international competitiveness, especially within the past five years. Fermented beverages have been one of the most researched industries to implement these technologies to assess product composition and improve production processes and product quality. Therefore, this Special Issue (SI) is focused on the latest research on the application of digital technologies on beverage fermentation monitoring and the improvement of processing performance, product quality and sensory acceptability. This SI consists of 13 publications with different applications related to product quality assessment using a low-cost electronic nose (e-nose) coupled with ML modelling as rapid methods to assess beer quality traits [1], for fault detection in beer [2] and to evaluate the success of amelioration techniques in smoke-tainted wines [3]. Furthermore, a study on the use of a low-cost ultrasonic sensor and ML to assess alcohol concentration in the beer fermentation process was published in [4]; the use of computer vision analysis to evaluate foam stability in beer using a three-dimensional (3D) camera was reported in [5], and morphological assessment of *Saccharomyces cerevisiae* yeast during fermentation was shown in [6]. Some methods based on liquid chromatography were reported to assess different chemical compounds in fermented beverages, such as the use of liquid chromatography coupled with electrospray ionization and quadrupole time of flight tandem mass spectrometry (LC-ESI-QTOF-MS/MS) for the characterization of phenolic compounds in herbal tea [7] and high-performance liquid chromatography with fluorescence detection (HPLC-FLD) to analyze biogenic amines in sparkling wine [8]. Different methods were used for the sensory analysis of fermented beverages, such as the use of focus groups with semi-trained panelists to assess the sensory descriptors of kombucha [9] or the use of AR to evaluate the acceptability of yogurt samples in different environments [10]. On the other hand, some applications for monitoring, optimization and improvements during processing have been presented, such as a robotic system, TeeBot, developed for fermentation and sampling [11]. Different techniques have been proposed for the optimization of beer fermentation accounting for parametric uncertainty [12], and a simulation of a closed system to was used to assess the effect of time and temperature on the respiration rate measured as consumed oxygen in olive fermentation used to extract oil [13].

Due to the expertise requirements and the time-consuming and costly nature of traditional methods to assess the quality traits of fermented beverages, novel rapid, reliable and cost-effective methods have been developed involving digital technologies. The use of different sensors coupled with ML and AI has shown to be effective predictive methods along the production line. A low-cost and portable e-nose has been developed and integrated with artificial neural networks (ANN) to accurately predict the type of fermentation

Citation: Gonzalez Viejo, C.; Fuentes, S. Editorial: Special Issue "Implementation of Digital Technologies on Beverage Fermentation". *Fermentation* **2022**, *8*, 127. https://doi.org/10.3390/fermentation8030127

Received: 7 March 2022
Accepted: 10 March 2022
Published: 15 March 2022

in beer (97%); consumer acceptability towards carbonation, bitterness, aroma, flavor and overall liking of beer (R = 0.95); different color and foam-related parameters (R = 0.93) [1]; and to assess the presence and level of faults (95%), as well as the specific off-flavors/off-aromas found in beer (>96%) [2]. The same e-nose showed to be highly accurate (>90%) at assessing the effectiveness of smoke taint amelioration techniques using activated carbon and enzymes [3]. A different approach for the use of sensors was presented using an ultrasonic sensor along with long short-term memory neural networks (LSTM) as a cost-effective method to predict alcohol during the fermentation process of beer with high accuracy (R^2 = 0.94) [4]. Computer vision is very convenient for developing non-invasive, contactless methods for assessing physical measurements. For instance, researchers have developed different approaches to determine foam in beer, one of the most important indicators of its quality. These consisted of measuring 13 different color- and foam-related parameters in beer using an automatic robotic pourer and computer vision algorithms [1], and foam stability in beer was assessed using a 3D camera based on measuring the distance between the camera and the beer [5]. On the other hand, Guadalupe-Daqui et al. [6] used automated computer vision to assess yeast (*S. cerevisiae*) morphological changes, which are related to yeast health and stress that, at the same time, has an impact on the quality of fermented products such as beer.

Sensory analysis of food and beverages is also important to determine the product's quality and consumers' acceptability. Therefore, researchers have developed novel methods applied to fermented beverages. Torrico et al. [10] designed sensory sessions to test different environments using mixed reality with AR headsets. Compared with the regular testing setting in individual booths, significant interactions between samples and the environment were found.

Monitoring and optimizing the production process of fermented beverages is also necessary to reduce time and improve product quality; researchers have developed a TeeBot automatic robotic sampler to monitor the fermentation of either 96 samples of 100 mL or 384 samples of 30 mL, which aids in minimizing the labor and time required for this part of the process [11]. Bhonsale et al. [12] presented three different approaches to optimize the temperature profile of beer fermentation, accounting for parametric uncertainty, which is usually not considered. The authors found that the second-order polynomial chaos was the best approach for high-quality beer manufacturing by maximizing the alcohol concentration and fermentation time.

Funding: This research received no external funding.

Institutional Review Board Statement: Not applicable.

Informed Consent Statement: Not applicable.

Data Availability Statement: Data availability is specified for every published paper.

Conflicts of Interest: The authors declared no conflict of interest.

References

1. Gonzalez Viejo, C.; Fuentes, S. Low-Cost Methods to Assess Beer Quality Using Artificial Intelligence Involving Robotics, an Electronic Nose, and Machine Learning. *Fermentation* **2020**, *6*, 104. [CrossRef]
2. Gonzalez Viejo, C.; Fuentes, S.; Hernandez-Brenes, C. Smart Detection of Faults in Beers Using Near-Infrared Spectroscopy, a Low-Cost Electronic Nose and Artificial Intelligence. *Fermentation* **2021**, *7*, 117. [CrossRef]
3. Summerson, V.; Gonzalez Viejo, C.; Torrico, D.D.; Pang, A.; Fuentes, S. Digital Smoke Taint Detection in Pinot Grigio Wines Using an E-Nose and Machine Learning Algorithms Following Treatment with Activated Carbon and a Cleaving Enzyme. *Fermentation* **2021**, *7*, 119. [CrossRef]
4. Bowler, A.; Escrig, J.; Pound, M.; Watson, N. Predicting alcohol concentration during beer fermentation using ultrasonic measurements and machine learning. *Fermentation* **2021**, *7*, 34. [CrossRef]
5. Nyarko, E.K.; Glavaš, H.; Habschied, K.; Mastanjević, K. Determination of foam stability in lager beers using digital image analysis of images obtained using RGB and 3D cameras. *Fermentation* **2021**, *7*, 46. [CrossRef]
6. Guadalupe-Daqui, M.; Chen, M.; Thompson-Witrick, K.A.; MacIntosh, A.J. Yeast Morphology Assessment through Automated Image Analysis during Fermentation. *Fermentation* **2021**, *7*, 44. [CrossRef]

7. Chou, O.; Ali, A.; Subbiah, V.; Barrow, C.J.; Dunshea, F.R.; Suleria, H.A. Lc-esi-qtof-ms/ms characterization of phenolics in herbal tea infusion and their antioxidant potential. *Fermentation* **2021**, *7*, 73. [CrossRef]
8. Mir-Cerdà, A.; Izquierdo-Llopart, A.; Saurina, J.; Sentellas, S. Oenological Processes and Product Qualities in the Elaboration of Sparkling Wines Determine the Biogenic Amine Content. *Fermentation* **2021**, *7*, 144. [CrossRef]
9. Alderson, H.; Liu, C.; Mehta, A.; Gala, H.S.; Mazive, N.R.; Chen, Y.; Zhang, Y.; Wang, S.; Serventi, L. Sensory profile of kombucha brewed with New Zealand ingredients by focus group and word clouds. *Fermentation* **2021**, *7*, 100. [CrossRef]
10. Dong, Y.; Sharma, C.; Mehta, A.; Torrico, D.D. Application of Augmented Reality in the Sensory Evaluation of Yogurts. *Fermentation* **2021**, *7*, 147. [CrossRef]
11. Van Holst Pellekaan, N.; Walker, M.E.; Watson, T.L.; Jiranek, V. 'TeeBot': A High Throughput Robotic Fermentation and Sampling System. *Fermentation* **2021**, *7*, 205. [CrossRef]
12. Bhonsale, S.; Mores, W.; Van Impe, J. Dynamic Optimisation of Beer Fermentation under Parametric Uncertainty. *Fermentation* **2021**, *7*, 285. [CrossRef]
13. Plasquy, E.; Florido, M.C.; Sola-Guirado, R.R.; García-Martos, J.M.; García-Martín, J.F. Effect of Temperature and Time on Oxygen Consumption by Olive Fruit: Empirical Study and Simulation in a Non-Ventilated Container. *Fermentation* **2021**, *7*, 200. [CrossRef]

Article

Low-Cost Methods to Assess Beer Quality Using Artificial Intelligence Involving Robotics, an Electronic Nose, and Machine Learning

Claudia Gonzalez Viejo * and Sigfredo Fuentes

Digital Agriculture, Food and Wine Sciences Group, School of Agriculture and Food, Faculty of Veterinary and Agricultural Sciences, University of Melbourne, Melbourne VIC 3010, Australia; sfuentes@unimelb.edu.au

* Correspondence: cgonzalez2@unimelb.edu.au

Received: 8 October 2020; Accepted: 30 October 2020; Published: 31 October 2020

Abstract: Beer quality is a difficult concept to describe and assess by physicochemical and sensory analysis due to the complexity of beer appreciation and acceptability by consumers, which can be dynamic and related to changes in climate affecting raw materials, consumer preference, and rising quality requirements. Artificial intelligence (AI) may offer unique capabilities based on the integration of sensor technology, robotics, and data analysis using machine learning (ML) to identify specific quality traits and process modifications to produce quality beers. This research presented the integration and implementation of AI technology based on low-cost sensor networks in the form of an electronic nose (e-nose), robotics, and ML. Results of ML showed high accuracy (97%) in the identification of fermentation type (Model 1) based on e-nose data; prediction of consumer acceptability from near-infrared (Model 2; R = 0.90) and e-nose data (Model 3; R = 0.95), and physicochemical and colorimetry of beers from e-nose data. The use of the RoboBEER coupled with the e-nose and AI could be used by brewers to assess the fermentation process, quality of beers, detection of faults, traceability, and authentication purposes in an affordable, user-friendly, and accurate manner.

Keywords: sensor networks; automation; beer acceptability; beer fermentation; RoboBEER

---

1. Introduction

Beer quality is usually regarded as a subjective concept that can be described as more objectively based on beer quality traits through physicochemical analysis and/or sensory analysis of beers using either trained panels or consumer tests, the latter for perceived quality. However, by themselves, parameters and tests may not describe specific quality traits related to the overall liking of beer and the complexity of consumer perception.

Based on the latest research, beers' physicochemical analysis has been mainly based on the bioactive compounds in craft beers [1], volatiles, and antioxidant properties when adding fruit using gas chromatography/mass-spectroscopy (GC-MS) and high-performance liquid chromatography (HPLC) [1–3]. Many other studies coupled these analyses of quality traits with sensory trials when analyzing processes to change beer characteristics, such as ultrasound-assisted thermal processing [4], antioxidant activity through natural additives [5–7], for sweet potato beers [8] and commercial beers [9]. In the case of traditional consumer sensory tests, they tend to be subjective, and require a laboratory with individual booths that meet specific requirements; it also involves the recruitment of a large number of participants, which often requires an incentive to participate, which leads to higher costs and time for conducting the sessions and analyzing the data [10–12]. Therefore, all these studies' translational results are not easy to implement in the broad brewing industry to be used in every single batch since they require laboratory instrumentation, specialized personnel, and skills for operation, data acquisition, and analysis and may be time-consuming and costly.

Many non-invasive technologies have been proposed for the analysis of beer quality traits, such as the implementation of robotics coupled with sensors and computer vision capabilities like the RoboBEER [13], with a similar approach applied for sparkling wines analysis though the FIZZEyeRobot [14] constructed with more ubiquitous components, making them more applicable to bigger brewing companies than to craft beer enterprises. Other non-invasive technologies based on near-infrared spectroscopy (NIR) have also been applied within the beer production process from fermentation analysis [15] and coupled with artificial intelligence (AI) to predict relative protein content of commercial beers [16], beer quality assessment [17], the effect of sonication on beer quality [18] and many other applications within AI [19]. More recently, an integrated gas sensor technology application has resulted in the development of a low-cost electronic nose (e-nose) that can be attached to robotics (i.e., RoboBEER) to assess aroma profiles [20] also implementing AI [21].

The implementation of e-noses for beer research is not new in the food science field as they have been used to assess aromas in products such as coffee acidity [22], and roasting level prediction [23], tea quality grading [24–26], classification of strawberry juices [24], assessment of wine smoke taint [27] and alcohol content [28], beer quality classification and aging [29–31], beer aroma prediction [20], meat quality and shelf life [32,33], olive oil origin and quality [33,34], milk spoilage [35], and rice infestation [36], among others. However, most of the e-noses used in those studies are expensive and need to be installed in a laboratory as they have a similar system to a gas chromatograph (GC), and/or are considered as low-cost compared to a GC, but still, cost ≥ USD 30,000 and require maintenance [37–39]. There have been some works using them for classification of beers using data-mining methods [40–43], to evaluate the water use in the brewing process [44], aroma discrimination [45], alcohol content [46], comparison of volatile compounds against gas chromatography methods [43]. However, there has not been an integration of these e-noses with low-cost robotic systems to uniform pouring conditions or comparing results with sensory analysis using machine-learning methods. This paper proposed an integrated low-cost AI system integrating robotics (RoboBEER), NIR spectroscopy, a newly developed e-nose, and sensory analysis to assess automatically the type of fermentation of beers (Model 1) based on e-nose data, sensory perception of beers based on NIR data (Model 2) and e-nose data (Model 3) and beer foamability parameters based on e-nose data (Model 4). All models presented high accuracy (R > 0.90) in the prediction of beer targets before they were mentioned.

The integrated AI system proposed would offer brewers critical information based on robotics and e-nose data related to the type of fermentation of beers that will help traceability and identification, analysis of consumer acceptance, and bubble and physical foam properties of beers. All this information would be possible with low-cost robotics and instrumentation, which could be affordable even to small brewing companies that may give them reliable and critical information to tailor their beers according to specific consumer expectations. Furthermore, the proposed e-nose and AI system offers the advantage of being versatile and may be adapted to be used for different purposes, such as the assessment of smoke taint in wines and berries [27].

2. Materials and Methods

2.1. Sample Description

A total of 20 different beer samples from the three types of fermentation (seven top, six bottom, and seven spontaneous) were analyzed in triplicates (three bottles of each; $n = 60$). The top fermentation samples consisted of beers from the abbey ale, porter, kölsch, red ale, steam ale, aged ale, and sparkling ale styles. The bottom fermentation samples involved beer styles such as pale lager, Vienna lager, and German pilsner. On the other hand, spontaneous fermentation samples consisted of different lambic flavors and a wild Saison style. More details on the specific samples are shown as supplementary material (Table S1).

2.2. Electronic Nose (E-Nose)

A low-cost and portable electronic nose (e-nose; Figure 1) developed by the Digital Agriculture Food and Wine (DAFW) Group from the Faculty of Veterinary and Agricultural Sciences (FVAS) of the University of Melbourne (UoM) was used. This e-nose, which is composed of nine different gas sensors (Table 1), is sensitive to different gases related to volatile aromatic compounds. It measures the signal from all sensors in volts to obtain all outputs in the same scale The readings were performed by placing the e-nose on a 500 mL beaker with each bottle's entire content; the e-nose started recording during the pouring of the sample and was placed on top of the beaker to collect readings for 3 min [20,27]. The e-nose was calibrated for 1–2 min between samples to ensure no carryover effects. All samples were measured at refrigeration temperature (4 °C), which is the consumption temperature and the same as used for sensory and physicochemical analysis.

(**a**) (**b**)

Figure 1. Electronic nose showing (**a**) the front side with the nine gas sensors attached to the printed circuit board (PCB) and the mini security digital (SD) card to store data, and (**b**) the back/side of the PCB.

Table 1. Sensors integrated into the electronic nose.

Sensor (Gas) *	Label/Model	Sensitivity
Alcohol	MQ3	0.5–10 mg L^{-1}
Methane	MQ4	200–10,000 ppm
Carbon monoxide	MQ7	20–2000 ppm
Hydrogen	MQ8	100–10,000 ppm
Ammonia/Alcohol/Benzene	MQ135	10–300 ppm/10–300 ppm/10–1000 ppm
Hydrogen Sulfide	MQ136	1–100 ppm
Ammonia	MQ137	5–200 ppm
Benzene/Alcohol/Ammonia	MQ138	10–1000 ppm/10–1000 ppm/10–3000 ppm
Carbon dioxide	MG811	350–10,000 ppm

* All sensors are from Henan Hanwei Electronics Co., Ltd., Henan, China.

2.3. Sensory: Consumer Acceptance Test

A sensory session was conducted with frequent beer consumers (n = 31) recruited from the staff and students at UoM, Australia. According to the Power analysis performed using SAS® v. 9.4 (SAS Institute, Cary, NC, USA), the number of participants was enough to find significant differences ($1–\beta > 0.99$). Participants were asked to sign a consent form approved by the FVAS–UoM Human Ethics Advisory Group (Ethics ID: 1545786.2). The session was conducted in individual sensory booths at the sensory laboratory of the FVAS–UoM. The samples were served at refrigeration temperature (4 °C) in standard wine-tasting glasses. Samples were served in a semi-randomized order (two groups of beers). Participants were provided with water and water crackers to cleanse their palates and were asked to rest for a few minutes between samples to avoid fatigue. The questionnaire was displayed in the BioSensory Application

(University of Melbourne, Parkville, Victoria, Australia, [47]). The sensory attributes evaluated for acceptability and the scale used are shown in Table 2.

Table 2. Sensory attributes evaluated in the beer consumer test.

Attribute	Label	Scale
Carbonation Mouthfeel	Mcarb	9-point hedonic
Bitterness	Tbitter	9-point hedonic
Aroma	Aroma Liking	9-point hedonic
Flavor	Flavor Liking	9-point hedonic
Overall Liking	Overall Liking	9-point hedonic

2.4. Near-Infrared Spectroscopy

Samples were analyzed using a NIR spectroscopy handheld device MicroPHAZIR™ RX Analyzer (Thermo Fisher Scientific, Waltham, MA, USA) able to read spectra from 1596–2396 nm. Gonzalez Viejo et al. [29] described a filter paper soaked with the beer sample and measured using the NIR device. The dry filter paper absorbance values were subtracted from the filter soaked in beer readings to obtain only the liquid's absorbance values.

2.5. Physical Parameters

The beer samples were analyzed using the robotic pourer RoboBEER (The University of Melbourne, Parkville, Vic, Australia), as Gonzalez Viejo et al. [10] described. Each beer bottle was poured using the pourer to record 5-min videos and further analyzed using computer vision algorithms in Matlab® R2020b (Mathworks, Inc., Natick, MA, USA) to obtain the physical parameters related to foam and color (Table 3).

Table 3. Physical parameters analyzed using the RoboBEER.

Parameter	Label
Maximum volume of foam	MaxVol
Total lifetime of foam	TLTF
Lifetime of foam	LTF
Foam drainage	FDrain
Color lab scale	L, a and b
Color RGB scale	R, G, and B
Small bubbles	SmBubb
Medium bubbles	MedBubb
Large bubbles	LgBubb

2.6. Statistical Analysis and Machine-Learning Modeling

A multivariate data analysis based on principal component analysis (PCA) was conducted using a code written in Matlab® R2020a to find relationships between the data obtained from the e-nose, RoboBEER, and the consumer test responses. Furthermore, a correlation matrix was developed using Matlab® R2020b to assess only the significant correlations ($p < 0.05$) between all parameters.

For machine-learning Model 1, pattern recognition, artificial neural networks (ANN) were used. This model consisted of the use of the maximum (Max), mean, and area under the curve (AUC) values of each sensor's outputs from the e-nose to classify the samples into the three types of fermentation (top, bottom, and spontaneous; Figure 2a). A total of 17 different training algorithms were tested in a loop using a code written in Matlab® R2020a to find the best model based on the lack of under- or overfitting, highest accuracy, and best performance (Levenberg–Marquardt). Data were divided randomly as 60% for training, 20% for validation with a mean squared error (MSE) performance algorithm, and 20% for training. A neuron trimming test was conducted using 3, 5, 7, and 10 neurons, with the latter resulting in the best performance with no signs of under- or overfitting.

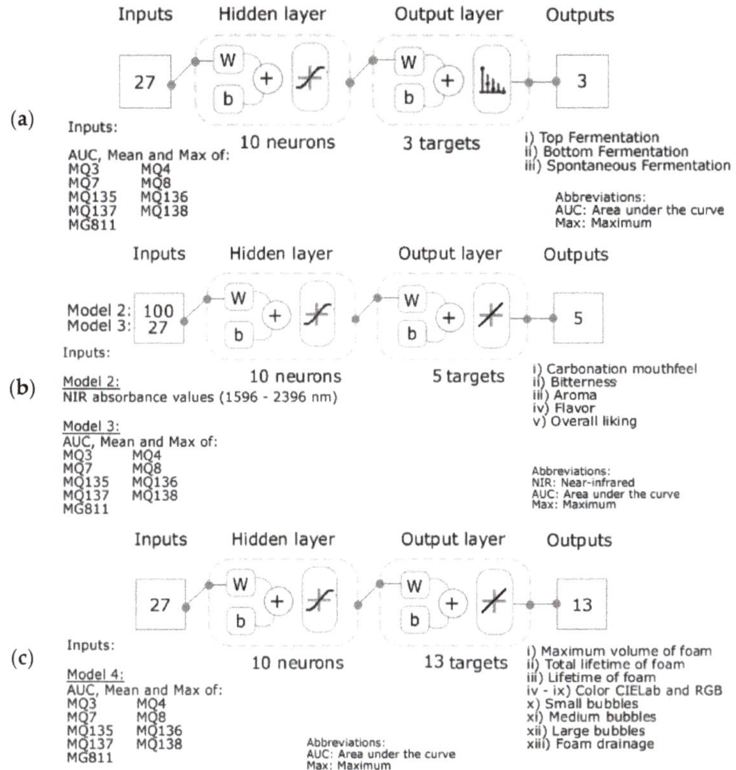

Figure 2. Diagrams of the two-layer feedforward artificial neural networks showing (**a**) pattern recognition Model 1 with a tan-sigmoid function in the hidden layer and Softmax neurons in the output layer, (**b**,**c**) regression Models 2–4 with a tan-sigmoid function in the hidden layer and a linear transfer function in the output layer.

Models 2–4 were developed using a regression ANN, similar to Model 1; 17 different training algorithms were tested, with Levenberg–Marquardt resulting in the best performance and accuracy for the three models. Model 2 was developed using the NIR absorbance values from the whole spectra (1596–2396 nm) as inputs, while Model 3 was created using the Max, mean, and AUC values of each sensor's outputs from the e-nose as inputs. For both models, the consumer test responses for the five attributes shown in Table 2 were used as targets (Figure 2b). On the other hand, Model 4 used the same e-nose inputs as Model 3, but with the RoboBEER outputs (Table 3) as targets (Figure 2c). A random data division was used as 60% for training, 20% for validation with an MSE performance algorithm, and 20% for testing. For these models, a trimming exercise was also performed using 3, 5, 7, and 10 neurons, with the latter resulting in the best performance with no signs of under- or over-fitting.

3. Results

3.1. Multivariate Data Analysis

As shown in Figure 3a, the PCA principal component one (PC1) explained 33.06% of total data variability, while PC2 accounted for 22.37% (total PC = 55.43%), which is close to the cut-off point of 60% [48]. According to the factor loadings (FL), the PC1 was mainly represented by the e-nose sensors MQ138 (FL = 0.32), MQ8 (FL = 0.31), and MQ137 (FL = 0.30) on the positive side of the axis, and by B (FL = −0.19) from RGB color scale on the negative side. On the other hand, PC2 was mainly represented by G (FL = 0.36) and R (FL = 0.34) from the RGB color scale, and L (FL = 0.35) from the CIELab color scale on the positive side of the axis, and by 'a' (FL = −0.32) from the CIELab color scale, and medium (FL = −0.22) and large (FL = −0.23) bubbles on the negative side. Positive relationships were observed between the e-nose sensors and physical parameters such as total lifetime of foam (TLTF) and MaxVol as well as liking of all sensory attributes, especially aroma. By contrast, the e-nose sensors had a negative correlation with foam drainage (FDrain). Samples from bottom fermentation were associated with lower voltage from the e-nose and with FDrain. In comparison, most top fermentation and three spontaneous fermentation beers were associated with higher voltage from the e-nose and the sensory attributes' liking. The other four spontaneous fermentation beers were more associated with the physical parameters such as bubble size and lifetime of foam (LTF) and 'a' and 'b' from the CIELab color scale.

Figure 3b shows the correlation matrix with all physical, sensory, and e-nose parameters. Positive and significant correlations ($p < 0.05$) were found between the sensory attributes' liking and e-nose sensors MQ8, MQ136, and MQ138. Furthermore, positive and significant correlations were found between physical parameters such as LTF, color ('a', 'b', and B), MedBubb, and the e-nose data. As expected, beer foam drainage (FDrain) was inversely correlated to maximum volume (MaxVol) and the total lifetime of foam (TLTF).

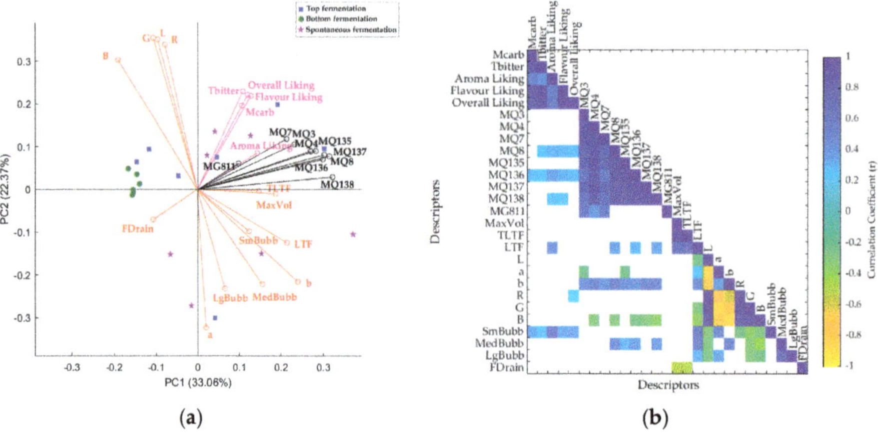

Figure 3. Multivariate data analysis results showing (**a**) the biplot from the principal components analysis (PCA) and (**b**) the correlation matrix depicting only the significant correlations ($p < 0.05$) of the data obtained from the electronic nose, RoboBEER physical parameters, and consumer sensory responses. Abbreviations are shown in Tables 1–3.

3.2. Machine-Learning Modeling

In Table 4, it can be observed that there was a high overall accuracy (97%) in predicting the type of beer fermentation using the Max, mean, and AUC data from the e-nose outputs as inputs. Furthermore, all stages had an accuracy of >90%. Compared to the validation and testing, the lower MSE value of the training stage and the latter two being the same, are indicators of no under- or overfitting. Figure 4 shows the receiver operating characteristic (ROC) curve depicting the true-positive (sensitivity) and false-positive (specificity) rates; as can be observed, the three categories had a sensitivity > 0.90.

Table 4. Statistical data from the pattern recognition artificial neural network Model 1 classify beers into the type of fermentation (top, bottom, and spontaneous) using the electronic nose outputs as inputs. The performance was based on means squared error (MSE).

Stage	Samples	Accuracy	Error	Performance (MSE)
Training	36	100%	0%	<0.01
Validation	12	92%	8%	0.10
Testing	12	92%	8%	0.10
Overall	60	97%	3%	N/A

N/A: Not applicable.

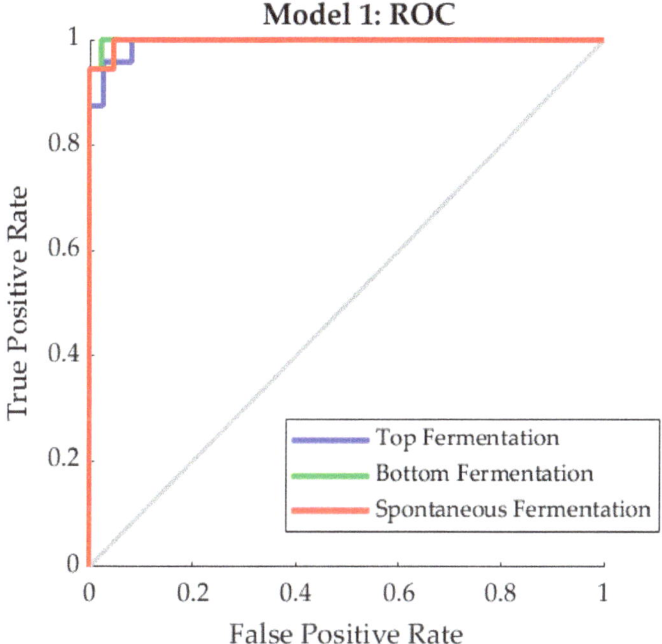

Figure 4. Receiver operating characteristics (ROC) curve depicting the true positive (y-axis) and false positive (x-axis) rates of Model 1 constructed to predict the type of fermentation of beers using the electronic nose outputs as inputs.

Table 5 shows the statistical data from the regression models. It can be observed that Model 2 had a high overall correlation coefficient ($r = 0.90$) and a high slope (>0.85) for all the stages. The training MSE value (MSE = 0.02) was lower than the validation and testing, which is a sign of no under- or overfitting; however, the validation (MSE = 0.12) and training (MSE = 0.30) were not as close, which may indicate some overfitting. Model 3, developed using the e-nose outputs to predict sensory descriptors' liking, had a higher overall correlation coefficient (R = 0.95) than Model 2, which was constructed using NIR inputs to predict the same sensory descriptors. Furthermore, Model 3 had high slope values for all stages. The training performance MSE value was lower than the validation and testing, and these two were close (validation MSE = 0.15; training MSE = 0.13); therefore, there were no signs of under- or overfitting. On the other hand, Model 3, developed using the e-nose outputs as inputs to predict the physical parameters obtained using RoboBEER, also had high overall accuracy ($r = 0.93$) with moderate-high slope values. This model had close MSE values for validation (MSE = 0.10) and testing

(MSE = 0.20) and with the training stage having a lower value (MSE = 0.02), therefore, no signs of under- or over-fitting.

Table 5. Statistical data from the regression artificial neural network Models 2–5. Performance was based on means squared error (MSE). Abbreviations: R: correlation coefficient.

Stage	Samples	Observations	R	Slope	Performance (MSE)
Model 2 (Near-infrared inputs/Sensory targets)					
Training	36	180	0.98	0.96	0.02
Validation	12	60	0.87	0.85	0.12
Testing	12	60	0.80	1.00	0.30
Overall	60	300	0.90	0.96	N/A
Model 3 (Electronic nose inputs/Sensory targets)					
Training	36	180	0.99	1.00	<0.01
Validation	12	60	0.95	0.94	0.15
Testing	12	60	0.85	0.94	0.13
Overall	60	300	0.95	0.97	N/A
Model 3 (Electronic nose inputs/RoboBEER targets)					
Training	36	468	0.98	0.93	0.02
Validation	12	156	0.90	0.80	0.10
Testing	12	156	0.82	0.87	0.20
Overall	60	780	0.93	0.89	N/A

N/A: Not applicable.

Figure 5 shows the overall regression models with their respective R values and regression equation. According to the 95% confidence bounds, Model 2 had 5% (15 out of 300 observations) of outliers, while Model 3 had 7% (21 out of 300 observations). On the other hand, Model 4 presented 5.3% (41 out of 780 observations) of outliers. It can be observed that in Model 2, flavor liking and carbonation mouthfeel were the attributes with the highest number of outliers (4 out of 60 each), while bitterness and aroma had the lowest (2 out of 60 each). In Model 3, bitterness was the highest in outliers (8 out of 60), while aroma was the lowest (1 out of 60). Conversely, in Model 4, 'a' and FDrain had the highest number of outliers (7 out of 60 each), while SmBubb did not present any outlier, and MedBubb only had 1 outlier out of 60.

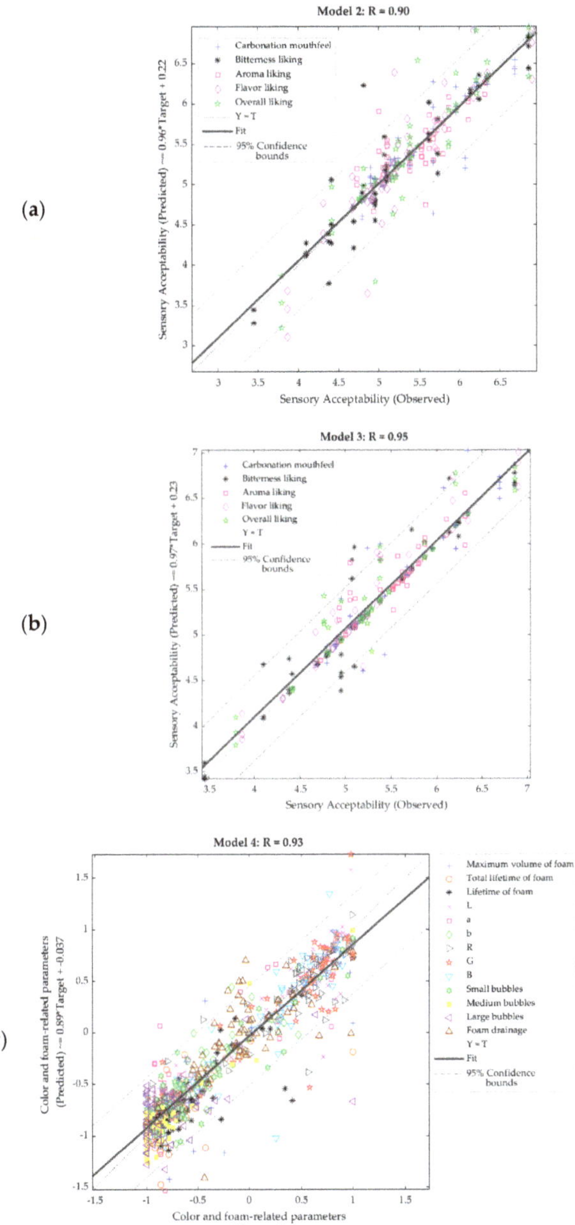

Figure 5. Overall regression models for (**a**) Model 2 to predict the consumer sensory responses using the near-infrared absorbance values as inputs, (**b**) Model 3 to predict the consumer sensory responses using the electronic nose outputs as inputs, and (**c**) Model 4 to predict the physical parameters measured with the RoboBEER using the electronic nose outputs as inputs.

4. Discussion

4.1. Relationships between E-Nose and Physicochemical Analysis

Most of the bottom fermentation beers were clustered close to vectors related to foam drainage (FDrain) of beers and color red, green and blue (RGB) and L; the latter may be explained due to the lighter color that these beers tend to have [49] compared to those form top and spontaneous fermentation. Furthermore, the spontaneous beers were grouped mostly close to bubble formation with the distribution of bubbles between small, medium, and large size (SmBubb, MedBubb, and LgBubb) contributing to a higher lifetime of foam (LTF), similar results have been previously reported for both physicochemical and sensory data [10,13,50,51]. On the other hand, top fermentation beers clustered mostly closer to all the gas sensors' sensitivity with a variation of foamability and bitterness which mainly influenced overall aroma, and flavor liking along with higher beer carbonation mouthfeel (Mcarb) (Figure 3a). The reason for top fermentation beers to be spread along the PCA may be mainly due to the variability in the bitterness and flavor due to the addition of different types and amounts of hops, which is greater than that in the bottom, and spontaneous fermentation beers [10,52,53].

Most of the gas sensors from the e-nose were related to each other as expected. Still, more importantly, there were statistically significant and positive correlations between sensors from the e-nose and carbonation sensation and bitterness, which can be directly related to the CO_2 release from beers in the pouring and aromas related to hops giving bitterness taste on beers [13,54] (Figures 2a and 3b), which directly influenced the flavor and overall liking from the sensory analysis. Furthermore, there were also significant and positive correlations between the e-nose sensors and the lifetime of foam and blue to yellow colors (b) from CIELab and blue (B) from the RGB color scales (Figure 2b). The latter may be explained since volatile compounds such as hydrogen sulfide (MQ136) are present in the malt and hops, which are responsible for beer color [49,55]. Furthermore, hydrogen sulfide (MQ136), alcohol (MQ3, MQ135, MQ138), and hydrogen are produced during fermentation in which a drop in pH is caused, provoking a change in color intensity [49,56,57].

There are significant correlations between small bubbles (SmBubb) and liking, which were directly related to the lifetime of foam (LTF) and retention of foam of beers, hence decreasing the release of gases after pouring, which can explain the absence of correlation between small bubbles and the e-nose gas sensors. Furthermore, correlations between LTF and SmBubb with 'a' from CIELab scale were found; this may be explained with the findings in other studies showing that beers and berries with more red color had higher sugar content [50,58], and at the same time sugars act as surfactant substances, which are responsible for increasing beer's viscosity and, therefore, increasing foam stability and reducing bubble size [13,17,59].

4.2. Artificial Intelligence Applied to Beer Quality Assessment

The machine learning-based Model 1 was able to classify all the beers studied into the top, bottom, and spontaneous fermentation with very high accuracy (97%) using data from the e-nose. The latter is consistent with the data from multivariate data analysis presented here (Figure 2a,b) and previous research using data from the RoboBEER as inputs to classify beers into the three types of fermentation with similarly reported accuracies of 92.4% [13]. These results are important since, for this specific type of application related to beer classification, data analysis is based on the signal analysis in the case of the e-nose compared to more complex implementation of computer vision algorithms in the case of RoboBEER. Furthermore, the e-nose could offer additional gas classification capabilities and identification of the specific gases that every sensor is sensitive to and aroma profiles from beers [20,27,29,31,60]. Further applications could be focused on identifying faults and proteins in beers [16] and traceability using new and emerging sensor technologies [61].

The use of the e-nose could also be more efficient, cost-effective, and user-friendly than NIR spectroscopy methods, as per the results shown here to assess consumer sensory perception and acceptability of beers, which resulted in Model 3 having higher accuracy (e-nose) compared to Model 2 (NIR). Finally, Model 4, based on e-nose inputs, was shown to be highly accurate at detecting physical parameters of foam formation, drainage, size, and distribution of bubbles and colorimetric assessment of beers compared to the RoboBEER. These results are very encouraging when aligned to Model 1 since this makes it possible for the RoboBEER to act just like a precise pouring device to be attached to the e-nose to obtain all the required physicochemical, colorimetric, and sensory perception by consumer parameters for a more accurate, objective, cost-effective and user-friendly beer quality assessment system. Furthermore, the e-nose may be installed as a stand-alone device at different stages of the production line to assess beer quality and detect any faults in real-time to take corrective actions in the processing line before the final product. These parameters obtained from the new system proposed here can be used as an integrative system coupled with blockchain for traceability and authentication of beers [62,63].

5. Conclusions

A novel system based on artificial intelligence (AI) has been proposed using robotics, electronic sensors (e-nose), and machine learning to assess beer quality more objectively. The system proposed here offers high accuracy in identifying beer fermentation, consumer preference, and acceptance, as well as physicochemical and colorimetric analysis. Furthermore, it potentially could be used for authentication and traceability coupled with blockchain technology for higher transparency within the brewing chain from raw materials, brewing, quality assurance, and sensory analysis, to commercialization. Finally, the integration of low-cost technology with AI makes it possible to apply these new and emerging technologies to the broad spectrum of brewing processes and brewing companies, from craft beer enterprises to the biggest brewers.

Supplementary Materials: The following are available online at http://www.mdpi.com/2311-5637/6/4/104/s1, Table S1. Style of the beer samples used for the study including their country of origin, fermentation type and alcohol content

Author Contributions: Conceptualization, C.G.V. and S.F.; methodology, C.G.V. and S.F.; validation, C.G.V. and S.F.; formal analysis, C.G.V.; investigation, C.G.V. and S.F.; data curation, C.G.V. and S.F.; writing—original draft preparation, C.G.V. and S.F.; writing—review and editing, C.G.V. and S.F.; software, C.G.V. and S.F.; visualization, C.G.V. and S.F.; supervision, S.F.; project administration, C.G.V. All authors have read and agreed to the published version of the manuscript.

Funding: This research received no external funding.

Acknowledgments: The authors would like to acknowledge Ranjith R. Unnithan, Bryce Widdicombe, Mimi Sun, and Jorge Gonzalez from the School of Engineering, Department of Electrical and Electronic Engineering of The University of Melbourne for their collaboration in the electronic nose development.

Conflicts of Interest: The authors declare no conflict of interest.

References

1. Gasiński, A.; Kawa-Rygielska, J.; Szumny, A.; Czubaszek, A.; Gąsior, J.; Pietrzak, W. Volatile Compounds Content, Physicochemical Parameters, and Antioxidant Activity of Beers with Addition of Mango Fruit (*Mangifera Indica*). *Molecules* **2020**, *25*, 3033. [CrossRef]
2. Gasiński, A.; Kawa-Rygielska, J.; Szumny, A.; Gąsior, J.; Głowacki, A. Assessment of Volatiles and Polyphenol Content, Physicochemical Parameters and Antioxidant Activity in Beers with Dotted Hawthorn (*Crataegus punctata*). *Foods* **2020**, *9*, 775. [CrossRef]
3. Kawa-Rygielska, J.; Adamenko, K.; Kucharska, A.Z.; Prorok, P.; Piórecki, N. Physicochemical and antioxidative properties of Cornelian cherry beer. *Food Chem.* **2019**, *281*, 147–153. [CrossRef]
4. Deng, Y.; Bi, H.; Yin, H.; Yu, J.; Dong, J.; Yang, M.; Ma, Y. Influence of ultrasound assisted thermal processing on the physicochemical and sensorial properties of beer. *Ultrason. Sonochem.* **2018**, *40*, 166–173. [CrossRef]
5. Pereira, I.M.C.; Neto, J.D.M.; Figueiredo, R.W.; Carvalho, J.D.G.; Figueiredo, E.A.T.D.; Menezes, N.V.S.D.; Gaban, S.V.F. Physicochemical characterization, antioxidant activity, and sensory analysis of beers brewed with cashew peduncle (*Anacardium occidentale*) and orange peel (*Citrus sinensis*). *Food Sci. Technol.* **2020**. [CrossRef]

6. Prestes, D.N.; Spessato, A.; Talhamento, A.; Gularte, M.A.; Schirmer, M.A.; Vanier, N.L.; Rombaldi, C.V. The addition of defatted rice bran to malted rice improves the quality of rice beer. *LWT* **2019**, *112*, 108262. [CrossRef]
7. Martínez, A.; Vegara, S.; Martí, N.; Valero, M.; Saura, D. Physicochemical characterization of special persimmon fruit beers using bohemian pilsner malt as a base. *J. Inst. Brew.* **2017**, *123*, 319–327. [CrossRef]
8. Humia, B.V.; Santos, K.S.; Schneider, J.K.; Leal, I.L.; de Abreu Barreto, G.; Batista, T.; Machado, B.A.S.; Druzian, J.I.; Krause, L.C.; da Costa Mendonça, M. Physicochemical and sensory profile of Beauregard sweet potato beer. *Food Chem.* **2020**, *312*, 126087. [CrossRef]
9. Sung, S.-A.; Lee, S.-J. Physicochemical and sensory characteristics of commercial top-fermented beers. *Korean J. Food Sci. Technol.* **2017**, *49*, 35–43. [CrossRef]
10. Viejo, C.G.; Fuentes, S.; Howell, K.; Torrico, D.; Dunshea, F. Integration of non-invasive biometrics with sensory analysis techniques to assess acceptability of beer by consumers. *Physiol. Behav.* **2019**, *200*, 139–147. [CrossRef]
11. Stone, H.; Bleibaum, R.; Thomas, H.A. *Sensory Evaluation Practices*; Elsevier/Academic Press: San Diego, CA, USA, 2012.
12. Kemp, S.; Hollowood, T.; Hort, J. *Sensory Evaluation: A Practical Handbook*; Wiley: Oxford, UK, 2011.
13. Viejo, C.G.; Fuentes, S.; Li, G.; Collmann, R.; Condé, B.; Torrico, D. Development of a robotic pourer constructed with ubiquitous materials, open hardware and sensors to assess beer foam quality using computer vision and pattern recognition algorithms: RoboBEER. *Food Res. Int.* **2016**, *89*, 504–513. [CrossRef] [PubMed]
14. Condé, B.C.; Fuentes, S.; Caron, M.; Xiao, D.; Collmann, R.; Howell, K.S. Development of a robotic and computer vision method to assess foam quality in sparkling wines. *Food Control* **2017**, *71*, 383–392. [CrossRef]
15. Vann, L.; Layfield, J.B.; Sheppard, J.D. The application of near-infrared spectroscopy in beer fermentation for online monitoring of critical process parameters and their integration into a novel feedforward control strategy. *J. Inst. Brew.* **2017**, *123*, 347–360. [CrossRef]
16. Viejo, C.G.; Caboche, C.H.; Kerr, E.D.; Pegg, C.L.; Schulz, B.L.; Howell, K.; Fuentes, S. Development of a Rapid Method to Assess Beer Foamability Based on Relative Protein Content Using RoboBEER and Machine Learning Modeling. *Beverages* **2020**, *6*, 28. [CrossRef]
17. Viejo, C.G.; Fuentes, S.; Torrico, D.; Howell, K.; Dunshea, F.R. Assessment of beer quality based on foamability and chemical composition using computer vision algorithms, near infrared spectroscopy and machine learning algorithms. *J. Sci. Food Agric.* **2018**, *98*, 618–627. [CrossRef]
18. Viejo, C.G.; Fuentes, S. A Digital Approach to Model Quality and Sensory Traits of Beers Fermented under Sonication Based on Chemical Fingerprinting. *Fermentation* **2020**, *6*, 73. [CrossRef]
19. Gonzalez Viejo, C.; Torrico, D.D.; Dunshea, F.R.; Fuentes, S. Emerging Technologies Based on Artificial Intelligence to Assess the Quality and Consumer Preference of Beverages. *Beverages* **2019**, *5*, 62. [CrossRef]
20. Viejo, C.G.; Fuentes, S.; Godbole, A.; Widdicombe, B.; Unnithan, R.R. Development of a low-cost e-nose to assess aroma profiles: An artificial intelligence application to assess beer quality. *Sens. Actuators B Chem.* **2020**, 127688. [CrossRef]
21. Viejo, C.G.; Fuentes, S. Beer Aroma and Quality Traits Assessment Using Artificial Intelligence. *Fermentation* **2020**, *6*, 56. [CrossRef]
22. Thazin, Y.; Pobkrut, T.; Kerdcharoen, T. Prediction of acidity levels of fresh roasted coffees using e-nose and artificial neural network. In Proceedings of the 2018 10th International Conference on Knowledge and Smart Technology (KST), Chiangmai, Thailand, 31 January–3 February 2018; pp. 210–215.
23. Romani, S.; Cevoli, C.; Fabbri, A.; Alessandrini, L.; Rosa, M.D. Evaluation of coffee roasting degree by using electronic nose and artificial neural network for off-line quality control. *J. Food Sci.* **2012**, *77*, C960–C965. [CrossRef]
24. Qiu, S.; Gao, L.; Wang, J. Classification and regression of ELM, LVQ and SVM for E-nose data of strawberry juice. *J. Food Eng.* **2015**, *144*, 77–85. [CrossRef]
25. Chen, Q.; Zhao, J.; Chen, Z.; Lin, H.; Zhao, D.-A. Discrimination of green tea quality using the electronic nose technique and the human panel test, comparison of linear and nonlinear classification tools. *Sens. Actuators B Chem.* **2011**, *159*, 294–300. [CrossRef]
26. Dutta, R.; Hines, E.; Gardner, J.; Kashwan, K.; Bhuyan, M. Tea quality prediction using a tin oxide-based electronic nose: An artificial intelligence approach. *Sens. Actuators B Chem.* **2003**, *94*, 228–237. [CrossRef]

27. Fuentes, S.; Summerson, V.; Viejo, C.G.; Tongson, E.; Lipovetzky, N.; Wilkinson, K.L.; Szeto, C.; Unnithan, R.R. Assessment of Smoke Contamination in Grapevine Berries and Taint in Wines Due to Bushfires Using a Low-Cost E-Nose and an Artificial Intelligence Approach. *Sensors* **2020**, *20*, 5108. [CrossRef]
28. Macías, M.; Agudo, J.; Manso, A.; Orellana, C.; Velasco, H.; Caballero, R. A compact and low cost electronic nose for aroma detection. *Sensors* **2013**, *13*, 5528–5541. [CrossRef]
29. Men, H.; Shi, Y.; Fu, S.; Jiao, Y.; Qiao, Y.; Liu, J. Mining Feature of Data Fusion in the Classification of Beer Flavor Information Using E-Tongue and E-Nose. *Sensors* **2017**, *17*, 1656.
30. Ghasemi-Varnamkhasti, M.; Mohtasebi, S.S.; Siadat, M.; Lozano, J.; Ahmadi, H.; Razavi, S.H.; Dicko, A. Aging fingerprint characterization of beer using electronic nose. *Sens. Actuators B Chem.* **2011**, *159*, 51–59. [CrossRef]
31. Pornpanomchai, C.; Suthamsmai, N. Beer classification by electronic nose. In Proceedings of the 2008 International Conference on Wavelet Analysis and Pattern Recognition, Hong Kong, China, 30–31 August 2008; pp. 333–338.
32. Wojnowski, W.; Majchrzak, T.; Dymerski, T.; Gębicki, J.; Namieśnik, J. Electronic noses: Powerful tools in meat quality assessment. *Meat Sci.* **2017**, *131*, 119–131. [CrossRef]
33. Wojnowski, W.; Majchrzak, T.; Dymerski, T.; Gębicki, J.; Namieśnik, J. Portable electronic nose based on electrochemical sensors for food quality assessment. *Sensors* **2017**, *17*, 2715. [CrossRef]
34. Deisingh, A.K.; Stone, D.C.; Thompson, M. Applications of electronic noses and tongues in food analysis. *Int. J. Food Sci. Technol.* **2004**, *39*, 587–604. [CrossRef]
35. Loutfi, A.; Coradeschi, S.; Mani, G.K.; Shankar, P.; Rayappan, J.B.B. Electronic noses for food quality: A review. *J. Food Eng.* **2015**, *144*, 103–111. [CrossRef]
36. Zhou, B.; Wang, J. Use of electronic nose technology for identifying rice infestation by Nilaparvata lugens. *Sens. Actuators B Chem.* **2011**, *160*, 15–21. [CrossRef]
37. Yang, W.; Yu, J.; Pei, F.; Mariga, A.M.; Ma, N.; Fang, Y.; Hu, Q. Effect of hot air drying on volatile compounds of Flammulina velutipes detected by HS-SPME–GC–MS and electronic nose. *Food Chem.* **2016**, *196*, 860–866. [CrossRef]
38. Di Rosa, A.R.; Leone, F.; Cheli, F.; Chiofalo, V. Fusion of electronic nose, electronic tongue and computer vision for animal source food authentication and quality assessment—A review. *J. Food Eng.* **2017**, *210*, 62–75. [CrossRef]
39. Zeiger, K. zNose Series. Available online: https://www.cbrnetechindex.com/Print/4362/electronic-sensor-technology-inc/znose-series (accessed on 26 October 2020).
40. Men, H.; Shi, Y.; Jiao, Y.; Gong, F.; Liu, J. Electronic nose sensors data feature mining: A synergetic strategy for the classification of beer. *Anal. Methods* **2018**, *10*, 2016–2025. [CrossRef]
41. Shi, Y.; Gong, F.; Wang, M.; Liu, J.; Wu, Y.; Men, H. A deep feature mining method of electronic nose sensor data for identifying beer olfactory information. *J. Food Eng.* **2019**, *263*, 437–445. [CrossRef]
42. Nimsuk, N. Improvement of accuracy in beer classification using transient features for electronic nose technology. *J. Food Meas. Charact.* **2019**, *13*, 656–662. [CrossRef]
43. Abdi, M.; Adebiyi, A.; Fasoli, A.; Mannari, A.; Labby, R.; Bozano, L. Model Comparison of Beer data classification using an electronic nose. In Proceedings of the Eighth International Conference on Learning Representations, Virtual Conference, Addis Ababa, Ethiopia, 26 April–1 May 2020.
44. Quarto, A.; Soldo, D.; Di Lecce, F.; Giove, A.; Di Lecce, V.; Castronovo, A. Electronic nose for evaluating water use in beer production. In Proceedings of the 2017 ISOCS/IEEE International Symposium on Olfaction and Electronic Nose (ISOEN), Montreal, QC, Canada, 28–31 May 2017; pp. 1–3.
45. Sànchez, C.; Lozano, J.; PedroSantos, J.; Azabal, A.; Ruiz-Valdepeñas, S. Discrimination of aromas in beer with electronic nose. In Proceedings of the 2018 Spanish Conference on Electron Devices (CDE), Salamanca, Spain, 14–16 November 2018; pp. 1–4.
46. Voss, H.G.J.; Mendes Júnior, J.J.A.; Farinelli, M.E.; Stevan, S.L. A Prototype to Detect the Alcohol Content of Beers Based on an Electronic Nose. *Sensors* **2019**, *19*, 2646. [CrossRef]
47. Fuentes, S.; Gonzalez Viejo, C.; Torrico, D.; Dunshea, F. Development of a biosensory computer application to assess physiological and emotional responses from sensory panelists. *Sensors* **2018**, *18*, 2958. [CrossRef]
48. Deep, K.; Jain, M.; Salhi, S. *Logistics, Supply Chain and Financial Predictive Analytics: Theory and Practices*; Springer: Berlin/Heidelberg, Germany, 2018.

49. Delcour, J.A.; Hoseney, R.C. *Principles of Cereal Science and Technology*; AACC International: Eagan, MN, USA, 2010.
50. Viejo, C.G.; Fuentes, S.; Torrico, D.; Howell, K.; Dunshea, F. Assessment of Beer Quality Based on a Robotic Pourer, Computer Vision, and Machine Learning Algorithms Using Commercial Beers. *J. Food Sci.* **2018**, *83*, 1381–1388. [CrossRef]
51. Viejo, C.G.; Fuentes, S.; Howell, K.; Torrico, D.; Dunshea, F.R. Robotics and computer vision techniques combined with non-invasive consumer biometrics to assess quality traits from beer foamability using machine learning: A potential for artificial intelligence applications. *Food Control* **2018**, *92*, 72–79. [CrossRef]
52. Perozzi, C.; Beaune, H. *The Naked Brewer: Fearless Homebrewing Tips, Tricks & Rule-breaking Recipes*; Penguin Publishing Group: New York, NY, USA, 2012.
53. Perozzi, C.; Beaune, H. *The Naked Pint: An Unadulterated Guide to Craft Beer*; Penguin Publishing Group: New York, NY, USA, 2012.
54. De Keukeleire, D. Fundamentals of beer and hop chemistry. *Quim. Nova* **2000**, *23*, 108–112. [CrossRef]
55. Anderson, R.; Howard, G. The production of hydrogen sulphide by yeast and by Zymomonas anaerobia. *J. Inst. Brew.* **1974**, *80*, 245–251. [CrossRef]
56. Bokulich, N.A.; Bamforth, C.W. The microbiology of malting and brewing. *Microbiol. Mol. Biol. Rev.* **2013**, *77*, 157–172. [CrossRef] [PubMed]
57. Stewart, G.G. The production of secondary metabolites with flavour potential during brewing and distilling wort fermentations. *Fermentation* **2017**, *3*, 63. [CrossRef]
58. Abeytilakarathna, P.; Fonseka, R.; Eswara, J.; Wijethunga, K. Relationship between total solid content and red, green and blue colour intensity of strawberry (Fragaria x ananassa Duch.) fruits. *J. Agric. Sci.* **2013**, *8*, 82–90. [CrossRef]
59. Badui, S. *Química de los Alimentos*; Pearson Education: Naucalpan de Juarez, México, 2006.
60. Yu, H.; Wang, J.; Yao, C.; Zhang, H.; Yu, Y. Quality grade identification of green tea using E-nose by CA and ANN. *LWT-Food Sci. Technol.* **2008**, *41*, 1268–1273. [CrossRef]
61. Violino, S.; Figorilli, S.; Costa, C.; Pallottino, F. Internet of Beer: A Review on Smart Technologies from Mash to Pint. *Foods* **2020**, *9*, 950. [CrossRef]
62. Patelli, N.; Mandrioli, M. Blockchain technology and traceability in the agrifood industry. *J. Food Sci.* **2020**. [CrossRef]
63. Kamiloglu, S. Authenticity and traceability in beverages. *Food Chem.* **2019**, *277*, 12–24. [CrossRef]

Article

Smart Detection of Faults in Beers Using Near-Infrared Spectroscopy, a Low-Cost Electronic Nose and Artificial Intelligence

Claudia Gonzalez Viejo [1,*], Sigfredo Fuentes [1] and Carmen Hernandez-Brenes [2]

[1] Digital Agriculture, Food and Wine Sciences Group, School of Agriculture and Food, Faculty of Veterinary and Agricultural Sciences, University of Melbourne, Melbourne, VIC 3010, Australia; sfuentes@unimelb.edu.au

[2] Tecnologico de Monterrey, Escuela de Ingeniería y Ciencias, Ave. Eugenio Garza Sada 2501, Monterrey 64849, Mexico; chbrenes@tec.mx

* Correspondence: cgonzalez2@unimelb.edu.au

Abstract: Early detection of beer faults is an important assessment in the brewing process to secure a high-quality product and consumer acceptability. This study proposed an integrated AI system for smart detection of beer faults based on the comparison of near-infrared spectroscopy (NIR) and a newly developed electronic nose (e-nose) using machine learning modelling. For these purposes, a commercial larger beer was used as a base prototype, which was spiked with 18 common beer faults plus the control aroma. The 19 aroma profiles were used as targets for classification ma-chine learning (ML) modelling. Six different ML models were developed; Model 1 (M1) and M2 were developed using the NIR absorbance values (100 inputs from 1596–2396 nm) and e-nose (nine sensor readings) as inputs, respectively, to classify the samples into control, low and high concentration of faults. Model 3 (M3) and M4 were based on NIR and M5 and M6 based on the e-nose readings as inputs with 19 aroma profiles as targets for all models. A customized code tested 17 artificial neural network (ANN) algorithms automatically testing performance and neu-ron trimming. Results showed that the Bayesian regularization algorithm was the most adequate for classification rendering precisions of M1 = 95.6%, M2 = 95.3%, M3 = 98.9%, M4 = 98.3%, M5 = 96.8%, and M6 = 96.2% without statistical signs of under- or overfitting. The proposed system can be added to robotic pourers and the brewing process at low cost, which can benefit craft and larger brewing companies.

Keywords: machine learning; off aromas; gas sensors; robotic pourer; aroma thresholds

Citation: Gonzalez Viejo, C.; Fuentes, S.; Hernandez-Brenes, C. Smart Detection of Faults in Beers Using Near-Infrared Spectroscopy, a Low-Cost Electronic Nose and Artificial Intelligence. *Fermentation* **2021**, 7, 117. https://doi.org/10.3390/fermentation7030117

Academic Editor: Daniel Cozzolino

Received: 29 June 2021
Accepted: 14 July 2021
Published: 15 July 2021

1. Introduction

In commercial settings, the assessment of beer faults is mainly the responsibility of the head brewer. They are usually determined from simple aroma profile assessment, sensitivity sensory tests such as absolute, recognition, differential, and/or terminal threshold using a trained panel [1–3] or utilizing specialized instrumentation such as gas chromatography-mass spectroscopy (GC-MS) [4]. Several types of faults (off-flavors/aromas) can develop in beers and with diverse origins and sensory perception thresholds, as shown in Table 1.

The drawbacks of common beer fault assessments are that they could be subjective. In the case of instrumentation or sensory sessions, they may require expensive equipment and special skills for usage, data handling, and analysis. Regarding sensory analysis, it requires a trained panel, which can also be cost-prohibitive and can assess only a few samples at any time to avoid increasing bias due to fatigue.

Table 1. Description and concentrations of typical beer faults (off-aromas/flavors).

Common Name	Chemical Compound	Aroma/Flavor	Origin	Contamination Stage	Detection Threshold in Water (mg L^{-1})	Typical Concentration in Beer (mg L^{-1})	Spoilage Concentration in Beer (mg L^{-1})	References
Diacetyl	2,3-butanediole	Butter	Low levels of valine in wort Microbial contamination	Wort	5×10^{-4}	8×10^{-3}–0.60	0.25	[5–8]
TCA *	2,4,6-trichloroanisole	Must taint/Moldy	Contaminated ingredients or other material (packaging)	Ageing Storage	3×10^{-8}–200×10^{-8}	Absent	0.02	[5,7,9]
Acetic acid	Acetic acid	Sour/Vinegar/Tangy	Spoilage bacteria Wild yeast	Fermentation Conditioning	0.10	30–200	60.0–120.0	[7,8,10]
Lactic acid	Lactic acid	Sour/Sour milk/Tart	Spoilage bacteria	Mashing Secondary fermentation	0.04	0.20–1.50	140	[8,10]
H_2S	Hydrogen sulfide	Rotten eggs	Raw material Yeast contamination	Fermentation	1×10^{-5}–10×10^{-5}	$\leq 1 \times 10^{-3}$	4×10^{-3}	[8,10–12]
DMS	Dimethyl sulfide	Sweet corn/Onion/Rotten vegetables	Microbial contamination	Wort boiling/cooling	3.3×10^{-7}	0.01–0.15	0.40	[8,10]
Papery	Trans-2-nonenal	Cardboard/Oxidized	Oxidation Staling	Fermentation Storage	8×10^{-8}	$< 5 \times 10^{-5}$	4×10^{-4}	[8,10]
Isovaleric acid	Isovaleric acid	Cheesy/Rancid/Sweaty feet	Old/Oxidized hops Process faults	Boiling Ageing	4.9×10^{-4}	≤ 0.20	1.00	[7,8,10]
Earthy	2-Ethyl fenchol	Soil/Compost/Moldy	Microbial contamination	Packaging	5×10^{-3} **	Absent	5×10^{-3}	[8]
Acetaldehyde	Acetaldehyde	Green apple/Bready/Grass	Staling Microbial contamination Poor yeast health	Fermentation Storage	2.5×10^{-5}–6.5×10^{-5}	2.00–15.0	20.0	[7,8]
Butyric	Butyric acid	Baby vomit/Putrid/Rancid butter	Microbial contamination Ageing	Wort production Packaging/Storage	2.4×10^{-3}	0.50–1.50	3.00	[7,8]
Caprylic	Caprylic acid	Goat/Soap/Sweaty	Microbial contamination Yeast breakdown	Maturation	0.013 **	2.00–8.00	10.0	[7,8]
Mercaptan	Ethanethiol	Drains/Sewer	Autolysis Poor yeast health	Fermentation Ageing	1.7×10^{-6} **	0.00–5×10^{-3}	1×10^{-3}	[7,8]
Spicy	Eugenol	Clove	Microbial contamination Wild yeast Oxidation	Ageing	7.1×10^{-7}	0.01–0.03	0.40	[7,8]
Metallic	Ferrous sulfate	Metal/Blood/Coin/Iron	Water sources Non-passivated vessels	Any brewing stage	1.00–1.50 **	≤ 0.50	1.00	[8]
Grainy	Isobutyraldehyde	Cereal husks/Green malt/Raw grain	Excessive run-off Insufficient boiling	Wort boiling	4.9×10^{-7}	1×10^{-3}–0.02	1.00–2.50	[7,8]
Indole	Indole	Farm/Barnyard/Fecal/Pig-like	Microbial contamination	Fermentation	5×10^{-3} **	$< 5 \times 10^{-3}$	0.01–0.02	[8]
Light-struck	2-Methyl-2-butene-1-thiol	Fecal/Skunky/Sulfury	Clear or green bottles	Storage	4×10^{-6} **	1×10^{-6}–5×10^{-6}	5.00–30.00	[8]
Bromophenol *	Bromophenol	Inky/Museum-like/Old electronics	Process/Equipment faults Contaminated raw material	Any brewing stage	3×10^{-9}	Absent	1.3×10^{-3}	[7,8]
Catty *	p-Methane-8-thiol-3-one	Oxidized/tomcat urine	Hops Contaminated raw material	Ageing Packaging	1.5×10^{-5} **	Absent	1.5×10^{-5}	[8]
Plastic *	Styrene	Burning plastic/Chemical	Brewing equipment and packaging material contamination	Any brewing stage	0.02	Absent	0.02	[8,13]

* Compounds not studied in this research. ** Detection threshold in beer as it has not been reported in water.

The implementation of new and emerging technologies for beer analysis [14], such as artificial intelligence (AI) [15–17] using robotics [18], near-infrared spectroscopy (NIR) [19–21], integrated gas sensors or low-cost electronic noses (e-noses) [22–24], and machine learning [25] is gaining traction recently for research and practical application purposes. One of those applications is the early detection of beer faults using e-noses [26,27] or beer classification [28].

This research focuses on implementing NIR spectroscopy and a recently developed low-cost e-nose using machine learning to create an integrated system for the smart detection of faults in beer. The integrated system proposed can become a big aid to brewing companies for the early assessment of faults in different manufacturing and processing stages to secure a high-quality product. These techniques can also be implemented for commercialization and authentication purposes to secure the provenance and consistency of quality for different markets.

2. Materials and Methods

2.1. Samples Description

A commercial Asahi Super Dry lager beer (Asahi Breweries, Sumida City, Tokyo, Japan) with 5% alcohol in 500 mL cans was used as the base samples for this study. This beer was selected because lager beers are less hoppy and less complex in aromas than other styles such as ales and lambics, which can be used as a prototype for testing purposes. The samples were spiked with 18 different flavor/aroma faults (Siebel Institute, Chicago, IL, USA) that are commonly found in beer (Table 2). A total of 1 L of beer (two cans) was used for each fault and was spiked with two different concentrations, as shown in Table 2. Besides the spiked samples, 1 L of the original beer was measured as a control, as two batches of beer were used, the control was taken as two samples (one per batch). All samples were measured in triplicates (replicates = 3), giving a total of n = 36 per concentration (n = 108) for the spiked samples, plus n = 6 control samples (n = 3 control per batch). Hence, giving a total of n = 114.

2.2. Near-Infrared Measurements

All samples were measured in triplicate for each of the three replicates (n = 9) using a handheld near-infrared (NIR) spectroscopy device (MicroPHAZIR™ RX Analyzer; Thermo Fisher Scientific, Waltham, MA, USA). This is able to measure the chemical fingerprinting within the 1596–2396 nm range. Samples were measured using the method described by Gonzalez Viejo et al. [19], which consists of using a filter paper Whatman® Grade 3 (Whatman plc., Maidstone, UK). The filter paper was first measured dry and then soaked in the sample using the white reference as background to avoid any noise from the environment. Then the absorbance values from the dry filter paper were subtracted from those with the sample to remove the paper components. For this study, both the raw absorbance values and the first derivative were used; the latter were obtained with the Savitzky-Golay method using the Unscrambler X ver. 10.3 (CAMO Software, Oslo, Norway) using the second polynomial order with the following smoothing parameters: number of left side points: 1, number of right-side points: 1, number of smoothing points: 3, and with symmetric kernel.

Table 2. Compounds used as off-flavors/aromas (faults) to spike beer samples and the concentrations used for training purposes according to human detection thresholds.

Number	Fault	Flavor/Aroma *	Concentration (Low; mg L^{-1})	Concentration (High; mg L^{-1})
1	Acetaldehyde	Green apple, cut grass	19.5	45.0
2	Acetic Acid	Vinegar	156	360
3	Butyric Acid	Putrid, baby vomit	3.25	7.50
4	Caprylic Acid	Soapy, wax, fatty	13.7	31.5
5	Contamination (Acetic Acid + Diacetyl)	Sour, buttery	156	361
6	Dimethyl Sulfide	Cooked vegetables	0.17	0.40
7	Diacetyl (2,3-Butanediol)	Butter, butterscotch	0.26	0.60
8	Earthy (2-Ethyl fenchol)	Soil	6.5×10^{-3}	0.02
9	Isobutyraldehyde	Grainy, husk, nut	1.63	3.75
10	Indole	Farm, barnyard	0.24	0.55
11	Isovaleric Acid	Cheese, sweaty socks, old hops	2.60	6.00
12	Lactic Acid	Sour milk	173.33	400
13	Light-struck (3-Methyl-2-butene-1-thiol)	Skunky, toffee, coffee	2.6×10^{-4}	$6x10^{-4}$
14	Mercaptan (Ethanethiol)	Sewer, drains	1.6×10^{-3}	3.8×10^{-3}
15	Ferrous Sulfate	Metallic, blood	1.63	3.75
16	Trans-2-nonenal	Papery, cardboard, oxidized	8.7×10^{-4}	2×10^{-3}
17	Eugenol	Cloves, spicy	0.05	0.12
18	Hydrogen Sulfide	Rotten Eggs	0.03	0.07

* As described by the beer faults kit supplier (Siebel Institute, Chicago, IL, USA).

2.3. Electronic Nose Measurements

A portable and low-cost electronic nose (e-nose) was used to measure the volatile compounds found in the samples. As described by Gonzalez Viejo et al. [29], the electronic nose consists of an array of nine different gas sensors (i) MQ3: alcohol, (ii) MQ4: methane (CH_4), (iii) MQ7: carbon monoxide (CO), (iv) MQ8: hydrogen (H_2), (v) MQ135, (vi) MQ136, (vii) MQ137, (viii) MQ138 and ix) MG811: carbon dioxide (CO_2). To measure the samples, the full amount of beer was poured into a clean 2 L jar, and the e-nose was placed on the top to acquire the volatile compound readings; all measurements were carried out in triplicates. The outputs were then analyzed using a supervised automatic code written in Matlab® R2020b (Mathworks, Inc., Natick, MA, USA). This code is able to automatically recognize features of curves (starting and end of stable signals) and create 10 subdivisions of each curve from the e-nose sensors from the stable signals to calculate 10 mean values [30]. This is done to increase variability of the data to further develop the ML models.

2.4. Alcohol and pH Measurements

Samples were analyzed for basic chemometrics in triplicates for pH and alcohol. The pH was measured in 50 mL samples of each replicate using a pH-meter (QM-1670, DigiTech, Sandy, UT, USA), the device was calibrated using a buffer solution (pH 7). Furthermore, an Alcolyzer Wine M with accuracy: <0.1% vv-1 (Anton Paar GmbH, Graz, Austria) was used to measure the alcohol content in 60 mL from each replicate.

2.5. Statistical Analysis and Machine Learning Modelling

The e-nose data were analyzed through ANOVA to assess significant differences ($p < 0.05$) among the samples with a Tukey honest significant difference (HSD) post hoc test ($\alpha = 0.05$) using XLSTAT 2020.3.1 (Addinsoft, New York, NY, USA). Furthermore, a code developed in Matlab® R2021a was used to conduct a correlation analysis and plot it in a matrix to assess only the significant correlations ($p < 0.05$) between the e-nose sensor outputs and the different faults.

Six supervised classification machine learning (ML) models were developed using artificial neural networks (ANN). All models were constructed using a customized code developed by the Digital Agriculture Food and Wine group from the University of Melbourne (DAFW; UoM) in Matlab® R2021a, which is able to test automatically 17 different ANN algorithms in a loop. The best models were selected based on the accuracy and performance, the Bayesian Regularization algorithm being the best for all four models.

The first two models were developed using the NIR absorbance raw values in the entire spectra (1596–2396 nm) (Model 1) and the 10 means (samples) from each sensor (inputs) of the e-nose outputs (Model 2) as inputs to classify the samples into (i) control, (ii) low concentration, and (iii) high concentration. Data were divided randomly as 70% of the samples used for training and 30% for testing. The performance was assessed from the means squared error (MSE), and a neuron trimming test was conducted to select the models with no under- or overfitting, being 10 the most optimal number for both models (Figure 1a).

For Models 3 and 4, the NIR absorbance raw values in the entire spectra (1596–2396 nm) were taken as inputs to predict the faults found in the sample for the low concentration (Model 3) and high concentration (Model 4). Data were divided using interleaved indices, which consists of cycling samples between the training (70%) and testing (30%) stages [31]. Performance was based on MSE; a neuron trimming test was conducted to select the models with no under- or overfitting, being 10 the most optimal number for both models (Figure 1b).

For the NIR models, the number of samples used was the number of beers with added faults plus control beers ($n = 19$), times the number of replicates (reps = 3; $n = 57$), multiplied by the number of measurements per replicate (measurements = 3; $n = 171$), giving a total of 180 samples considering the control as six replicates (+9).

The other two models were developed using the e-nose outputs as inputs to predict the fault found in the sample for the low concentration (Model 5) and high concentration (Model 6). Data were divided randomly into training (70%) and testing (30%) stages. Similar to Models 1 and 2, the performance was based on MSE, and 10 neurons (Figure 1c) were used for both models as they provided the best models with no under- or overfitting after conducting a neuron trimming test.

For the e-nose models, the number of samples used was the number of beer faults plus control ($n = 19$), times the number of replicates (reps = 3; $n = 57$), multiplied by the number of mean values obtained from the e-nose curves per beer sample (values = 10; $n = 570$), giving a total of 600 considering the control as six replicates (+30).

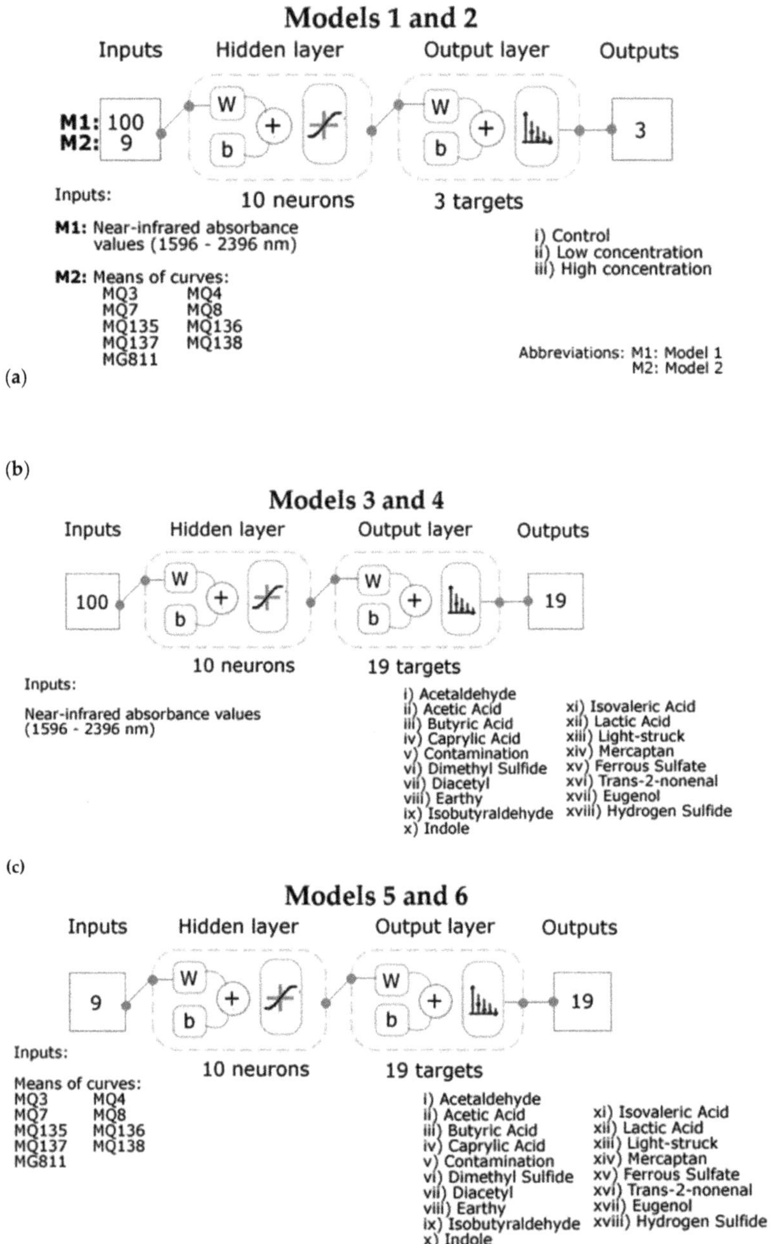

Figure 1. Diagrams of the two-layer feedforward models with a tan-sigmoid function in the hidden layer and a Softmax function in the output layer for (**a**) Models 1 (Near-infrared inputs), and 2 (electronic nose inputs), (**b**) Models 3 (low concentration) and 4 (high concentration), and (**c**) Models 5 (low concentration) and 6 (high concentration). Abbreviations: W: weights; b: bias.

3. Results

Figure 2 shows the curves of the NIR raw values for the control and each fault tested for low and high concentrations. It can be observed that the overtones were similar for both low and high concentrations; however, the absorbance values were different. The overtones found for all samples are within the 1900 and 2000 nm and >2250 nm. On the other hand, in Figure 3, which shows the absorbance values for the first derivative transformation, the overtones were enhanced within the 1850–1905 nm and 1950–2140 nm ranges. In both Figures, it could be observed that the sample with hydrogen sulfide at low concentration had higher absorbance values than with the high concentration.

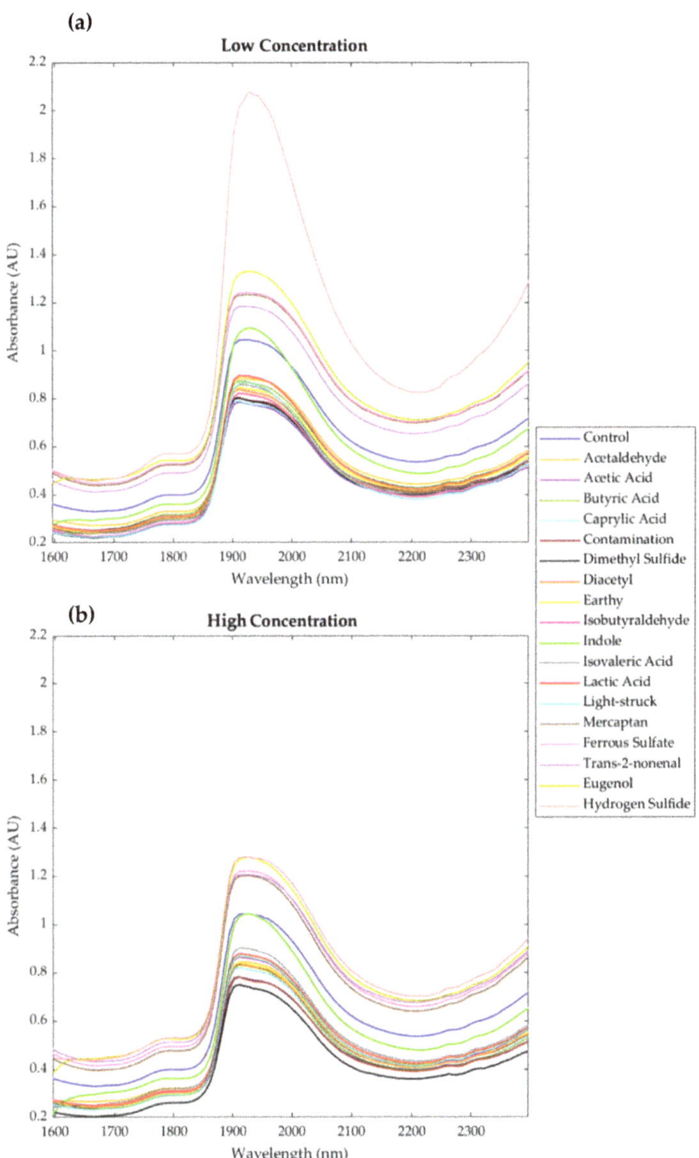

Figure 2. Near-infrared curves for all beer samples using the raw absorbance values at (**a**) low and (**b**) high concentrations.

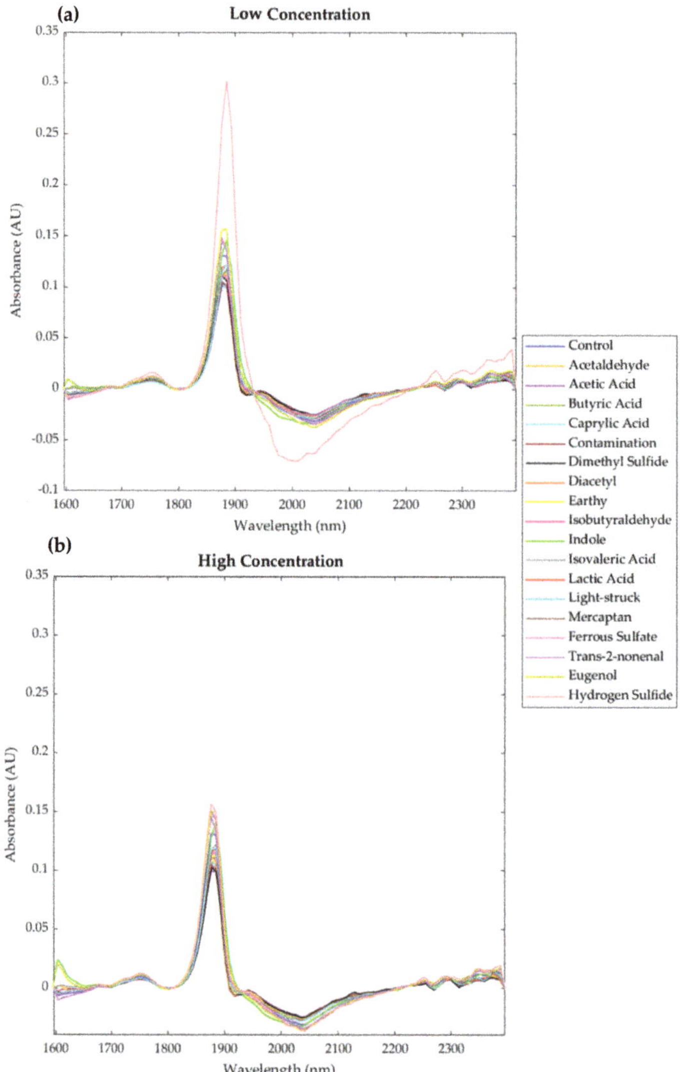

Figure 3. Near-infrared curves for all beer samples using the first derivative of the absorbance values at (**a**) low and (**b**) high concentrations.

Figure 4 shows the stacked bar graphs depicting the ANOVA results, mean values, and error bars based on the standard error of each e-nose sensor and each beer sample for low and high concentrations. It can be observed that there were significant differences ($p < 0.05$) between samples for all sensors. The MQ3 (alcohol) presented the highest voltage values for all samples, followed by the MQ4 (CH_4). The Mercaptan sample was the highest in alcohol gas release for the low concentration, while acetic acid presented the lowest signal. Lactic acid and light-struck had the highest signal for MQ4 (CH_4) and MQ137 (ammonia). On the other hand, for the high concentration samples, Eugenol was the highest in voltage for MQ3, MQ4, MQ135, and MQ137, while Indole was the lowest for MQ3. Even though a beer sample with H_2S was evaluated, it did not have the highest signal for the MQ136 sensor (H_2S) in both concentrations.

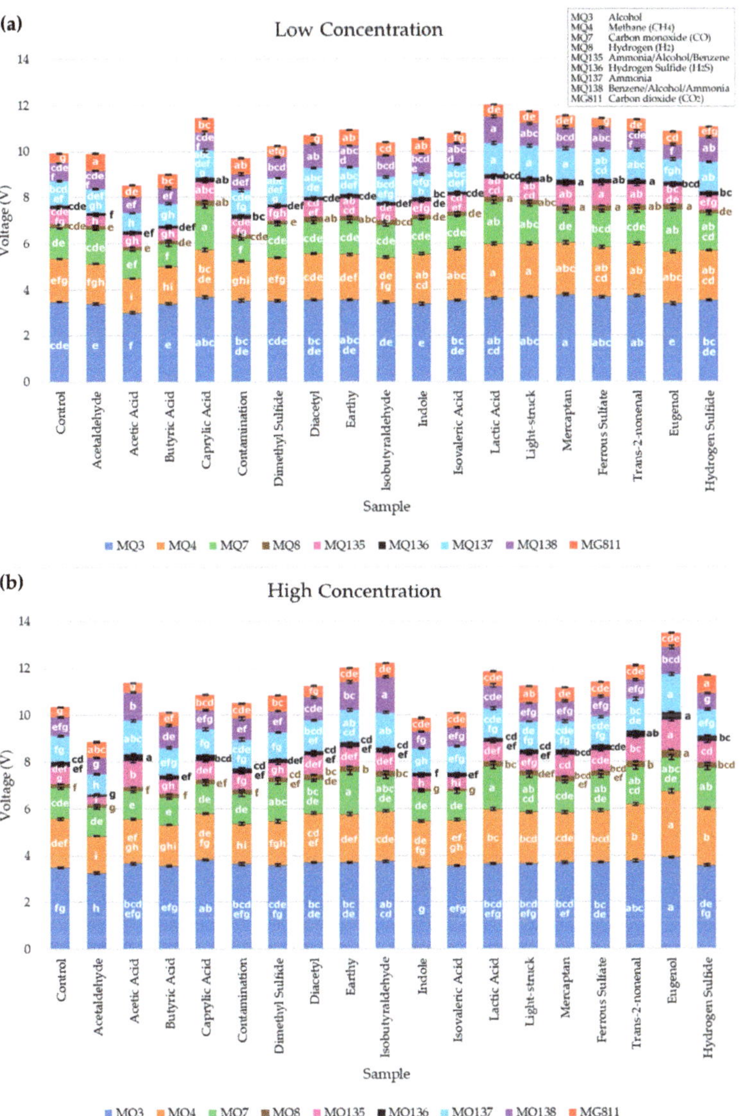

Figure 4. Stacked bar graphs for (**a**) low and (**b**) high concentrations showing the mean voltage values obtained from the electronic nose sensors. Error bars are based on standard error. Different letters depict significant differences ($p < 0.05$) between samples for each sensor based on the Tukey honest significant difference (HSD) post-hoc test.

Figure 5 shows the correlation matrix for the e-nose outputs and beer spiked with faults. It can be observed that MQ3 had positive and low but significant ($p < 0.05$) differences with caprylic acid (r = 0.16), mercaptan (r = 0.11), trans-2-nonenal and Eugenol (r = 0.14), and negative correlations with acetaldehyde (r = −0.23) and indole (r = −0.10). The MQ136 (H_2S) sensor had a positive correlation with trans-2-nonenla (r = 0.23), eugenol (r = 0.28), and hydrogen sulfide (r = 0.10), and a negative correlation with acetaldehyde (r = −0.28). Furthermore, Eugenol presented positive correlations with MQ4 (r = 0.35),

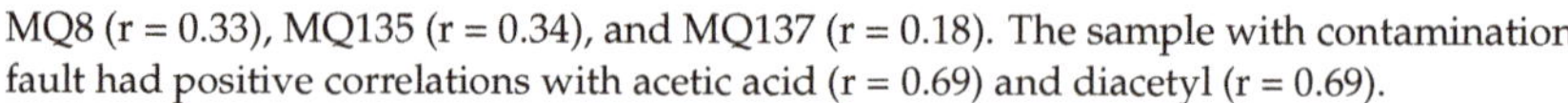
MQ8 (r = 0.33), MQ135 (r = 0.34), and MQ137 (r = 0.18). The sample with contamination fault had positive correlations with acetic acid (r = 0.69) and diacetyl (r = 0.69).

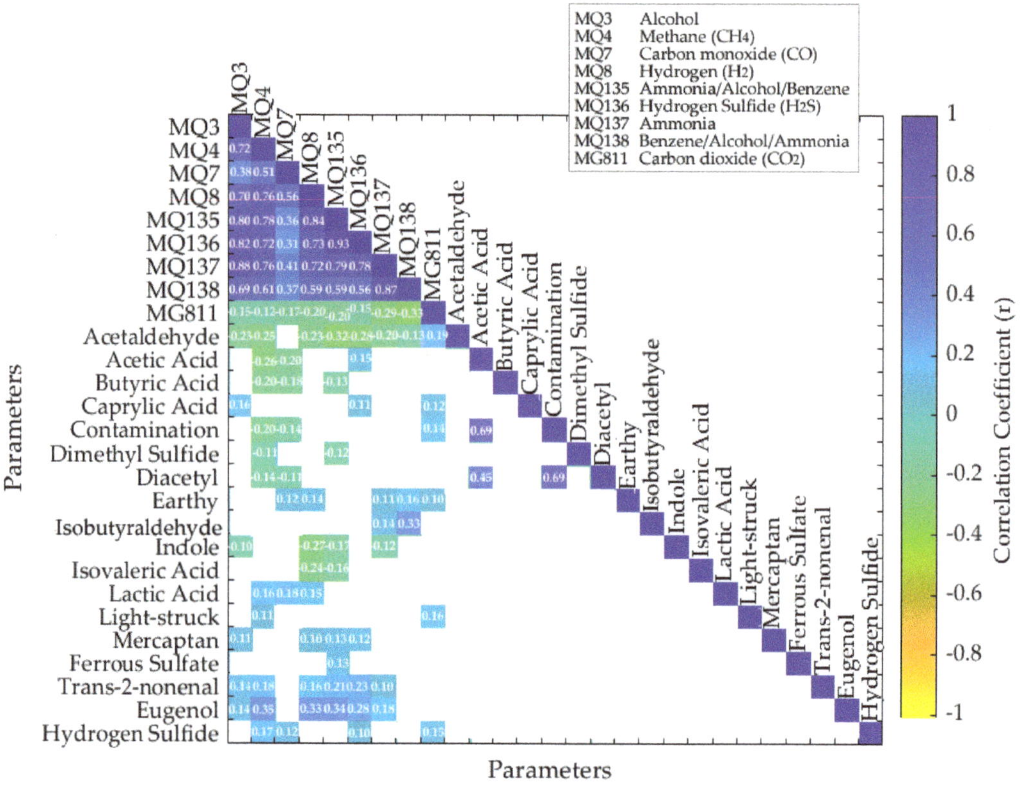

Figure 5. Matrix showing the significant ($p < 0.05$) correlations between the electronic nose sensors and the beer samples spiked with faults. Color bar represents the positive correlations on the blue side and negative correlations on the yellow side.

Figure 6 shows significant differences ($p < 0.05$) between samples for pH and alcohol content for low and high concentration treatments. It can be observed that at both low and high concentrations, the beer with eugenol had the highest pH (Low: 4.22; High: 4.33), while the sample with lactic acid presented the lowest pH (Low: 4.03; High: 3.90), which differed from the control (4.13). On the other hand, the sample with Eugenol and low and high concentration had the highest alcohol content (Low: 4.73; High: 4.77), while the light-stuck sample had the lowest alcohol content (Low and High: 4.71%), which was similar to the control (4.71%).

Table 3 shows that both Model 1 (NIR inputs) and Model 2 (e-nose inputs) had high overall accuracy (>95%) to classify the beer samples into control, low and high concentrations of faulty aromas. None of these models was under- or overfitted because their training MSE values were lower than the testing stage, which indicates they gave a high performance. Furthermore, in their receiver operating characteristic (ROC) curves (Figure 7), it can be seen that the three classifications in the two models had very high sensitivity (true positive rate; >0.94). It can be observed that the lowest sensitivity for Models 1 and 2 was the low concentration (0.94 and 0.95, respectively).

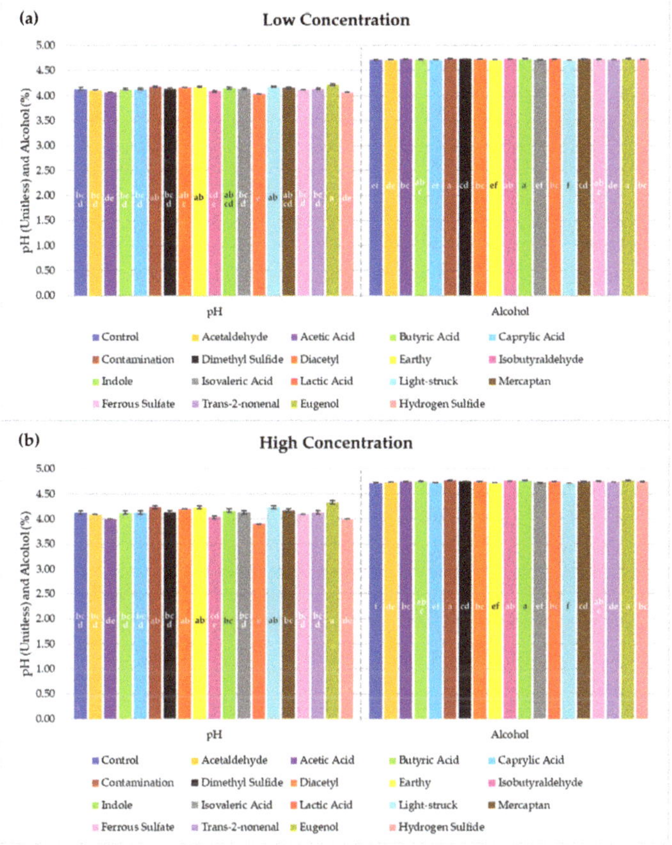

Figure 6. Bar graphs of the pH and alcohol mean values for (**a**) low and (**b**) high concentration treatment. Error bars are based on standard error. Different letters depict significant differences ($p < 0.05$) between samples for each sensor based on the Tukey honest significant difference (HSD) post-hoc test.

Table 3. Statistical results of the artificial neural network classification models developed using the near-infrared absorbance values Model 1 and electronic nose outputs (Model 2) as inputs to predict the concentration level. Abbreviations: MSE: means squared error.

Stage	Samples	Accuracy	Error	Performance (MSE)
		Model 1: Near-infrared inputs		
Training	239	100%	0.0%	<0.001
Testing	103	85.4%	14.6%	0.08
Overall	342	95.6%	4.4%	-
		Model 2: Electronic nose inputs		
Training	239	98.5%	1.5%	0.01
Testing	103	87.7%	12.3%	0.08
Overall	342	95.3%	4.7%	-

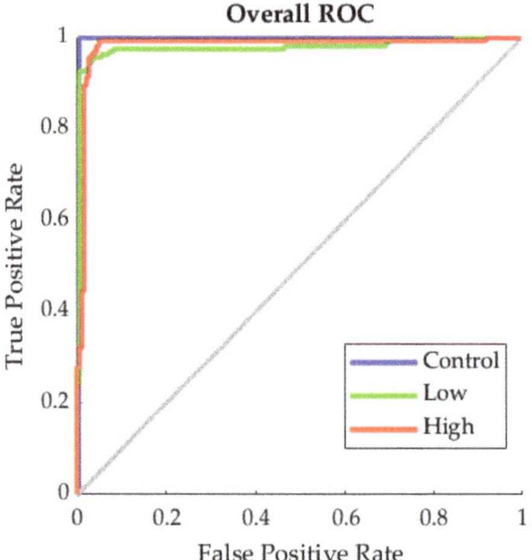

(a)

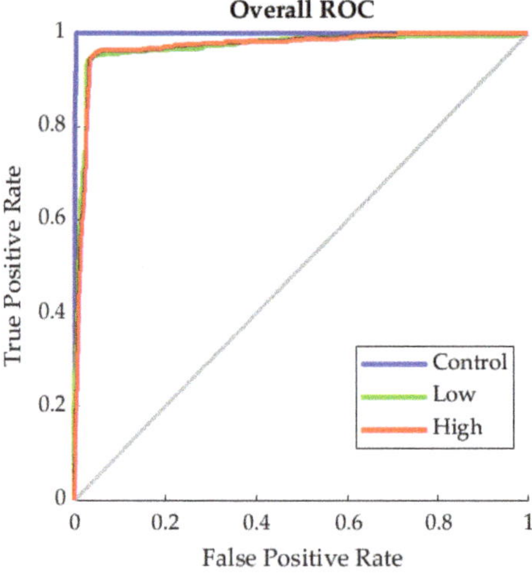

(b)

Figure 7. Receiver operating characteristic (ROC) curves of the models developed using (**a**) near-infrared (Model 1) and (**b**) electronic nose (Model 2) values as inputs to classify beers according to the concentration of faults present in them (control, low and high concentration).

Table 4 shows that Models 3 (low concentration) and 4 (high concentration) developed using the NIR absorbance values as inputs had high overall accuracy with 99% and 98%, respectively. None of the models presented any signs of under- or overfitting as the training performance (Models 3 and 4: MSE < 0.001) was lower than the testing (Model 3: MSE = 0.003; Model 4: MSE = 0.005). Figure 8 depicts the ROC curves for Models 3 and 4, in which it can be observed that all of the classification categories had high sensitivity (>0.89). In Model 3, acetic acid and H_2S were the lowest in sensitivity (0.89), while in Model 4, the lowest were samples earthy, light-struck, and H_2S (0.89).

Table 4. Statistical results of the artificial neural network classification models developed using the near-infrared absorbance values as inputs. Abbreviations: MSE: means squared error.

Stage	Samples	Accuracy	Error	Performance (MSE)
		Model 3: Near-infrared inputs—Low concentration		
Training	126	100%	0.0%	<0.001
Testing	54	96.3%	3.7%	0.003
Overall	180	98.9%	1.1%	-
		Model 4: Near-infrared inputs—High concentration		
Training	126	100%	0.0%	<0.001
Testing	54	94.4%	5.6%	0.005
Overall	180	98.3%	1.7%	-

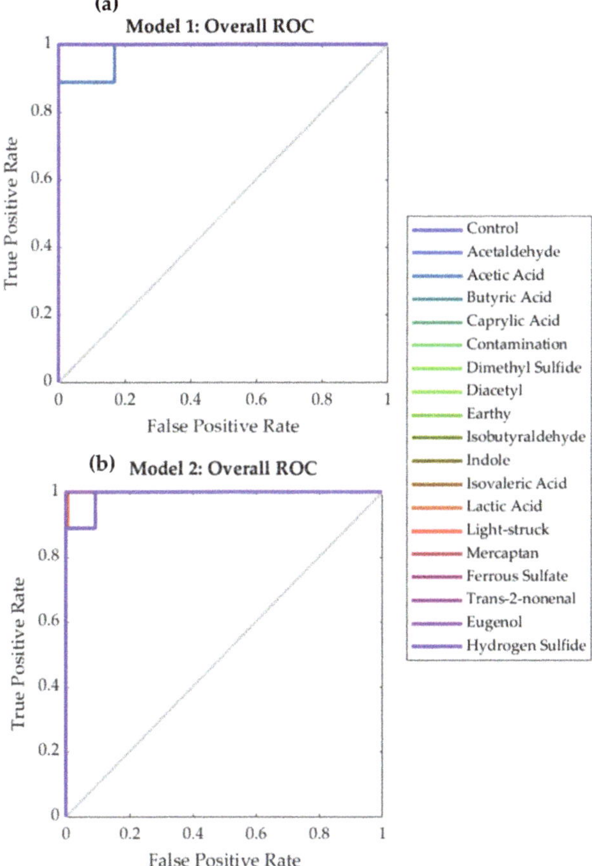

Figure 8. Receiver operating characteristic (ROC) curves of the (**a**) low (Model 3) and (**b**) high (Model 4) concentration samples to classify beers according to the faults present in them using the near-infrared absorbance values as inputs.

Table 5 shows that Models 5 (low concentration) and 6 (high concentration) developed using the e-nose outputs as inputs had high overall accuracy (97% and 96%, respectively).

In both, Models 5 and 6, the training performance (Models 5 and 6: MSE < 0.001) was lower than the testing (Model 5: MSE = 0.009; Model 6: MSE = 0.011). Based on the ROC curves (Figure 9), all classification categories had high sensitivity (>0.87), with caprylic acid being the lowest sensitivity in Model 5 (0.90) and diacetyl and ferrous sulfate in Model 6 (0.87).

Table 5. Statistical results of the artificial neural network classification models developed using the electronic nose outputs as inputs. Abbreviations: MSE: means squared error.

Stage	Samples	Accuracy	Error	Performance (MSE)
Model 5: Electronic nose inputs—Low concentration				
Training	420	99.8%	0.2%	<0.001
Testing	180	90.0%	10.0%	0.009
Overall	600	96.8%	3.2%	-
Model 6: Electronic nose inputs—High concentration				
Training	420	100%	0.0%	<0.001
Testing	180	87.2%	12.8%	0.011
Overall	600	96.2%	3.8%	-

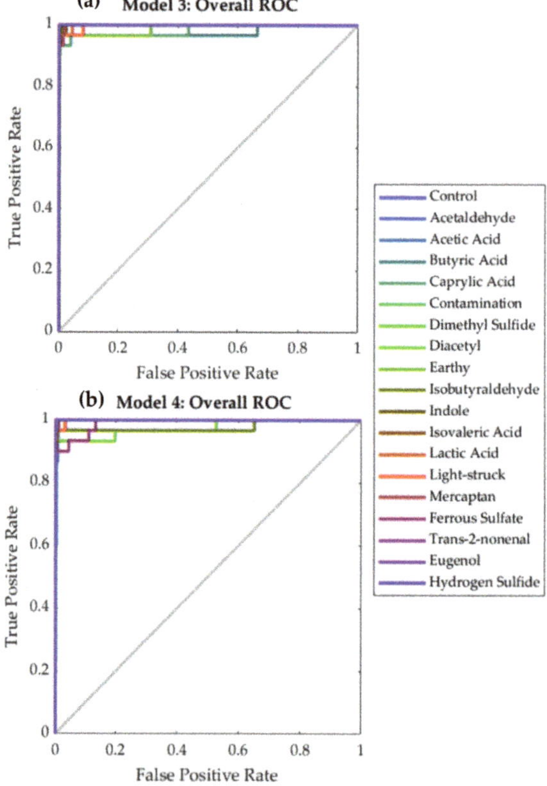

Figure 9. Receiver operating characteristic (ROC) curves for the (**a**) low (Model 5) and (**b**) high (Model 6) concentration samples to classify beers according to the faults present in them using the electronic nose outputs as inputs.

4. Discussion

4.1. Near-Infrared Spectroscopy (NIR)

The NIR range used in this study includes the aromatic overtones, which corresponded to 1596–2396 nm. This NIR range has been used in previous studies to characterize and model different volatile aromatic compounds in beers [19,25,32] and wine [33,34]. The major raw NIR peaks shown for all samples were in 1900–2000 nm and >2300 nm (Figure 2), which correspond to overtones of compounds such as carboxylic acid, pOH, water, amides, alcohol, proteins, and carbohydrates [35]. After a first derivative transformation (Figure 3), several other peaks and valleys were enhanced, such as other water overtones, thiols, and starch, all of which are found in beer [35]. The major differences in the latter case were observed between 1600 and 1700 nm, where aromatic compounds and alkyls can be found, at 1890 nm as the major peak, which corresponds to carboxylic acid, which may be present in beer samples in the form of different compounds such as acetic and lactic acids, and may derivate into esters, which are responsible for different aromas in beer [29,35], and between 1950 and 2150 nm, where proteins, amides, alcohol, and sucrose are found [35,36]. Interestingly, one off-aroma addition was significantly different compared to the rest (hydrogen sulfide) for low concentrations (Figure 3a) but not for high concentrations used (Figure 3b). The latter effect may be related to the initial interaction between the H_2S compound and other minerals in the beer, increasing the chemical fingerprint in the specific overtones, which decrease at higher H_2S concentrations [37]. These patterns and differences between the chemical fingerprint for samples with different fault additions are assessed using machine learning for discrimination purposes and potential identification and classification using ANN modeling techniques.

4.2. Low-Cost e-Nose and Beer Chemometrics

The low-cost e-nose used was preliminarily tested on different commercial beers for ML classification purposes and determination of aroma profiles compared to gas chromatography [29]. The same e-nose was successfully used to assess and quantify smoke taint compounds in wine [38]. The variation of the e-nose sensors response after the addition of fault compounds is significant for most of the sensors (Figure 4) and with differences between variations for low concentrations (Figure 4a) compared to high concentrations (Figure 4b) of faults, which helps to justify the classifications performed by the machine learning modeling techniques.

The latter is also supported by the general correlation matrix analysis (Figure 5) between all the fault compound additions and the different e-nose sensors. More negative and statistically significant correlations ($p < 0.05$) were found between acetaldehyde and most sensors except for MQ7 and with a positive correlation with MG811. The negative correlations between acetaldehyde and sensors sensitive to alcohol may be associated with the fact that acetaldehyde is often produced from the oxidation of ethanol [8,39]. The same trends were shown for sensor MQ4 (CH_4) and acetic acid; this negative correlation was found in breweries, where CH_4 production decreased acetic acid formation [40]. As expected, contamination was positively correlated with diacetyl, and butyric acid was formed when the latter two are mixed. Furthermore, Indole was negatively correlated to MQ3, MQ137, MQ8, and MQ135. The negative correlation between Indole and the sensors sensitive to alcohol may be since Indole is formed by coliform bacteria, which is eliminated with the presence of alcohol; therefore, at higher alcohol, lower indole production [8]. Furthermore, as expected, hydrogen sulfide was positively correlated with MQ136 (H_2S) sensor; however, the correlation, although positive, was weak due to the sensitivity (1–100 mg L^{-1}) of the sensor [29], which is higher than the concentrations used in the samples for this study (0.03 and 0.07 mg L^{-1}). The rest of the fault compounds were positively correlated at different levels with most of the sensors.

There are fewer variations concerning basic chemometrics, such as pH and alcohol levels, which can be explained through the interactions of fault compounds (Figure 6), especially at high concentrations (Figure 6b), which were more significant for the case of pH compared to alcohol concentration. As expected, an increase in alcohol and a decrease in pH can be observed in samples with compounds that have an alcohol group and acidic faults, respectively.

4.3. Supervised Machine Learning Classification Models and Deployment

Models 1 and 2 were developed to determine whether the beers are a control (no faults), low, or high concentration of faults. This will indicate which model should be used to further predict the specific fault present in the sample: Models 3 (low concentration) or 4 (high concentration) for NIR and Models 5 (low concentration) or 6 (high concentration) for e-nose.

All six models based on NIR and e-nose resulted in high accuracies (>95% for NIR and e-nose, Models 1 and 2; >98% for NIR Models 3 and 4, and >96% for e-nose, Models 5 and 6). These accuracies and performances are consistent with those presented in previous studies using NIR and e-nose for beer to assess aroma compounds and for the classification of commercial beers [22,29].

In terms of practicality, even though the e-nose models have lower performance than the NIR models, the low-cost and integrability nature of the e-nose makes it more flexible for deployment to be added to different brewing processes with automated data acquisition and interpretation through AI. The e-nose processed data can be readily available to brewers for decision-making using wireless data transmission through the Internet of Things (IoT) [41,42]. On the contrary, the commercial NIR instrument used in this study cannot be integrated with AI as it can only incorporate models based on partial least squares (PLS), which have the limitation of assessing regression levels of single compounds per model for manual and punctual measurements, also requiring a trained operator for the instrument usage, data acquisition, handling, and interpretation. Furthermore, additional software is required for PLS modeling and integration with the NIR at a considerable cost.

Another advantage of the e-nose and AI method proposed compared to NIR is the implementation and deployment costs using the beforementioned data transmission and cloud processing using AI since it is based on numeric data transmission from the different voltage readings of sensors. The NIR instrument can be cost-prohibitive for craft brewing companies compared to the cost of the e-nose hardware, which corresponds to 2.5% of the NIR instrument.

5. Conclusions

The comparative accuracies of ML models developed for NIR and e-nose make the latter method cost-effective, reliable, and easy to deploy for craft, medium, and big brewing companies. The latter also allows the implementation and deployment of this method with the option of replication to assess multiple batches simultaneously. Due to the portability of new versions of e-noses considering local data storage and analysis using local and inexpensive microprocessors (i.e., Jetson® from NVIDIA, Arduino® or Raspberry Pi ®), these could be added to robots such as the RoboBEER for quality traits and consumer perception assessment using AI. Further studies are required to model fault assessment on different beer styles with more complex aroma profiles, such as lambic and within different stages of the brewing process. The results presented here are from a preliminary study on the integration of low-cost sensor technology and AI, and the models developed can be enriched with further data making the most of the learning capabilities of the ANN models considered.

Author Contributions: Conceptualization, C.G.V. and S.F.; data curation, C.G.V. and S.F.; formal analysis, C.G.V.; funding acquisition, C.H.-B.; investigation, C.G.V. and S.F.; methodology, C.G.V. and S.F.; project administration, C.G.V. and S.F.; resources, C.G.V. and S.F.; software, C.G.V. and S.F.; validation, C.G.V. and S.F.; visualization, C.G.V. and S.F.; writing—original draft, C.G.V. and S.F.; writing—review and editing, C.G.V., S.F., and C.H.-B. All authors have read and agreed to the published version of the manuscript.

Funding: This research was funded by the Mexican Beer and Health Council (Consejo de Investigación sobre Salud y Cerveza de México).

Institutional Review Board Statement: Not applicable.

Informed Consent Statement: Not applicable.

Data Availability Statement: Data and intellectual property belong to The University of Melbourne; any sharing needs to be evaluated and approved by the university.

Acknowledgments: The authors would like to acknowledge Ranjith R. Unnithan, and Bryce Widdicombe from the School of Engineering, Department of Electrical and Electronic Engineering of The University of Melbourne for their collaboration in the electronic nose development.

Conflicts of Interest: The authors declare no conflict of interest.

References

1. Spedding, G.; Aiken, T. Sensory analysis as a tool for beer quality assessment with an emphasis on its use for microbial control in the brewery. In *Brewing Microbiology*; Elsevier: Amsterdam, The Netherlands, 2015; pp. 375–404.
2. Amerine, M.A.; Pangborn, R.M.; Roessler, E.B. *Principles of Sensory Evaluation of Food*; Elsevier: Amsterdam, The Netherlands, 2013.
3. Lawless, H.T. *Quantitative Sensory Analysis: Psychophysics, Models and Intelligent Design*; Wiley: Hoboken, NJ, USA, 2013.
4. Ristic, R.; Danner, L.; Johnson, T.; Meiselman, H.; Hoek, A.; Jiranek, V.; Bastian, S. Wine-related aromas for different seasons and occasions: Hedonic and emotional responses of wine consumers from Australia, UK and USA. *Food Qual. Prefer.* **2019**, *71*, 250–260. [CrossRef]
5. Ragazzo-Sanchez, J.; Chalier, P.; Chevalier-Lucia, D.; Calderon-Santoyo, M.; Ghommidh, C. Off-flavours detection in alcoholic beverages by electronic nose coupled to GC. *Sens. Actuators B Chem.* **2009**, *140*, 29–34. [CrossRef]
6. Díaz, A.; Ventura, F.; Galceran, M.T. Identification of 2, 3-butanedione (diacetyl) as the compound causing odor events at trace levels in the Llobregat River and Barcelona's treated water (Spain). *J. Chromatogr. A* **2004**, *1034*, 175–182. [CrossRef] [PubMed]
7. Miller, G.H. *Whisky Science: A Condensed Distillation*; Springer: Berlin/Heidelberg, Germany, 2019.
8. Barnes, T. The complete beer fault guide v. 1.4. January 2011. Available online: https://londonamateurbrewers.co.uk/wp-content/uploads/2015/05/Complete_Beer_Fault_Guide.pdf (accessed on 10 June 2021).
9. Cravero, M.C.; Bonello, F.; Pazo Alvarez, M.d.C.; Tsolakis, C.; Borsa, D. The sensory evaluation of 2, 4, 6-trichloroanisole in wines. *J. Inst. Brew.* **2015**, *121*, 411–417. [CrossRef]
10. Wray, E. Common faults in beer. In *The Craft Brewing Handbook*; Elsevier: Amsterdam, The Netherlands, 2020; pp. 217–246.
11. Mander, L.; Liu, H.-W. *Comprehensive Natural Products II: Chemistry and Biology*; Elsevier: Amsterdam, The Netherlands, 2010; Volume 1.
12. Pomeroy, R.D.; Cruse, H. Hydrogen sulfide odor threshold. *J. Am. Water Work. Assoc.* **1969**, *61*, 677. [CrossRef]
13. World Health Organization. *WHO Guidelines for Drinking Water Quality*, 2nd ed.; WHO: Geneva, Switzerland, 1996.
14. Viejo, C.G.; Fuentes, S.; Howell, K.; Torrico, D.; Dunshea, F.R. Robotics and computer vision techniques combined with non-invasive consumer biometrics to assess quality traits from beer foamability using machine learning: A potential for artificial intelligence applications. *Food Control* **2018**, *92*, 72–79. [CrossRef]
15. Viejo, C.G.; Fuentes, S. Beer Aroma and Quality Traits Assessment Using Artificial Intelligence. *Fermentation* **2020**, *6*, 56. [CrossRef]
16. Wilson, C.; Threapleton, L. Application of artificial intelligence for predicting beer flavours from chemical analysis. In Proceedings of the 29th European Brewery Convention Congress, Dublin, Ireland, 17–22 May 2003; pp. 17–22.
17. Geilings, B. Using Artificial Intelligence to Positively Impact the Beer Brewing Process. Bachelor's Thesis, Haaga-Helia University of Applied Sciences, Helsinki, Finland, 2021.
18. Viejo, C.G.; Fuentes, S.; Li, G.; Collmann, R.; Condé, B.; Torrico, D. Development of a robotic pourer constructed with ubiquitous materials, open hardware and sensors to assess beer foam quality using computer vision and pattern recognition algorithms: RoboBEER. *Food Res. Int.* **2016**, *89*, 504–513. [CrossRef]
19. Gonzalez Viejo, C.; Fuentes, S.; Torrico, D.; Howell, K.; Dunshea, F.R. Assessment of beer quality based on foamability and chemical composition using computer vision algorithms, near infrared spectroscopy and machine learning algorithms. *J. Sci. Food Agric.* **2018**, *98*, 618–627. [CrossRef]
20. Vann, L.; Layfield, J.B.; Sheppard, J.D. The application of near-infrared spectroscopy in beer fermentation for online monitoring of critical process parameters and their integration into a novel feedforward control strategy. *J. Inst. Brew.* **2017**, *123*, 347–360. [CrossRef]
21. Fox, G. The brewing industry and the opportunities for real-time quality analysis using infrared spectroscopy. *Appl. Sci.* **2020**, *10*, 616. [CrossRef]
22. Gonzalez Viejo, C.; Fuentes, S. Low-Cost Methods to Assess Beer Quality Using Artificial Intelligence Involving Robotics, an Electronic Nose, and Machine Learning. *Fermentation* **2020**, *6*, 104. [CrossRef]
23. Voss, H.G.J.; Mendes Júnior, J.J.A.; Farinelli, M.E.; Stevan, S.L. A Prototype to Detect the Alcohol Content of Beers Based on an Electronic Nose. *Sensors* **2019**, *19*, 2646. [CrossRef] [PubMed]

24. Ghasemi-Varnamkhasti, M.; Mohtasebi, S.; Rodriguez-Mendez, M.; Lozano, J.; Razavi, S.; Ahmadi, H. Potential application of electronic nose technology in brewery. *Trends Food Sci. Technol.* **2011**, *22*, 165–174. [CrossRef]
25. Viejo, C.G.; Torrico, D.D.; Dunshea, F.R.; Fuentes, S. Development of Artificial Neural Network Models to Assess Beer Acceptability Based on Sensory Properties Using a Robotic Pourer: A Comparative Model Approach to Achieve an Artificial Intelligence System. *Beverages* **2019**, *5*, 33. [CrossRef]
26. Santos, J.P.; Lozano, J. Real time detection of beer defects with a hand held electronic nose. In Proceedings of the 2015 10th Spanish Conference on Electron Devices (CDE), Madrid, Spain, 11–13 February 2015; pp. 1–4.
27. Sanaeifar, A.; ZakiDizaji, H.; Jafari, A.; de la Guardia, M. Early detection of contamination and defect in foodstuffs by electronic nose: A review. *TrAC Trends Anal. Chem.* **2017**, *97*, 257–271. [CrossRef]
28. Pornpanomchai, C.; Suthamsmai, N. Beer classification by electronic nose. In Proceedings of the 2008 International Conference on Wavelet Analysis and Pattern Recognition, Hong Kong, China, 30–31 August 2008; pp. 333–338.
29. Gonzalez Viejo, C.; Fuentes, S.; Godbole, A.; Widdicombe, B.; Unnithan, R.R. Development of a low-cost e-nose to assess aroma profiles: An artificial intelligence application to assess beer quality. *Sens. Actuators B Chem.* **2020**, *308*, 127688. [CrossRef]
30. Gonzalez Viejo, C.; Tongson, E.; Fuentes, S. Integrating a Low-Cost Electronic Nose and Machine Learning Modelling to Assess Coffee Aroma Profile and Intensity. *Sensors* **2021**, *21*, 2016. [CrossRef] [PubMed]
31. Perez, C. *Big Data and Deep Learning. Examples with Matlab*; Lulu Press, Inc: Morrisville, NC, USA, 2020.
32. Gonzalez Viejo, C.; Torrico, D.D.; Dunshea, F.R.; Fuentes, S. Emerging Technologies Based on Artificial Intelligence to Assess the Quality and Consumer Preference of Beverages. *Beverages* **2019**, *5*, 62. [CrossRef]
33. Smyth, H.; Cozzolino, D.; Cynkar, W.; Dambergs, R.; Sefton, M.; Gishen, M. Near infrared spectroscopy as a rapid tool to measure volatile aroma compounds in Riesling wine: Possibilities and limits. *Anal. Bioanal. Chem.* **2008**, *390*, 1911–1916. [CrossRef]
34. Ferrer-Gallego, R.; Hernández-Hierro, J.M.; Rivas-Gonzalo, J.C.; Escribano-Bailón, M.T. Evaluation of sensory parameters of grapes using near infrared spectroscopy. *J. Food Eng.* **2013**, *118*, 333–339. [CrossRef]
35. Burns, D.A.; Ciurczak, E.W. *Handbook of Near-Infrared Analysis*; CRC press: Boca Raton, FL, USA, 2007.
36. Osborne, B.G.; Fearn, T.; Hindle, P.H. *Practical NIR Spectroscopy with Applications in Food and Beverage Analysis*; Longman Scientific and Technical: London, UK, 1993.
37. Brenner, M.; Khan, A.; Bernstein, L. Formation and Retention of Free H2S and of Free Volatile Thiols during Beer Fermentation. In Proceedings of the Annual Meeting-American Society of Brewing Chemists, St. Paul, MN, USA, 1 December 1975; pp. 82–93.
38. Fuentes, S.; Summerson, V.; Gonzalez Viejo, C.; Tongson, E.; Lipovetzky, N.; Wilkinson, K.L.; Szeto, C.; Unnithan, R.R. Assessment of Smoke Contamination in Grapevine Berries and Taint in Wines Due to Bushfires Using a Low-Cost E-Nose and an Artificial Intelligence Approach. *Sensors* **2020**, *20*, 5108. [CrossRef] [PubMed]
39. Pesselman, R.L.; Meshbesher, T.M.; Floyd, S.D.; Langer, S.H. Electrogenerative oxidation of ethanol to acetaldehyde. *Chem. Eng. Commun.* **1985**, *38*, 265–273. [CrossRef]
40. Gomes, M.M.; Sakamoto, I.K.; Rabelo, C.A.B.S.; Silva, E.L.; Varesche, M.B.A. Statistical optimization of methane production from brewery spent grain: Interaction effects of temperature and substrate concentration. *J. Environ. Manag.* **2021**, *288*, 112363. [CrossRef] [PubMed]
41. Violino, S.; Figorilli, S.; Costa, C.; Pallottino, F. Internet of beer: A review on smart technologies from mash to pint. *Foods* **2020**, *9*, 950. [CrossRef]
42. Betts, B. Brewing up: A technology revolution. *Eng. Technol.* **2016**, *11*, 54–57. [CrossRef]

Article

Digital Smoke Taint Detection in Pinot Grigio Wines Using an E-Nose and Machine Learning Algorithms Following Treatment with Activated Carbon and a Cleaving Enzyme

Vasiliki Summerson [1], Claudia Gonzalez Viejo [1], Damir D. Torrico [2], Alexis Pang [1] and Sigfredo Fuentes [1,*]

[1] Digital Agriculture, Food and Wine Group, School of Agriculture and Food, Faculty of Veterinary and Agricultural Sciences, The University of Melbourne, Building 142, Parkville, VIC 3010, Australia; vsummerson@student.unimelb.edu.au (V.S.); cgonzalez2@unimelb.edu.au (C.G.V.); alexis.pang@unimelb.edu.au (A.P.)

[2] Department of Wine, Food and Molecular Biosciences, Faculty of Agriculture and Life Sciences, Lincoln University, Lincoln 7647, Canterbury, New Zealand; Damir.Torrico@lincoln.ac.nz

* Correspondence: sfuentes@unimelb.edu.au

Abstract: The incidence and intensity of bushfires is increasing due to climate change, resulting in a greater risk of smoke taint development in wine. In this study, smoke-tainted and non-smoke-tainted wines were subjected to treatments using activated carbon with/without the addition of a cleaving enzyme treatment to hydrolyze glycoconjugates. Chemical measurements and volatile aroma compounds were assessed for each treatment, with the two smoke taint amelioration treatments exhibiting lower mean values for volatile aroma compounds exhibiting positive 'fruit' aromas. Furthermore, a low-cost electronic nose (e-nose) was used to assess the wines. A machine learning model based on artificial neural networks (ANN) was developed using the e-nose outputs from the unsmoked control wine, unsmoked wine with activated carbon treatment, unsmoked wine with a cleaving enzyme plus activated carbon treatment, and smoke-tainted control wine samples as inputs to classify the wines according to the smoke taint amelioration treatment. The model displayed a high overall accuracy of 98% in classifying the e-nose readings, illustrating it may be a rapid, cost-effective tool for winemakers to assess the effectiveness of smoke taint amelioration treatment by activated carbon with/without the use of a cleaving enzyme. Furthermore, the use of a cleaving enzyme coupled with activated carbon was found to be effective in ameliorating smoke taint in wine and may help delay the resurgence of smoke aromas in wine following the aging and hydrolysis of glycoconjugates.

Keywords: climate change; artificial neural networks; volatile phenols; glycoconjugates; bushfires

Citation: Summerson, V.; Gonzalez Viejo, C.; Torrico, D.D.; Pang, A.; Fuentes, S. Digital Smoke Taint Detection in Pinot Grigio Wines Using an E-Nose and Machine Learning Algorithms Following Treatment with Activated Carbon and a Cleaving Enzyme. *Fermentation* **2021**, *7*, 119. https://doi.org/10.3390/fermentation7030119

Academic Editor: Antonio Morata

Received: 29 June 2021
Accepted: 14 July 2021
Published: 16 July 2021

Publisher's Note: MDPI stays neutral with regard to jurisdictional claims in published maps and institutional affiliations.

1. Introduction

The exposure of grapevines to smoke during the critical period between veraison and harvest can result in the uptake and accumulation of volatile phenols and their glycoconjugates in grape berries, which can negatively affect the composition and sensory properties of wines [1–3]. Once absorbed, volatile phenols from smoke are rapidly glycosylated and stored in the skin and pulp of grape berries [4,5]. During winemaking, many of these glycoconjugates are hydrolyzed back into their free active forms where they can impart their unpleasant 'smoky' aromas, with a large proportion of the glycoconjugate pool remaining in the final wine [5–8]. Furthermore, the in-mouth hydrolysis of volatile phenol glycoconjugates has also been shown to occur, which may further impact the flavor and aftertaste of smoke-tainted wines [9,10].

The degree of smoke taint in the final wine depends on several factors, including the timing and duration of smoke exposure and winemaking practices such as yeast selection and skin contact time during fermentation [3,6,11,12]. Some techniques have been investigated to mitigate the effects of grapevine smoke exposure in-field and to find methods for ameliorating smoke taint in the final wine. The research by van der Hulst et al. [2] found that a foliar application of kaolin resulted in significantly lower glycoconjugate levels in Merlot grapes following smoke exposure compared to the control; however, no significant differences were found for Sauvignon Blanc and Chardonnay grapes. Further research by Favell et al. [13] found that applying an artificial grape cuticle one week prior to the application of smoke can significantly impede the uptake of smoke-derived volatile phenols. In other research, Szeto et al. [12] found that in-canopy misting partially reduced the uptake of smoke-derived volatile phenols by grapes; however, it was not significant enough to readily influence the concentration of volatile phenols and the negative sensory characteristics associated with smoke taint in the final wine. For the amelioration of smoke taint in wine, Fudge et al. [14] found that reverse osmosis and solid-phase adsorption effectively reduced the concentrations of smoke-derived volatile phenols. Hence, the negative sensory attributes of smoke taint. Further research also found that the treatment of smoke-affected wines with activated carbon was also effective in reducing volatile phenol concentrations and the intensity of smoke-related sensory attributes [15]. However, both activated carbon and reverse osmosis treatments were ineffective at reducing the levels of volatile phenol glycoconjugates. Consequently, over time, the concentration of free volatile phenols and, hence, smoke-related sensory characteristics may increase due to the gradual hydrolysis of glycoconjugates [14,15]. Therefore, to effectively ameliorate smoke taint in wine, it is necessary to reduce both free volatile phenols and their bound glycoconjugates to prevent the gradual return of negative 'smoky' characteristics.

Aroma is one of the most important quality attributes for wine. While thousands of volatile compounds have been identified, which contribute to the complexity and varietal character of wine, in the case of smoke taint, they can attribute undesirable characteristics such as 'smoky', 'ash' and 'medicinal' aromas and flavors [8,14,16,17]. To assess the levels of the smoke compounds (both volatile phenols and their glycoconjugates), growers and winemakers are required to send samples of grapes and/or wines to commercial laboratories or conduct mini harvests to perform a sensory analysis. However, the high cost associated with laboratory testing may prevent this form of analysis for many producers, and a sensory analysis is time-consuming and may not allow for timely actions within the time constraints of a vintage [4,18]. There is, therefore, a need for a rapid and affordable alternative method for assessing smoke contamination.

Electronic noses (e-noses) have been developed for olfactory analyses for use in many industries, such as in agriculture and forestry [19] and medical diagnostics [20,21], as well as in the food industry for various applications, including quality control and the assessment of food authentication and adulteration, as well as freshness and shelf-life prediction [22–26]. In most cases, e-noses are comprised of an array of gas sensors with high sensitivity to detect volatile compounds coupled with a data processing unit and pattern recognition methods for identifying aroma profiles [18,24,25]. The portability, ease of use, and nondestructive nature of e-noses have increased interest in their use [23,24], particularly for smoke taint analyses in wine. The research by Antolini et al. [27] found that an e-nose was effective at discriminating between different smoke-treated wines, and it could serve as a useful tool for the early detection of smoke taint before it is perceived by a sensory analysis. Furthermore, the research by Fuentes et al. [18] for the assessment of smoke-tainted grape berries and wine used the readings from a low-cost e-nose as inputs for machine learning algorithms to create different artificial neural network (ANN) models with high accuracy to (1) classify wines according to different smoke and/or misting treatments, (2) predict the levels of 20 glycoconjugates and 10 volatile phenols in grapes at one hour post-smoke exposure and at harvest, as well as in the final wine, and (3) to predict the intensity of 12 wine descriptors based on a consumer sensory evaluation.

This study investigated the effectiveness of enzyme preparation with glycosidase activity in cleaving volatile phenol glycoconjugates prior to treatment with activated carbon for a more effective smoke taint amelioration. In addition to this, the use of a low-cost e-nose to distinguish smoke-tainted and non-smoke-tainted wines was also assessed.

Measurements from the e-nose were used as inputs for machine learning modeling to classify wine samples according to the type of smoke taint amelioration treatment applied (activated carbon with/without enzyme treatment) and smoke taint status (smoke-tainted or non-smoke-tainted). This may provide winemakers with an alternative method for assessing the effectiveness of smoke taint amelioration treatment by activated carbon with/without the use of a cleaving enzyme. Furthermore, the use of a glycosidase enzyme treatment before the addition of activated carbon may offer an improved method for ameliorating smoke taint in wine.

2. Materials and Methods

2.1. Wine Samples and Smoke Taint Amelioration Treatments

Commercial Pinot Grigio wines supplied by a winery were used in this study. The smoke-tainted (ST) wines were produced from grapes harvested from a vineyard located in the King Valley, Victoria, Australia exposed to moderate levels of smoke during the 2019/2020 harvest period. The non-smoke-tainted (NS) wines were produced from grapes harvested from a vineyard located in the Murray Valley, Victoria, Australia. From each type of wine, 1.5-L samples were taken and used for two different smoke taint amelioration treatments, as well as for a control treatment (CO, i.e., no treatments applied). The first treatment involved the use of activated carbon (Smartvin® FPS, Enologica Vason, Verona, Italy) (activated carbon treatment; AC) applied at a concentration of 2 g L^{-1} and left for 48 h. The second treatment involved the use of an enzyme preparation (ZIMAROM®, Enologica Vason) to cleave volatile phenol glycoconjugates at a concentration prior to the addition of activated carbon. The enzyme preparation was applied at a rate of 3 g hL^{-1} and left to work for four days; after which, the wine underwent treatment with activated carbon, as described above (carbon enzyme treatment; CE). Following treatment with activated carbon, bentonite was added at a rate of 150 g hL^{-1} to clarify the wines (Flottobent®, Enologica Vason). The bentonite mixture was first allowed to swell in water (ratio of bentonite to water was 1:15) and left to settle for 48 h; after which, the wines were racked and, because of the small volumes of the samples, were filtered using a size four filter paper designed for filtering coffee (E. H. Harris & Co. Pty. Ltd., Kingsgrove, NSW, Australia). During the entire process, wines were stored in a temperature-controlled environment between 20 and 23 °C.

2.2. Chemical Measurements

A sample of 200 mL of each replicate of each treatment was measured for pH using a Fisherbrand Accumet® AB15 pH meter (Fisher Scientific, Hampton, NH, USA) that was calibrated with a buffer solution at pH 7.0. Total dissolved solids (TDS) and electrical conductance (EC) were also assessed using a Yuelong YL-TDS2-A digital water quality tester (Zhengzhou Yuelong Electronic Technology Co., Ltd., Zhengzhou, China). Furthermore, the total soluble solids (TSS) were measured in °Brix (Brix) using a portable Alla France REFBX010 optical refractometer (Alla France Sarl, Chemillé-Melay, France) that was rinsed with distilled water and dried between each measurement. Lastly, the alcohol content of each sample was determined using 20 mL of sample injected into an Alcolyzer Wine M alcohol meter (Anton Paar GmbH, Graz, Austria) that was set to use the wine extension method located in the equipment settings. All measurements were performed in triplicates and at room temperatures between 20 and 23 °C.

2.3. Electronic Nose

A low-cost, portable e-nose comprised of nine different sensors that are sensitive to different gases (Table 1) was used to assess the wine samples in triplicates. The e-nose was developed by the Digital Agriculture, Food and Wine (DAFW) Group from the Faculty of Veterinary and Agricultural Sciences (FVAS) of the University of Melbourne (UoM), and the details have been described previously by Gonzalez Viejo et al. [28]. This e-nose has previously demonstrated great success for smoke taint assessment and the prediction of levels of smoke-derived volatile phenols and their glycoconjugates in Cabernet Sauvignon wines, as illustrated by Fuentes et al. [18]. Measurements were performed by pouring 200 mL of wine sample into a 500-mL beaker with the e-nose placed on top for approximately 3 min to collect the gas readings. The e-nose was calibrated for 20–30 s between samples to prevent any carryover effects and the signal readings monitored to ensure they reached baseline values prior to testing the next sample. Readings were monitored prior to placing the e-nose over the sample, during measurement and after the sample was removed to ensure stability and accuracy of the readings.

Table 1. Sensors integrated in the electronic nose and the gases to which they are sensitive.

Sensor *	Gases
MQ3	Ethanol
MQ4	Methane
MQ7	Carbon monoxide
MQ8	Hydrogen
MQ135	Ammonia, alcohol, and benzene
MQ136	Hydrogen sulfide
MQ137	Ammonia
MQ138	Benzene, alcohol, and ammonia
MG811	Carbon dioxide

* All sensors were manufactured by Henan Hanwei Electronics Co., Ltd., Zhengzhou, China.

Data acquisition was performed using a customized code written in MATLAB® R2020a (Mathworks, Inc., Natick, MA, USA) to identify the stable e-nose signals from the moment the e-nose was placed on the beaker containing the wine sample until prior to its removal. As described by Gonzalez Viejo et al. [29], the e-nose data was automatically divided into ten subdivisions to extract the average values per sensor, which were then used as inputs for the machine learning modeling.

2.4. GC-MS Analysis of Volatile Aroma Compounds

A gas-chromatograph with a mass-selective detector 5977B (GC-MSD; Agilent Technologies, Inc., Santa Clara, CA, USA) with an HP-5MS column (length 30 m, inner diameter 0.25 mm and film 0.25 μ; Agilent Technologies, Inc.) with helium as the carrier gas (flow rate of 1 mL min^{-1}) and an integrated autosampler system PAL3 (CTC Analytics AG, Zwingen, Switzerland) was used to evaluate the aroma compounds present in each of the wine samples (done in triplicates). A total of 5 mL of each wine sample was placed into a 20-mL vial and assessed using the headspace method with a solid-phase microextraction (SPME) divinylbenzene–carboxen–polydimethylsiloxane (DVB–CAR–PDMS) 1.1 mm grey fiber (Agilent Technologies, Inc.). A blank was used at the start to ensure no carryover effects from any previous analyses. Further details about the method used were described by Gonzalez Viejo et al. [30]. Furthermore, the National Institute of Standards and Technology (NIST; National Institute of Standards and Technology, Gaithersburg, MD, USA) library was used to identify the compounds observed, and only compounds with greater than 70% certainty were used.

2.5. Statistical Analysis and Machine Learning Modeling

Results from the chemical measurements and the relative peak areas of the volatile compounds identified by GC-MS were analyzed by analysis of variance (ANOVA) using Minitab® version 19.2020.1 (Minitab Inc., State College, PA, USA), with mean comparisons performed using Fisher's least significant difference (LSD) *post hoc* test at $\alpha = 0.05$ to assess if there were significant differences among the wine samples. A principal components analysis (PCA) was also conducted for the chemical measurements and volatile compounds using a customized code written in MATLAB® R2020a. In addition to this, a matrix was also developed using MATLAB® R2020a to assess the significant correlations ($p < 0.05$) between the chemical measurements, volatile aroma compounds and e-nose readings.

The ten mean values of each of the e-nose outputs for the NSCO, NSAC, NSCE and STCO were used as inputs for machine learning modeling based on artificial neural networks (ANN) to create a pattern recognition model that classifies the wine samples according to the smoke taint status (i) non-smoked without amelioration treatment (NSCO), (ii) non-smoked with carbon treatment (NSAC), (iii) non-smoked with carbon and enzyme treatment and (iv) smoked (SMCO). The model was developed using a customized code written in MATLAB® R2020a that tested 17 different training algorithms, with the optimal selected based on the highest accuracy and performance, as described by Gonzalez Viejo et al. [31] (Figure 1). In this instance, the Bayesian Regularization algorithm was found to be the best algorithm, and further training was performed to develop a more accurate ANN model with no signs of under- or overfitting. The inputs were divided randomly, with 60% used for the training stage and 40% for testing, with performance tested based on the mean squared error (MSE). A total of five neurons were used following a trimming test with 3, 5, 7 and 10 neurons (data not shown). The ANN model was then used to classify the remaining wine samples (smoked treatments: STAC and STCE) to assess the effectiveness of the smoke taint amelioration treatments.

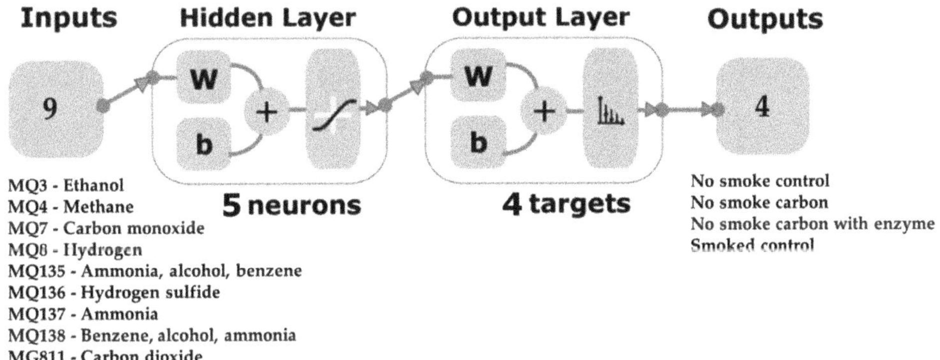

Figure 1. Two-layer feed-forward network with five hidden neurons and sigmoid function to classify wine samples according to the type of smoke taint amelioration treatment applied and smoke taint status (smoked or unsmoked). Abbreviations: w: weights; b: bias.

3. Results

3.1. Chemical Measurements

The results for the chemical measurements are shown in Table 2. Significant differences ($p < 0.05$) were found amongst the different wine treatments for all the parameters, with the STAC and STCE treatments exhibiting the highest mean values for TDS (802.50 and 832.00 ppm, respectively) and EC (1706.80 and 1769.80 µs cm^{-1}), with the NSCE (665.70 ppm and 1415.70 µs cm^{-1}) and STCO (670.30 ppm and 1425.50 µs cm^{-1}) treatments exhibiting the lowest values. The NSCO treatment had the highest mean °Brix and alcohol content (6.00 Brix and 12.18%), while the STCE treatment had the lowest °Brix (5.09 Brix) and the STAC (9.64%) and STCE (9.41%) treatments had the lowest alcohol contents. Lastly, the NSCE treatment had the highest pH (3.87), while the STAC, STCE and STCO treatments had the lowest (3.48, 3.48 and 3.43, respectively).

Table 2. Results from the chemical measurements.

Sample	TDS (ppm)		EC (μs cm^{-1})		°Brix		pH		Alcohol (%)	
	Mean	SE	Mean	SE	Mean	SE	Mean	SE	Mean	SE
NSAC	700.00 ^{bc}	±24.80	1489.00 ^{bc}	±52.70	5.23 ^{c}	±0.03	3.73 ^{b}	±0.03	10.70 ^{c}	±0.04
NSCE	665.70 ^{c}	±16.80	1415.70 ^{c}	±35.60	5.80 ^{b}	±0.00	3.87 ^{a}	±0.07	11.34 ^{b}	±0.01
NSCO	727.00 ^{b}	±8.08	1546.30 ^{b}	±17.40	6.00 ^{a}	±0.00	3.63 ^{b}	±0.03	12.18 ^{a}	±0.02
STAC	802.50 ^{a}	±1.43	1706.80 ^{a}	±2.99	5.15 ^{cd}	±0.03	3.48 ^{c}	±0.03	9.64 ^{d}	±0.10
STCE	832.00 ^{a}	±6.43	1769.80 ^{a}	±13.70	5.09 ^{d}	±0.05	3.48 ^{c}	±0.02	9.41 ^{d}	±0.13
STCO	670.30 ^{c}	±16.50	1425.50 ^{c}	±35.20	5.90 ^{ab}	±0.03	3.43 ^{c}	±0.02	10.76 ^{c}	±0.08

Abbreviations: NSAC = non-smoke-tainted wine with activated carbon treatment, NSCE = non-smoke-tainted wine with enzyme and activated carbon treatment, NSCO = control non-smoke-tainted wine, STAC = smoke-tainted wine with activated carbon treatment, STCE = smoke-tainted wine with enzyme and activated carbon treatment, STCO = control smoke-tainted wine and SE = standard error. Means followed by different letters are significantly different based on Fisher's least significant difference (LSD) post hoc test ($\alpha = 0.05$).

3.2. GC-MS Analysis

Table 3 shows the mean values for the peak areas of the volatile aroma compounds and their standard deviations with their aroma descriptions. Significant differences ($p < 0.05$) were found between the wine samples for all the aroma compounds. Both CO treatments had the highest mean levels of aroma compounds within their (NS or ST) groups, with the NSCO treatment having the highest mean values for 1-butanol, 3-methyl- acetate, butanedioic acid, hexanoic acid, ethyl ester and phenylethyl alcohol. These aroma compounds are associated primarily with positive 'fruit' aromas, such as banana, pear, apple, pineapple, rose and honey aromas (Table 3). The STCO sample illustrated high levels of dodecanoic acid, ethyl ester, which is associated with sweet, waxy, soapy and floral aromas, as well as decanoic acid, ethyl ester and octanoic acid, ethyl ester, which are associated with aromas such as apple, sweet, waxy, apricot and banana. The AC and CE treatments appeared to have the lowest mean levels of aroma compounds, with the STAC and STCE treatments having the lowest values for 1-butanol, 3-methyl- acetate and hexanoic acid, ethyl ester.

Table 3. Mean values of the peak areas for the volatile aroma compounds detected from the GC-MS analysis (top) and their standard error (bottom), as well as their aroma descriptions and retention times. All values are in scientific notation 10^5.

Volatile Aromatic Compound	Odor Description	RT (min)	NSAC	NSCE	NSCO	STAC	STCE	STCO
1-Butanol, 3-methyl-, acetate	Banana, pear, alcohol	13.67	3.97 ^{c} ±0.03	4.03 ^{c} ±0.01	23.88 ^{a} ±0.20	1.81 ^{d} ±0.63	2.51 ^{cd} ±0.62	18.04^{b} ±0.42
Butanedioic acid, diethyl ester	Fruity, grape, wine [32,33]	19.18	1.71 ^{b} ±0.09	1.37 ^{b} ±0.02	4.74 ^{a} ±2.65	0 0	0.30 ^{b} ±0.10	0 0
Decanoic acid, ethyl ester	Apple, grape, sweet, brandy, waxy [32–35]	23.02	75.95 ^{b} ±35.12	10.33 ^{b} ±2.70	119.21 ^{b} ±119.21	35.68 ^{b} ±18.30	1.74 ^{b} ±0.24	326.15 ^{a} ±68.10
Dodecanoic acid, ethyl ester	floral, waxy, soap [32,35]	26.33	7.61 ^{b} ±2.71	2.01 ^{b} ±0.31	49.33 ^{a} ±12.87	8.36 ^{b} ±2.56	1.96 ^{b} ±0.18	63.87 ^{a} ±6.17
Hexanoic acid, ethyl ester	Sweet, fruity, wine [17,32–34]	16.4	14.90 ^{c} ±0.08	12.21 ^{c} ±0.12	90.43 ^{a} ±0.78	7.74 ^{d} ±0.57	7.35 ^{d} ±1.00	83.60 ^{b} ±0.81
Octanoic acid, ethyl ester	Fruity, banana, sweet, apple, pineapple [17,34,35]	19.72	15.88 ^{b} ±1.87	9.47 ^{b} ±0.36	347.84 ^{a} ±32.08	13.08^{b} ±6.11	6.35 ^{b} ±0.89	361.81 ^{a} ±26.41
Phenylethyl alcohol	Rose, honey, floral [33,35]	18.93	3.11 ^{c} ±0.53	1.25 ^{cd} ±0.63	16.27 ^{a} ±1.88	0.25 ^{d} ±0.25	0 0	6.21^{b} ±0.19

Abbreviations: NSAC: non-smoke-tainted wine with activated carbon treatment, NSCE = non-smoke-tainted wine with enzyme and activated carbon treatment, NSCO = control non-smoke-tainted wine, STAC = smoke-tainted wine with activated carbon treatment, STCE = smoke-tainted wine with enzyme and activated carbon treatment, STCO = control smoke-tainted wine and RT = retention time. Means followed by different letters are significantly different based on Fisher's least significant difference (LSD) post hoc test.

3.3. Electronic Nose

Figure 2 shows the mean values and standard errors for each gas sensor that makes up the e-nose for each sample. Again, significant differences ($p < 0.05$) were observed between wine samples for each of the gas sensors, indicating that the e-nose is able to differentiate the samples and, hence, smoke taint amelioration treatments. The highest readings for all wine samples were seen for ethanol gas release (sensor MQ3), with the NSAC treatment demonstrating the highest mean value (4.42 V), while STCE exhibited the lowest (4.29 V). Hydrogen sulfide gas (sensor 136) presented the lowest values for all wine samples, with the NSAC and NSCO treatments exhibiting the highest values (0.48 and 0.47, respectively), while the STCO treatment exhibited the lowest (0.39 V). For the CO_2 gas readings (sensor MG811), the values were inversed, and hence, a higher voltage corresponded to a lower concentration. The STCE and STAC treatments exhibited the lowest mean CO_2 readings (1.16, and 1.15 V, respectively), while the NSAC and NSCO treatments exhibited the highest (1.03 and 1.00 V, respectively).

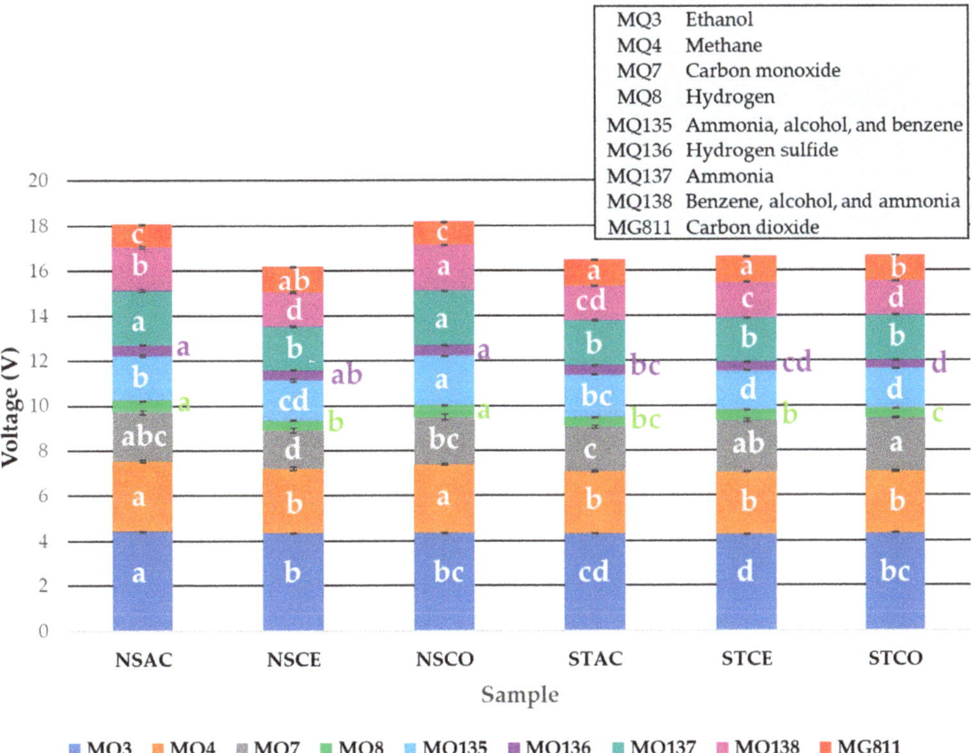

Figure 2. Stacked mean values of the electronic nose outputs showing the letters of significance from the ANOVA and Fisher's least significant difference (LSD) post hoc test ($p < 0.05$; $\alpha = 0.05$). Differences between samples are compared for each sensor (bar colors). Abbreviations: NSAC: non-smoke-tainted wine with activated carbon treatment, NSCE = non-smoke-tainted wine with enzyme and activated carbon treatment, NSCO = control non-smoke-tainted wine, STAC = smoke-tainted wine with activated carbon treatment, STCE = smoke-tainted wine with enzyme and activated carbon treatment and STCO = control smoke-tainted wine.

3.4. Multivariate Data Analysis

Figure 3 shows the PCA with data from the chemical measurements, volatile aromatic compounds and e-nose readings. Principal component one (PC1) accounted for 49.9% of the data variability, while principal component two (PC2) accounted for 26.7%. Therefore, the PCA accounted for a total of 76.6% of the data variability. From the factor loadings (FL), PC1 was primarily described by MG811 gas sensors (FL = 0.29), TDS (FL = 0.17) and EC (FL = 0.17) on the positive side of the axis and phenylethyl alcohol (FL = −0.28), alcohol content (FL = −0.28) and butanedioic acid, diethyl ester (FL = −0.27) on the negative side. On the other hand, PC2 was predominantly described by dodecanoic acid, ethyl ester (FL = 0.34), decanoic acid, ethyl ester (FL = 0.34) and octanoic acid, ethyl ester (FL = 0.30) on the positive side of the axis and MQ136 (FL = −0.31), pH (FL = −0.29) and MQ8 gas sensors (FL = −0.25) on the negative side. It can be observed that the smoked samples with amelioration treatments (STAC and STCE) were grouped with sample NSCE and associated with parameters such as CO_2 (MG811), TDS and EC. Sample NSAC was associated with pH and sensors MQ136, MQ3, MQ8 and MQ4, while the non-smoked control sample NSCO was related with alcohol, phenylethyl alcohol and butanedioic acid, diethyl ester. On the other hand, the smoked control sample (STCO) was associated with sensor MQ7 and decanoic acid, ethyl ester.

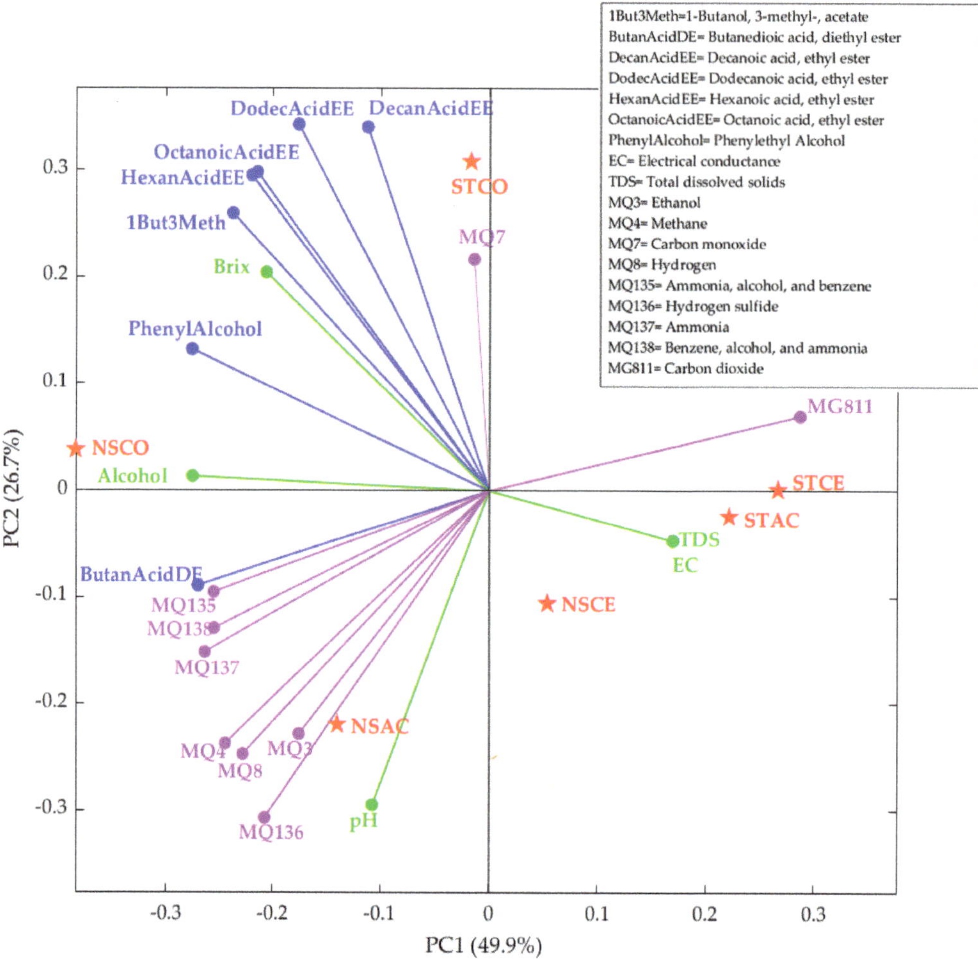

Figure 3. Principal component analysis showing the chemical measurements (blue), e-nose readings (purple) and volatile aromatic compounds (green). Abbreviations: NSAC = non-smoke-tainted wine with activated carbon treatment, NSCE = non-smoke-tainted wine with enzyme and activated carbon treatment, NSCO = control non-smoke-tainted wine, STAC = smoke-tainted wine with activated carbon treatment, STCE = smoke-tainted wine with enzyme and activated carbon treatment, STCO = control smoke-tainted wine and PC = principal component. Red stars in the figure depict the samples.

Figure 4 shows the significant correlations ($p < 0.05$) between the chemical measurements, volatile aroma compounds and e-nose outputs. From the matrix, positive correlations could be seen between 1-butanol, 3-methyl-, acetate and octanoic acid, ethyl ester ($r = 0.9$); hexanoic acid, ethyl ester ($r = 0.99$); phenylethyl alcohol ($r = 0.93$) and °Brix ($r = 0.82$), as well as between butanedioic acid, diethyl ester and the alcohol content ($r = 0.82$) and MQ135 gas sensor ($r = 0.90$). Negative correlations could be observed between the MG811 gas sensor and butanedioic acid, diethyl ester ($r = -0.82$) and MQ4 ($r = -0.90$), MQ8 ($r = -0.88$), MQ135 ($r = -0.88$), MQ137 ($r = -0.98$) and MQ138 ($r = -0.95$) gas sensors.

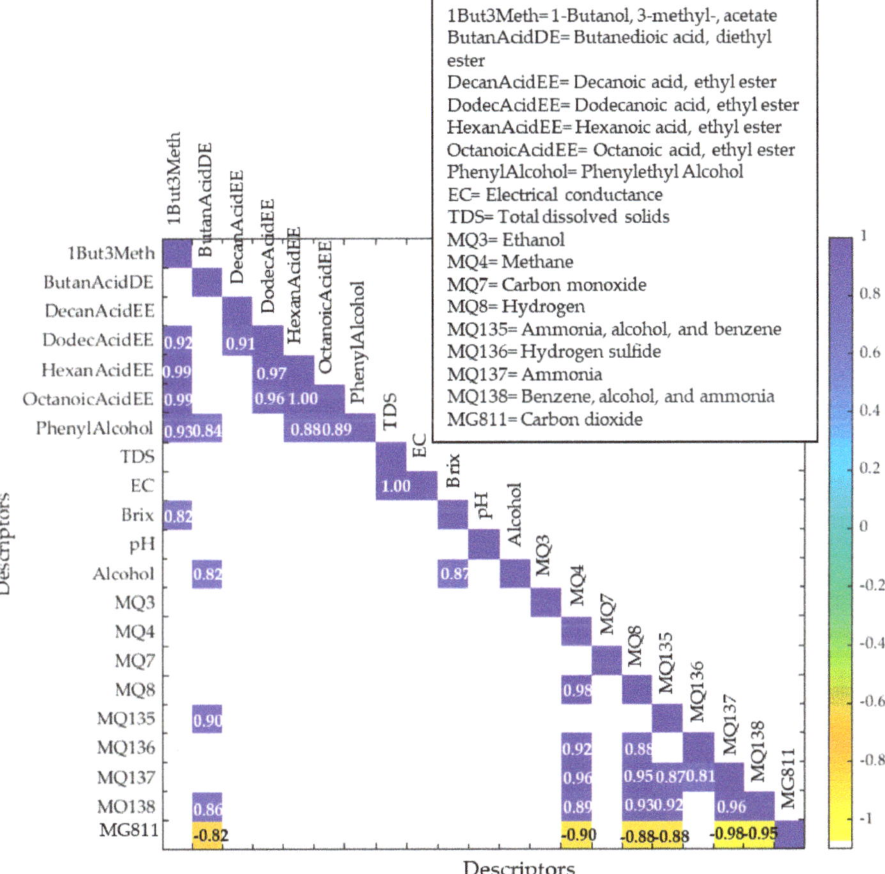

Figure 4. Matrix showing the significant ($p < 0.05$) correlations between the chemical measurements, volatile aroma compounds and e-nose readings. Color bar: the blue side depicts the positive correlations, while the yellow side depicts the negative correlations. Darker blue and yellow colors denote higher correlations.

3.5. *Artificial Neural Network Model*

The statistical data for the machine learning model developed using the e-nose readings as inputs and the smoke taint amelioration treatments as targets are shown in Table 4. The model displayed a very high overall accuracy of 98% in classifying the e-nose readings according to the experimental treatments. There were no signs of under- or overfitting, as shown by the lower performance value for the training stage (MSE < 0.01) than that for the testing stage (MSE = 0.02).

Table 4. Statistical results for the artificial neural network pattern recognition model developed. Performance is based on the mean squared error (MSE).

Stage	Number of Samples	Accuracy (%)	Error (%)	Performance (MSE)
Training	90	100	0	<0.01
Testing	60	95	5	0.02
Overall	150	98	2	-

- Not applicable.

The receiver operating characteristic (ROC) curve shown in Figure 5 also showed high true-positive rates (sensitivity) and low false-positive rates (specificity) in classifying the e-nose readings according to the smoke taint amelioration treatments. The NSCO treatment had the highest sensitivity (100%), while the NSCE treatment had the lowest (93.3%).

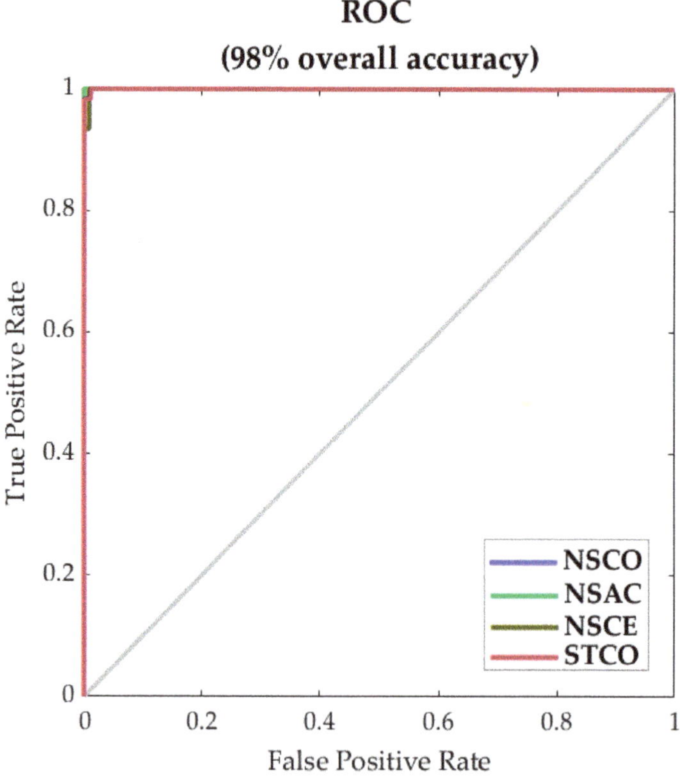

Figure 5. Receiver operating characteristic (ROC) curve for the machine learning model developed. Abbreviations: NSAC = non-smoke-tainted wine with activated carbon treatment, NSCE = non-smoke-tainted wine with enzyme and activated carbon treatment, NSCO = control non-smoke-tainted wine and STCO = control smoke-tainted wine.

The developed model was also used to classify the two remaining smoke taint amelioration treatments (STAC and STCE). The model showed that the amelioration treatments had 90% success, as they were classified within the non-smoked treatments (Table 5).

Table 5. Classification rates for the STCE and STAC treatments according to the machine learning model developed, with an overall success of 90%.

		Amelioration Treatments	
		STCE	**STAC**
Classification rates	NSCO	20	18
	NSAC	22	17
	NSCE	3	1
	STCO	5	4

Abbreviations: NSAC = non-smoke-tainted wine with activated carbon treatment, NSCE = non-smoke-tainted wine with enzyme and activated carbon treatment, NSCO = control non-smoke-tainted wine, STAC = smoke-tainted wine with activated carbon treatment, STCE = smoke-tainted wine with enzyme and activated carbon treatment and STCO = control smoke-tainted wine.

4. Discussion

The STAC and STCE wine samples had the highest mean values for TDS and EC, which may be a result of the bentonite treatment releasing minerals into the wine. Interestingly, the NSCO wine had the highest °Brix and alcohol contents. This is similar to the findings by Kennison et al. [3], who found that smoke exposure had an adverse effect on grape berry ripening and, hence, sugar accumulation, irrespective of the timing and duration of smoke exposure. It is, therefore, not surprising that the ST wine samples had lower °Brix and alcohol contents due to the long periods of smoke exposure associated with wildfire events that may have impacted the ripening of the grape berries. The activated carbon treatment did not appear to affect the pH of the smoke-tainted wines; however, it did appear to lower the pH of the non-smoke-tainted wines, potentially due to the adsorption of tartaric acid and malic acid [15].

As expected, the STCO and NSCO treatments exhibited higher mean values for the peak areas of the volatile aroma compounds (Table 3), and this was further highlighted by the PCA (Figure 3), which showed that these wine samples were associated with volatile aroma compounds, including phenylethyl alcohol and butanedioic acid, diethyl ester (NSCO) and decanoic acid, ethyl ester (STCO), as well as the correlation matrix illustrating the positive correlations between the aroma compounds (Figure 4). Activated carbon is a nonspecific fining agent and is, therefore, capable of removing positive aroma compounds in addition to those responsible for smoke taint [36–38]. However, research by Fudge et al. [15] found that a significant reduction of 'smoke' characters through treatment with activated carbon resulted in an increase in 'fruit' characters. Previous studies have found a negative correlation between the intensity of 'smoke' and 'fruit' aromas; thus, by reducing the negative 'smoke' characters, the positive 'fruit' characters may be enhanced [3,6,15]. Therefore, while treatment with activated carbon with/without the addition of a cleaving enzyme reduced the concentration of volatile aromatic compounds in smoke-tainted wines, the reduction in smoke-derived volatiles may help lead to a resurgence in the positive 'fruit' characters that were previously suppressed.

To date, traditional methods for assessing wine quality and the degree of smoke taint have involved the use of chromatographic techniques for the identification of aroma volatiles and trained panels [4,39–41]. However, there are several drawbacks to these techniques, as they can be time-consuming in terms of sample preparation, as well as training sensory experts, which is expensive, as chromatographic techniques require costly reagents and training, and maintaining trained panels can also be expensive, as well as being destructive in their forms of assessment. Furthermore, the results from sensory evaluations using human panels can be affected by physiological and psychological issues within the individuals, such as fatigue and decreased sensitivity to samples due to prolonged exposure [4,39–41]. Thus, the use of low-cost e-noses may offer a more cost-effective and accurate form of quality and smoke taint assessments. Previous research by Fuentes et al. [18] developed ANN models using readings from a low-cost e-nose as inputs to accurately predict the levels of volatile phenols and their glycoconjugates in grapes and wine, as well as a model predicting 12 consumer sensory responses towards the wine.

The models developed in their study may offer grape growers a rapid and cost-effective technique for assessing smoke-tainted wines; however, the models have yet to be tested on wines that have undergone smoke taint amelioration techniques to assess the accuracy of the models following treatment with activated carbon.

The ANN model developed in this study showed high accuracy in classifying the e-nose readings according to the different smoke taint amelioration techniques applied. It may offer winemakers a rapid, cost-effective tool to assess the effectiveness of smoke taint amelioration by activated carbon with/without the addition of a cleaving enzyme. Furthermore, as this method is nondestructive, repeated measurements can be made to assess wine samples as they age over time. In addition to this, when the STCE enzyme e-nose readings were assessed using the model, most of the readings were classified as either NSCO or NSAC, illustrating that the enzyme treatment together with activated carbon was effective in ameliorating smoke taint. Previous research by Fudge et al. [15] found that treatment with activated carbon alone was effective in removing smoke-derived volatile phenols; however, it did not remove their glycoconjugates. Therefore, the hydrolysis of these glycoconjugates over time may result in a resurgence of smoke aromas. The work by Herderich and Krstic [42] found that the treatment of smoke-affected Pinot Noir wine with a glucosidase enzyme reduced the concentrations of six glycosides by approximately 50%. Thus, combining the use of cleaving enzymes with activated carbon may enhance smoke taint amelioration by removing a significant amount of glycoconjugates and, therefore, delay the resurgence of smoke aromas as the wine ages. Further research is required to assess the exact levels of the glycoconjugates in the final wine samples.

5. Conclusions

The use of an electronic nose coupled with machine learning modeling can offer winemakers a cost-effective and rapid technique for assessing the effectiveness of smoke taint amelioration treatment by activated carbon with/without the addition of a cleaving enzyme without requiring the use of a trained sensory panel. Related to the latter, further advantages are in the repeatability of the integrated hardware (e-nose) and software (AI), which allows the monitoring of multiple batches at the same time by integrating local analytical microprocessors (e.g., Jetson ® from NVIDIA®). Furthermore, the use of a cleaving enzyme can enhance smoke taint amelioration by removing a significant proportion of glycoconjugates, thereby delaying the resurgence of smoke aromas and maintaining the value of the wine for a longer period. The integrated system proposed could offer winemakers a near-real-time system to assess the effect of treatments on microbrewing for decision-making purposes to manage the smoke taint in wines.

Author Contributions: Conceptualization, V.S., S.F. and C.G.V.; methodology, V.S.; formal analysis, V.S. and C.G.V.; investigation, V.S. and C.G.V.; software, C.G.V. and S.F.; resources, S.F; data curation, V.S and C.G.V.; writing—original draft preparation, V.S.; writing—review and editing, C.G.V., A.P., D.D.T., V.S. and S.F.; visualization, V.S.; supervision, S.F., A.P. and D.D.T.; validation, C.G.V., V.S. and S.F.; project administration, S.F. and funding acquisition, S.F. All authors have read and agreed to the published version of the manuscript.

Funding: This research was supported through the Australian Government Research Training Program Scholarship.

Institutional Review Board Statement: Not applicable.

Informed Consent Statement: Not applicable.

Data Availability Statement: The data and intellectual property belong to The University of Melbourne; any sharing needs to be evaluated and approved by the university.

Acknowledgments: The authors greatly acknowledge Brown Brothers Winemakers and Wendy Cameron for organizing and supplying the wine samples and Enologica Vason for supplying the compounds for this research. The authors would like to acknowledge Ranjith R. Unnithan and Bryce

Widdicombe from the School of Engineering, Department of Electrical and Electronic Engineering of The University of Melbourne for their collaboration in the electronic nose development.

Conflicts of Interest: The authors declare no conflict of interest.

References

1. Pardo-Garcia, A.; Wilkinson, K.; Culbert, J.; Lloyd, N.; Alonso, G.; Salinas, M.R. Accumulation of guaiacol glycoconjugates in fruit, leaves and shoots of vitis vinifera cv. Monastrell following foliar applications of guaiacol or oak extract to grapevines. *Food Chem.* **2017**, *217*, 782–789. [CrossRef]
2. van der Hulst, L.; Munguia, P.; Culbert, J.A.; Ford, C.M.; Burton, R.A.; Wilkinson, K.L. Accumulation of volatile phenol glycoconjugates in grapes following grapevine exposure to smoke and potential mitigation of smoke taint by foliar application of kaolin. *Planta* **2019**, *249*, 941–952. [CrossRef] [PubMed]
3. Kennison, K.; Wilkinson, K.; Pollnitz, A.; Williams, H.; Gibberd, M. Effect of timing and duration of grapevine exposure to smoke on the composition and sensory properties of wine. *Aust. J. Grape Wine Res.* **2009**, *15*, 228–237. [CrossRef]
4. Summerson, V.; Gonzalez Viejo, C.; Torrico, D.D.; Pang, A.; Fuentes, S. Detection of smoke-derived compounds from bushfires in cabernet-sauvignon grapes, must, and wine using near-infrared spectroscopy and machine learning algorithms. *OENO One* **2020**, *54*, 1105–1119. [CrossRef]
5. Hayasaka, Y.; Baldock, G.; Pardon, K.; Jeffery, D.; Herderich, M. Investigation into the formation of guaiacol conjugates in berries and leaves of grapevine vitis vinifera l. Cv. Cabernet sauvignon using stable isotope tracers combined with hplc-ms and ms/ms analysis. *J. Agric. Food Chem.* **2010**, *58*, 2076–2081. [CrossRef]
6. Ristic, R.; Osidacz, P.; Pinchbeck, K.; Hayasaka, Y.; Fudge, A.; Wilkinson, K. The effect of winemaking techniques on the intensity of smoke taint in wine. *Aust. J. Grape Wine Res.* **2011**, *17*, S29–S40. [CrossRef]
7. Ristic, R.; van der Hulst, L.; Capone, D.; Wilkinson, K. Impact of bottle aging on smoke-tainted wines from different grape cultivars. *J. Agric. Food Chem.* **2017**, *65*, 4146–4152. [CrossRef]
8. Kennison, K.; Gibberd, M.; Pollnitz, A.; Wilkinson, K. Smoke-derived taint in wine: The release of smoke-derived volatile phenols during fermentation of merlot juice following grapevine exposure to smoke. *J. Agric. Food Chem.* **2008**, *56*, 7379–7383. [CrossRef]
9. Mayr, C.M.; Parker, M.; Baldock, G.A.; Black, C.A.; Pardon, K.H.; Williamson, P.O.; Herderich, M.J.; Francis, I.L. Determination of the importance of in-mouth release of volatile phenol glycoconjugates to the flavor of smoke-tainted wines. *J. Agric. Food Chem.* **2014**, *62*, 2327–2336. [CrossRef]
10. Parker, M.; Osidacz, P.; Baldock, G.A.; Hayasaka, Y.; Black, C.A.; Pardon, K.H.; Jeffery, D.W.; Geue, J.P.; Herderich, M.J.; Francis, I.L. Contribution of several volatile phenols and their glycoconjugates to smoke-related sensory properties of red wine. *J. Agric. Food Chem.* **2012**, *60*, 2629–2637. [CrossRef]
11. Summerson, V.; Gonzalez Viejo, C.; Pang, A.; Torrico, D.D.; Fuentes, S. Review of the effects of grapevine smoke exposure and technologies to assess smoke contamination and taint in grapes and wine. *Beverages* **2021**, *7*, 7. [CrossRef]
12. Szeto, C.; Ristic, R.; Capone, D.; Puglisi, C.; Pagay, V.; Culbert, J.; Jiang, W.; Herderich, M.; Tuke, J.; Wilkinson, K. Uptake and glycosylation of smoke-derived volatile phenols by cabernet sauvignon grapes and their subsequent fate during winemaking. *Molecules* **2020**, *25*, 3720. [CrossRef]
13. Favell, J.W.; Noestheden, M.; Lyons, S.-M.; Zandberg, W.F. Development and evaluation of a vineyard-based strategy to mitigate smoke-taint in wine grapes. *J. Agric. Food Chem.* **2019**, *67*. [CrossRef]
14. Fudge, A.; Ristic, R.; Wollan, D.; Wilkinson, K. Amelioration of smoke taint in wine by reverse osmosis and solid phase adsorption. *Aust. J. Grape Wine Res.* **2011**, *17*, S41–S48. [CrossRef]
15. Fudge, A.; Schiettecatte, M.; Ristic, R.; Hayasaka, Y.; Wilkinson, K. Amelioration of smoke taint in wine by treatment with commercial fining agents. *Aust. J. Grape Wine Res.* **2012**, *18*, 302–307. [CrossRef]
16. Kelly, D.; Zerihun, A. The effect of phenol composition on the sensory profile of smoke affected wines. *Molecules* **2015**, *20*, 9536–9549. [CrossRef] [PubMed]
17. Zhao, P.; Qian, Y.; He, F.; Li, H.; Qian, M. Comparative characterization of aroma compounds in merlot wine by lichrolut-en-based aroma extract dilution analysis and odor activity value. *Chemosens. Percept.* **2017**, *10*, 149–160. [CrossRef]
18. Fuentes, S.; Summerson, V.; Gonzalez Viejo, C.; Tongson, E.; Lipovetzky, N.; Wilkinson, K.L.; Szeto, C.; Unnithan, R.R. Assessment of smoke contamination in grapevine berries and taint in wines due to bushfires using a low-cost e-nose and an artificial intelligence approach. *Sensors* **2020**, *20*, 5108. [CrossRef]
19. Wilson, A.D. Diverse applications of electronic-nose technologies in agriculture and forestry. *Sensors* **2013**, *13*, 2295–2348. [CrossRef] [PubMed]
20. Chen, Q.; Chen, Z.; Liu, D.; He, Z.; Wu, J. Constructing an e-nose using metal-ion-induced assembly of graphene oxide for diagnosis of lung cancer via exhaled breath. *ACS Appl. Mater. Interfaces* **2020**, *12*, 17713–17724. [CrossRef]
21. Lekha, S.; Suchetha, M. Recent advancements and future prospects on e-nose sensors technology and machine learning approaches for non-invasive diabetes diagnosis: A review. *IEEE Rev. Biomed. Eng.* **2020**, *14*. [CrossRef]
22. Gamboa, J.C.R.; da Silva, A.J.; de Andrade Lima, L.L.; Ferreira, T.A. Wine quality rapid detection using a compact electronic nose system: Application focused on spoilage thresholds by acetic acid. *LWT* **2019**, *108*, 377–384. [CrossRef]
23. Mu, F.; Gu, Y.; Zhang, J.; Zhang, L. Milk source identification and milk quality estimation using an electronic nose and machine learning techniques. *Sensors* **2020**, *20*, 4238. [CrossRef] [PubMed]

24. Nimsuk, N. Improvement of accuracy in beer classification using transient features for electronic nose technology. *J. Food Meas. Charact.* **2019**, *13*, 656–662. [CrossRef]
25. Xu, J.; Liu, K.; Zhang, C. Electronic nose for volatile organic compounds analysis in rice aging. *Trends Food Sci. Technol.* **2021**, 109. [CrossRef]
26. Zarezadeh, M.R.; Aboonajmi, M.; Varnamkhasti, M.G.; Azarikia, F. Olive oil classification and fraud detection using e-nose and ultrasonic system. *Food Anal. Methods* **2021**, 1–12. [CrossRef]
27. Antolini, A.; Forniti, R.; Modesti, M.; Bellincontro, A.; Catelli, C.; Mencarelli, F. First application of ozone postharvest fumigation to remove smoke taint from grapes. *Ozone Sci. Eng.* **2020**, 1–9. [CrossRef]
28. Viejo, C.G.; Fuentes, S.; Godbole, A.; Widdicombe, B.; Unnithan, R.R. Development of a low-cost e-nose to assess aroma profiles: An artificial intelligence application to assess beer quality. *Sens. Actuators B Chem.* **2020**, *308*, 127688. [CrossRef]
29. Gonzalez Viejo, C.; Tongson, E.; Fuentes, S. Integrating a low-cost electronic nose and machine learning modelling to assess coffee aroma profile and intensity. *Sensors* **2021**, *21*, 2016. [CrossRef]
30. Gonzalez Viejo, C.; Fuentes, S.; Torrico, D.D.; Godbole, A.; Dunshea, F.R. Chemical characterization of aromas in beer and their effect on consumers liking. *Food Chem.* **2019**, *293*, 479–485. [CrossRef] [PubMed]
31. Gonzalez Viejo, C.; Torrico, D.D.; Dunshea, F.R.; Fuentes, S. Development of artificial neural network models to assess beer acceptability based on sensory properties using a robotic pourer: A comparative model approach to achieve an artificial intelligence system. *Beverages* **2019**, *5*, 33. [CrossRef]
32. Arcari, S.G.; Caliari, V.; Sganzerla, M.; Godoy, H.T. Volatile composition of merlot red wine and its contribution to the aroma: Optimization and validation of analytical method. *Talanta* **2017**, *174*, 752–766. [CrossRef]
33. Issa-Issa, H.; Noguera-Artiaga, L.; Sendra, E.; Pérez-López, A.J.; Burló, F.; Carbonell-Barrachina, Á.A.; López-Lluch, D. Volatile composition, sensory profile, and consumers' acceptance of fondillón. *J. Food Qual.* **2019**, *2019*. [CrossRef]
34. Ayestarán, B.; Martínez-Lapuente, L.; Guadalupe, Z.; Canals, C.; Adell, E.; Vilanova, M. Effect of the winemaking process on the volatile composition and aromatic profile of tempranillo blanco wines. *Food Chem.* **2019**, *276*, 187–194. [CrossRef] [PubMed]
35. Viejo, C.G.; Fuentes, S. Beer aroma and quality traits assessment using artificial intelligence. *Fermentation* **2020**, *6*, 56. [CrossRef]
36. Filipe-Ribeiro, L.; Milheiro, J.; Matos, C.C.; Cosme, F.; Nunes, F.M. Reduction of 4-ethylphenol and 4-ethylguaiacol in red wine by activated carbons with different physicochemical characteristics: Impact on wine quality. *Food Chem.* **2017**, *229*, 242–251. [CrossRef]
37. Krstic, M.; Johnson, D.; Herderich, M. Review of smoke taint in wine: Smoke-derived volatile phenols and their glycosidic metabolites in grapes and vines as biomarkers for smoke exposure and their role in the sensory perception of smoke taint. *Aust. J. Grape Wine Res.* **2015**, *21*, 537–553. [CrossRef]
38. Mirabelli-Montan, Y.A.; Marangon, M.; Graça, A.; Mayr Marangon, C.M.; Wilkinson, K.L. Techniques for mitigating the effects of smoke taint while maintaining quality in wine production: A review. *Molecules* **2021**, *26*, 1672. [CrossRef]
39. Han, F.; Zhang, D.; Aheto, J.H.; Feng, F.; Duan, T. Integration of a low-cost electronic nose and a voltammetric electronic tongue for red wines identification. *Food Sci. Nutr.* **2020**, *8*, 4330–4339. [CrossRef] [PubMed]
40. Sherman, E.; Harbertson, J.F.; Greenwood, D.R.; Villas-Bôas, S.G.; Fiehn, O.; Heymann, H. Reference samples guide variable selection for correlation of wine sensory and volatile profiling data. *Food Chem.* **2018**, *267*, 344–354. [CrossRef] [PubMed]
41. Summerson, V.; Gonzalez Viejo, C.; Szeto, C.; Wilkinson, K.L.; Torrico, D.D.; Pang, A.; Bei, R.D.; Fuentes, S. Classification of smoke contaminated cabernet sauvignon berries and leaves based on chemical fingerprinting and machine learning algorithms. *Sensors* **2020**, *20*, 5099. [CrossRef] [PubMed]
42. Herderich, M.; Krstic, M. Mitigation of Climate Change Impacts on the National Wine Industry by Reduction in Losses from Controlled Burns and Wildfires and Improvement in Public Land Management. Available online: https://www.awri.com.au/research_and_development/2017-2025-rde-plan-projects/project-3-4-3/ (accessed on 5 June 2021).

Article

Predicting Alcohol Concentration during Beer Fermentation Using Ultrasonic Measurements and Machine Learning

Alexander Bowler [1], Josep Escrig [2], Michael Pound [3] and Nicholas Watson [1,*]

1 Food, Water, Waste Research Group, Faculty of Engineering, University Park, University of Nottingham, Nottingham NG7 2RD, UK; alexander.bowler@nottingham.ac.uk
2 i2CAT Foundation, Calle Gran Capita, 2-4 Edifici Nexus (Campus Nord Upc), 08034 Barcelona, Spain; josep.escrig@i2cat.net
3 School of Computer Science, Jubilee Campus, University of Nottingham, Nottingham NG8 1BB, UK; michael.pound@nottingham.ac.uk
* Correspondence: nicholas.watson@nottingham.ac.uk

Abstract: Beer fermentation is typically monitored by periodic sampling and off-line analysis. In-line sensors would remove the need for time-consuming manual operation and provide real-time evaluation of the fermenting media. This work uses a low-cost ultrasonic sensor combined with machine learning to predict the alcohol concentration during beer fermentation. The highest accuracy model (R^2 = 0.952, mean absolute error (MAE) = 0.265, mean squared error (MSE) = 0.136) used a transmission-based ultrasonic sensing technique along with the measured temperature. However, the second most accurate model (R^2 = 0.948, MAE = 0.283, MSE = 0.146) used a reflection-based technique without the temperature. Both the reflection-based technique and the omission of the temperature data are novel to this research and demonstrate the potential for a non-invasive sensor to monitor beer fermentation.

Keywords: machine learning; ultrasonic measurements; long short-term memory; industrial digital technologies

Citation: Bowler, A.; Escrig, J.; Pound, M.; Watson, N. Predicting Alcohol Concentration during Beer Fermentation Using Ultrasonic Measurements and Machine Learning. *Fermentation* **2021**, *7*, 34. https://doi.org/10.3390/fermentation7010034

Received: 16 February 2021
Accepted: 2 March 2021
Published: 4 March 2021

1. Introduction

During beer fermentation, yeast metabolism produces ethanol and carbon dioxide from a sugar-water mixture called wort [1,2]. The fermentation is conventionally monitored through off-line wort density measurements until a predetermined ethanol concentration is reached [3], after which the process is continued for a predefined time for development of flavour compounds [4]. This requires manual sampling, takes time, and wastes resources by disposing of the measured sample. In-line measurement techniques directly measure the process material and on-line methods use bypasses to automatically collect, analyse, and return samples to the process [5]. By providing real-time, automatic alcohol concentration measurements, in-line and on-line techniques would ensure product quality through early detection of anomalous batches, allow effective scheduling of production equipment by predicting fermentation endpoint, and reduce the burden of manual sampling by operators. Furthermore, real-time data is key to the Fourth Industrial Revolution, which will implement industrial digital technologies such as the Internet of Things, cloud computing, and machine learning (ML) to integrate entire processes, automatically make decisions, and improve manufacturing productivity, efficiency, and sustainability [6].

Several in-line and on-line methods to monitor alcoholic fermentation have been investigated, including in-situ transflectance near-infrared spectroscopy [7,8], and Raman spectroscopy probes [9]; automated flow-through mid-infrared spectroscopy [10], Fourier transform infrared spectroscopy [11], and piezoelectric MEMS resonators [12]; non-invasive Raman spectroscopy through transparent vessel walls [13]; and CO_2 emission monitoring [14]. Ultrasonic (US) sensors are an attractive monitoring technique owing to their low cost and have previously been used to study fermentation, including as in-line methods on circulation lines [15], in-situ in tanks [16], and using non-invasive, through-transmission of the fermenting media [17,18]. US monitoring techniques use high frequency (>1 MHz) and low power (<1 Wcm^{-2}) pressure waves to characterise material properties whilst causing no alterations to the material in which they propagate [19]. However, US properties vary with temperature and the presence of gas bubbles causes attenuation of the sound wave [20]. Previous in-line, on-line, and off-line studies to monitor fermentation using US measurements have developed empirical or semi-empirical models from the speed of sound or acoustic impedance to determine alcohol content [16]. These methods require extensive calibration procedures to compensate for the effects of temperature, dissolved CO_2 [16,18,21], and yeast cell concentration [18]. Supervised ML uses data to train predictive algorithms for classification or regression problems. Through ML, compensation procedures are not required as the complexities caused by varying process parameters imbedded in the sensor data can be unravelled. Furthermore, procedures for accurate determination of the speed of sound are not necessary [15,16,22].

This work presents three novel contributions to US monitoring of alcoholic fermentations: Firstly, ML is used to predict alcohol concentration during lab-scale beer fermentations from US measurements. Secondly, although an in-situ sensor probe is used, the potential for non-invasive monitoring of fermentation is investigated by only using the US wave reflected from the interface between the probe and the wort. This technique is similar to previous work by our group [23–26]. Implementation of this technique would provide in-line, non-invasive process monitoring without the need for circulation or bypass lines. This method would also not require transmission through the total vessel contents, which would be impossible at industrial scale. Therefore, this technique could be inexpensively fitted to the outside of existing vessels. Finally, exclusion of the temperature as a feature in the ML models is evaluated. Effective monitoring without the need for an invasive temperature sensor would further reduce the cost and complexity of industrial implementation.

2. Materials and Methods

The fermentation was conducted in a 30 L cylindrical plastic vessel (Figure 1). A lid sealed the vessel to protect the wort from contamination. The lid contained an air lock to release the CO_2 produced during fermentation. A belt heater increased the temperature of the wort to facilitate fermentation. The wort was prepared in the vessel by dissolving and mixing 1.5 kg of malt (Coopers Real Ale, UK) and 1 kg of sugar (brewing sugar, the Home Brew Shop, UK) in 22 L of water. Once the ingredients were mixed, a US probe was installed, consisting of a US transducer (Sonatest, 2 MHz central frequency, UK) and a temperature sensor (RTD, PT1000, UK). The US transducer was connected to a Lecouer Electronique US Box (France) that excited the transducer and digitised the received US signal. The temperature sensor was connected to a Pico electronic box (PT-104 Data Logger, UK). The two electronic boxes were connected to a laptop that controlled the data acquisition. Coupling gel was applied between the US transducer and the probe, and a spring was used to maintain the contact pressure. A Tilt hydrometer was installed to provide real-time density measurements. The real-time density measurements were required as the ground truth data of the wort alcohol concentration to train the ML models. This device was a small cylinder that floats in the liquid with its centre of gravity different from its centre of buoyancy. This causes an inclination of the device that is dependent on the specific gravity of the fermenting media. The inclination of the hydrometer was measured by a self-contained accelerometer and was transmitted by radio to a smartphone located outside of the vessel. A calibration procedure related the inclination to the specific gravity. It should be noted hydrometers are not suitable for in-line monitoring of industrial fermentations. Firstly, the balance of the device can be easily distorted by foam or solids floating on the surface, or by bubbles produced during fermentation. Secondly, as the hydrometer floats on the wort surface, it would need manual removal at the end of each fermentation batch. The most accurate method of specific gravity measurement is to extract samples and use a portable density meter. However, this would require manual sample withdrawal at least every 2 h and would decrease the volume of liquid in the vessel, affecting the fermentation process. Furthermore, this would only produce sparse ground truth measurements of the density to train the ML models.

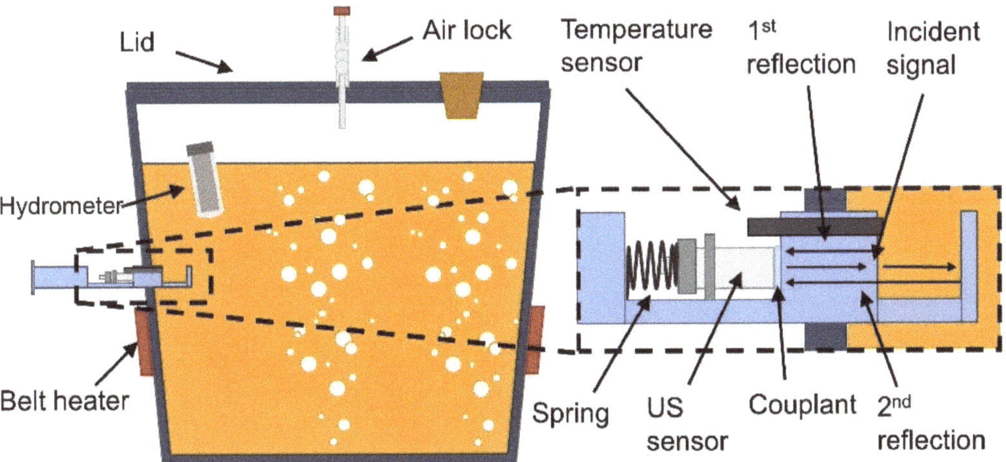

Figure 1. Experimental apparatus and measured US wave reflections.

The yeast (Coopers Real Ale, UK) was distributed on the surface and the vessel sealed. The mixture was left for 4 to 7 days while the fermentation occurred. After this time, the fermentation equipment was cleaned and a new batch was prepared. In total, 13 batches were completed. During fermentation, data was collected from the three different sensors: the US sensor, the temperature sensor, and the hydrometer. The time of each measurement was also recorded. The fermentation batches were conducted over a period of approximately 3 months. This meant that the environmental and water temperature in the laboratory changed during this time. Furthermore, the belt heater was only in contact with the lower section of the vessel. This produced temperature variations from around 20 to 30 °C. However, this temperature variation is beneficial to our ML evaluation as each model must be able to generalise across a wide range of process temperatures.

Sets of US and temperature data were collected periodically. Each of the sets consisted of 36 US waves and 36 temperature readings. For the US signal, 7000 sampling points were collected at 80 MHz sampling frequency. The time between each wave acquisition was 0.55 s. Between each set of data collection, 200 s elapsed. As depicted in Figure 1, the US transducers emitted sound waves which travelled along a PMMA buffer. At the interface between the buffer material and the wort, part of the sound wave is reflected back to the transducer (the 1st reflection). The rest of the sound wave continues through the wort, reflects at the opposite probe wall, and travels back to the transducer to be recorded (the 2nd reflection). An example of the signal recorded by the transducer is presented in Figure 2a. Close-ups of each reflection are presented in Figure 2b,c. The first section of the waveform (sample points < 500) is reflected back to the transducer before contacting the buffer material and wort interface and therefore contains no useful information about the process. The 1st reflection is identified between sample point 900 and 1500, and the 2nd reflection between 6000 and 6500, as shown in Figure 2.

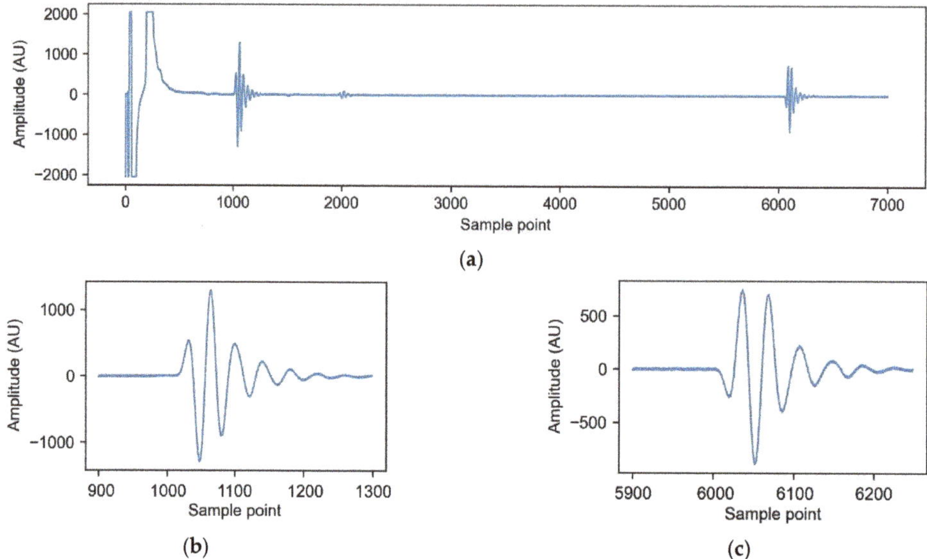

Figure 2. Example ultrasonic waveform obtained: (**a**) The 1st reflection is located around sample point 1000, the 2nd reflection is located around sample point 6000; (**b**) a close-up of the 1st reflection; (**c**) a close-up of the 2nd reflection.

2.1. Volume of Alcohol Calculation

The volume of alcohol (%) can be calculated from the specific gravity of the fermenting media using Equation (1) [27].

$$ABV = (SG_{in} - SG) \times 131.25, \tag{1}$$

where ABV is the alcohol by volume (%), SG_{in} is the starting specific gravity of the liquid before the yeast was added, and SG is the current specific gravity of the fermenting liquid. The multiplier of this equation is based on the stoichiometric relationship of the fermentation reaction, where the decreasing density is due to CO_2 production and escape through the air lock [28].

2.2. Ultrasonic Wave Features

The following features were calculated from the obtained US waveform to use in the ML models. These are common features extracted from US waveforms [29]. The theory behind the selection of each feature is presented in their respective sections. Different combinations of these features were tested during ML model optimisation. The optimal feature combinations are presented in Table 1, Section 3.1.

2.2.1. Energy

The waveform energy is a measure of the size of the waveform received by the transducer. For the 1st reflection, this is a measure of the proportion of the sound wave reflected from the interface between the buffer material and the wort. This is dependent on the change in acoustic impedance between these two materials [30]. Monitoring the waveform energy of the 2nd reflection offers additional information on the level of sound wave attenuation in the wort. This is caused by viscous losses in the media and scattering due to heterogeneities such as bubbles and yeast cells [30].

$$E = \sum_{i=start}^{i=end} A_i^{\,2}, \tag{2}$$

where E is the waveform energy, A_i is the waveform amplitude at sample point i, and *start* and *end* denote the range of samples points for the reflection of interest [29].

Table 1. Results for the long short-term memory neural network (LSTM) models. MAE: mean absolute error; MSE: mean squared error; ABV: alcohol by volume. The regression metrics for the models evaluated on the test set are highlighted in bold at the bottom of the table.

Model	1	2	3	4
Reflections	1st and 2nd	1st and 2nd	1st	1st
Temperature	Yes	No	Yes	No
Optimal features	• 1st reflection energy • 2nd reflection energy • 1st reflection energy standard deviation • 2nd reflection energy standard deviation • Time of flight • Temperature	• 1st reflection energy • 2nd reflection energy • 1st reflection energy standard deviation • 2nd reflection energy standard deviation • Time of flight	• 1st reflection energy • 1st reflection energy standard deviation • Temperature	• 1st reflection energy • 1st reflection energy standard deviation • 1st reflection peak-to-peak amplitude • 1st reflection maximum amplitude • 1st reflection minimum amplitude
Feature gradients	Yes	Yes	Yes	Yes
Batch size	2	2	4	4
Learning rate	0.01	0.01	0.033	0.033
LSTM units	2	2	4	4
L2 regularisation	0.0001	0.0001	0.00001	0.0001
Dropout rate	0	0	0	0
Epochs	100	100	100	100
Clip norm value	1	1	1	1
R^2	**0.952**	**0.939**	**0.878**	**0.948**
MAE (% ABV)	**0.265**	**0.355**	**0.426**	**0.283**
MSE (% ABV)	**0.136**	**0.173**	**0.345**	**0.146**

2.2.2. Peak-to-Peak Amplitude, Maximum Amplitude, and Minimum Amplitude

The peak-to-peak amplitude, maximum amplitude, and minimum amplitude provide additional information as to how the energy is distributed in the waveform. Changes in wort composition or temperature may affect how the sound wave travels and reflects from boundaries, presenting differences in the shape of the received waveform. These three features were calculated for both the 1st and 2nd reflections.

$$PPA = \max(A_{start:end}) - \min(A_{start:end}), \tag{3}$$

$$A_{max} = \max(A_{start:end}), \tag{4}$$

$$A_{min} = \min(A_{start:end}), \tag{5}$$

where PPA is the peak-to-peak amplitude, A_{max} is the maximum amplitude, and A_{min} is the minimum amplitude.

2.2.3. Energy Standard Deviation

A total of 36 US waves were collected during each acquisition block. Phenomena in the process—e.g., the presence of bubbles at different times during fermentation—may cause fluctuations in the energy of the received waveforms. Therefore, the standard deviation of the energy in a block of acquired waveforms was investigated as a feature. The standard deviation of the energy was calculated for both the 1st and 2nd reflections.

$$STD = \sqrt{\frac{1}{W}\sum_{i=1}^{i=W}\left(E_i - \overline{E}\right)^2} \tag{6}$$

where *STD* is the standard deviation, *W* is the number of waveforms collected in the block, *i* is an individual waveform, and $\overline{E}$ is the mean waveform energy in the block.

2.2.4. Time of Flight

The time of flight was calculated using a thresholding method, i.e., the waveform sample point where the second reflection amplitude rises above the signal noise. This is a measure of the speed of sound in the wort that is dependent on its density and compressibility [20].

2.2.5. Feature Gradients

A one-sided, backwards moving mean was applied to obtain lagged feature representations over the previous 5 h. For the artificial neural networks (ANNs), this allows the use of past process information. For the long short-term memory neural networks (LSTMs), this allows for a way of storing past process information in some features, reducing the burden on the LSTM units to remember all feature trajectories.

2.3. *Machine Learning*

The ground truth data for the percentage volume of alcohol during fermentation was calculated from the portable density meter and hydrometer measurements. In total, 13 fermentation batches were monitored. The final two batches were selected as the test set to provide an unbiased assessment of the experimental methodology used. In an industrial setting, the final ML models would be deployed after collecting the training set runs. The remaining 11 batches were used in a 5-fold cross-validation procedure to optimize the ML models' hyperparameters. Long short-term memory neural networks (LSTMs) are able to retain information from previous time-steps in a sequence. LSTMs are a type of recurrent neural network that reduces the likelihood of vanishing or exploding gradients by using gate units. This enables their use over much longer sequences [31]. To evaluate the utility of using LSTMs to predict alcohol concentration, they were compared with artificial neural networks (ANNs) which are unable to store past process information. ANNs combine input features to produce new features which can approximate the relationship with the target variable given enough neurons in the hidden layer [32,33].

For the LSTMs, zero-padding was applied to the US features to make every fermentation batch sequence an equal length. A masking layer specified that the LSTM units ignore this padding. Each sequence consisted of 4646 timesteps. All timesteps for each batch were used as a single sequence rather than being split into multiple sequences of shorter length. While long LSTM sequences (250–500 timesteps) are prone to produce vanishing gradients when predicting a single output, this is not a problem when predicting an output at every timestep as used in this task [34].

For the ANNs, a single hidden layer and the Adam optimisation algorithm was used. Cross-validation determined the optimal batch size, number of neurons in the hidden layer, learning rate, drop-out rate, L2 regularisation penalty, and number of epochs for training. For the LSTMs, the Adam optimisation algorithm was used and the cross-validation procedure determined the optimal batch size, number of LSTM units, learning rate, drop-out rate, L2 regularisation penalty, gradient norm clipping value, and number of epochs. After cross-validation, the set of hyperparameters which resulted in the lowest average validation error were used to train a final model using all of the training set. The networks were trained using TensorFlow 2.3.0. The coefficient of determination (R^2), mean squared error (MSE) and mean absolute error (MAE) were used as performance metrics to evaluate the ML models. Multiple metrics produce a comprehensive assessment of a model's ability to fit to the test set and improve comparison between models.

3. Results

Figure 3 displays selected features from all the fermentation batches. It is shown that the energy of 1st reflection (Figure 3a), energy of the 2nd reflection (Figure 3b), and the time of flight of the sound wave through the wort (Figure 3c) start at different values for each batch. There are several explanations for this. Firstly, as presented in Figure 3d, the process temperature is not the same at the start of each batch. As the speed of sound is highly dependent on temperature, the US properties begin from different magnitudes. Secondly, the US probe required manual removal and repositioning when disposing each batch after fermentation. This disturbed the spring maintaining the contact pressure of the US transducer, which affects the sound energy transferred through the materials from the sensor.

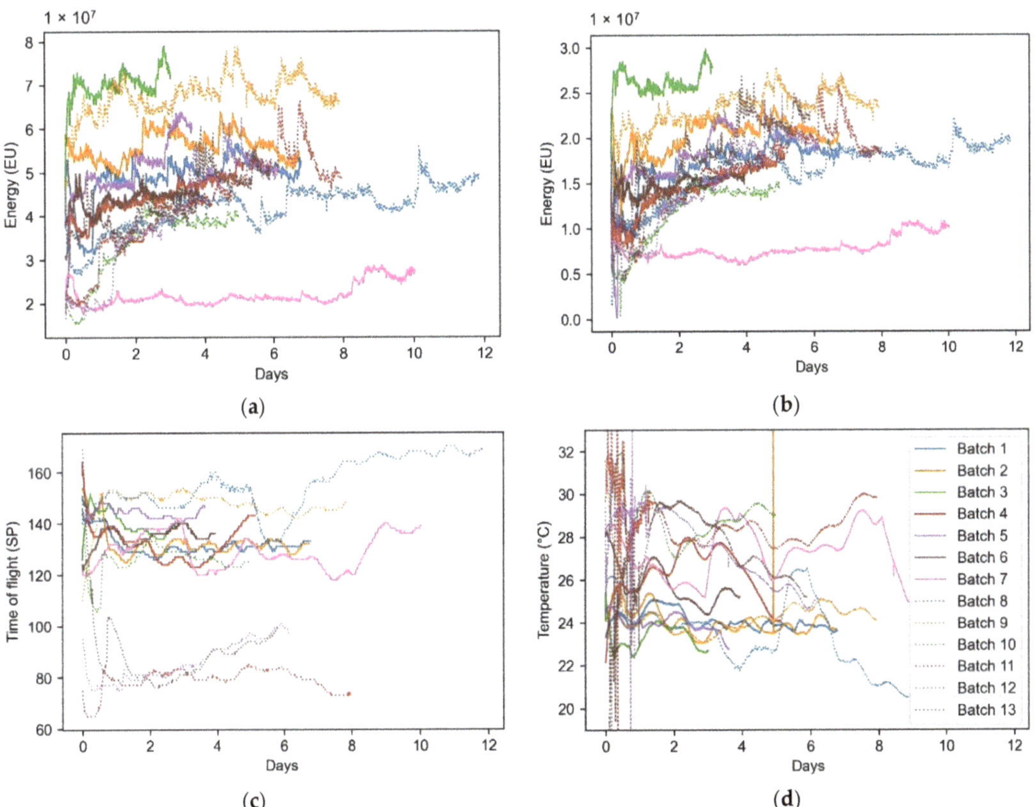

Figure 3. US waveform features for all fermentation batches: (**a**) The energy of the 1st reflection; (**b**) the energy of the 2nd reflection; (**c**) the time of flight for the 2nd reflection; (**d**) the process temperature.

The trajectories of the waveform features are also not smooth. Again, this is partly due to the oscillating process temperature. In addition, bubbles of CO_2 produced during the fermentation were observed to attach to the surface of the probe material, which would cause scattering and reflection of the sound wave. During the fermentation, as further CO_2 bubbles were produced, the new bubbles would replace the previous ones on the surface. This is likely to cause fluctuations in the waveform energy transferring through the interface between the probe and the wort.

The energy of the 1st reflection increases throughout the fermentation (Figure 3a). The energy of the 1st reflection is proportional to the change in acoustic impedance at the buffer-wort interface, with the acoustic impedance being a product of the material density and speed of sound [20]. As the density of the wort decreases during fermentation, the speed of sound also decreases as found in [17,18,35]. As the solid buffer material has a greater density and speed of sound than the starting wort, the proportion of sound wave reflected at the buffer-wort material increases throughout the fermentation. However, the time of flight (Figure 3c), the inverse of the speed of sound, shows no general trend. This contrasts with the results obtained in [17,18,35], which suggests that it should increase. This is likely due to the changing process temperature masking an increasing time of flight. The results in [17,18,35] were all obtained at a constant temperature. The reduced time of flight for the last three batches (Batches 11, 12, and 13 in Figure 3c) is most likely due to a disturbance of the sensor positioning after Batch 10. These batches were kept in order to provide an unbiased assessment of the experimental methodology used. In an industrial setting because the test set data (batches 12 and 13) would not be available to analyse prior to the ML model training. The energy of the 2nd reflection is diminished compared with the 1st reflection at the beginning of the fermentation until approximately Day 3, as presented in Figure 4. A similar result was found in [17] and is due to the fermentation being most vigorous at the start of the process. This causes more CO_2 bubbles to be produced and therefore greater attenuation of the sound wave travelling through the wort.

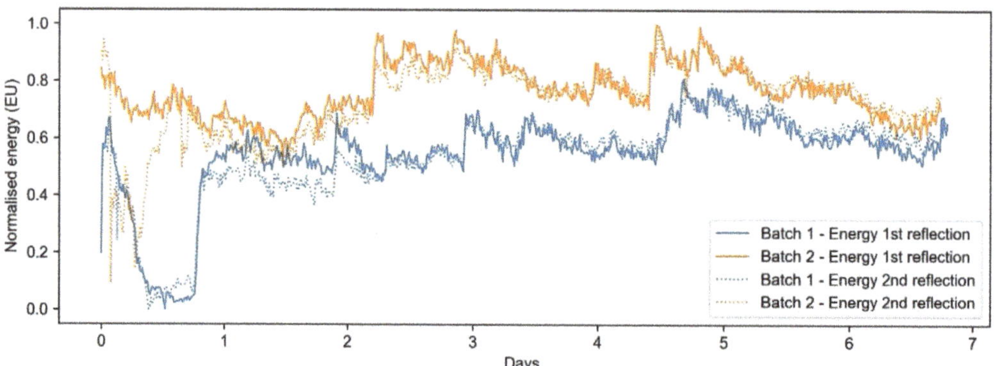

Figure 4. A comparison between the waveform energy of the 1st and 2nd reflections for batches 1 and 2. The energy of the 2nd reflection is diminished compared to the 1st until approximately Day 3. This is due to sound wave attenuation because of CO_2 bubbles being produced.

3.1. Machine Learning

The ANN model with the highest accuracy only achieved an R^2 of 0.398 (MAE = 1.010% ABV, MSE = 1.942% ABV). As such, only results from the LSTM models are included in Table 1. This shows that the gradients of the features, as provided to the ANNs, is insufficient, and the enhanced memory of the feature history provided by the LSTM units is required for this process. The results of four final LSTM models are presented in Table 1, which either use the 1st reflection or both reflections, and either use the temperature as a feature or not. The optimal features, optimal hyperparameters, and performance metrics are included. The most accurate LSTM model (Model 1) used features from both the 1st and 2nd reflections and the process temperature. Interestingly, the second most accurate model (Model 4) only used features from the 1st reflection, excluding the process temperature. The third most accurate model (Model 2) used features from the 1st and 2nd reflections without the process temperature. Finally, the least accurate model (Model 3) combined features from the 1st reflection and the process temperature. Graphical representations of these predictions are shown in Figure 5a–h.

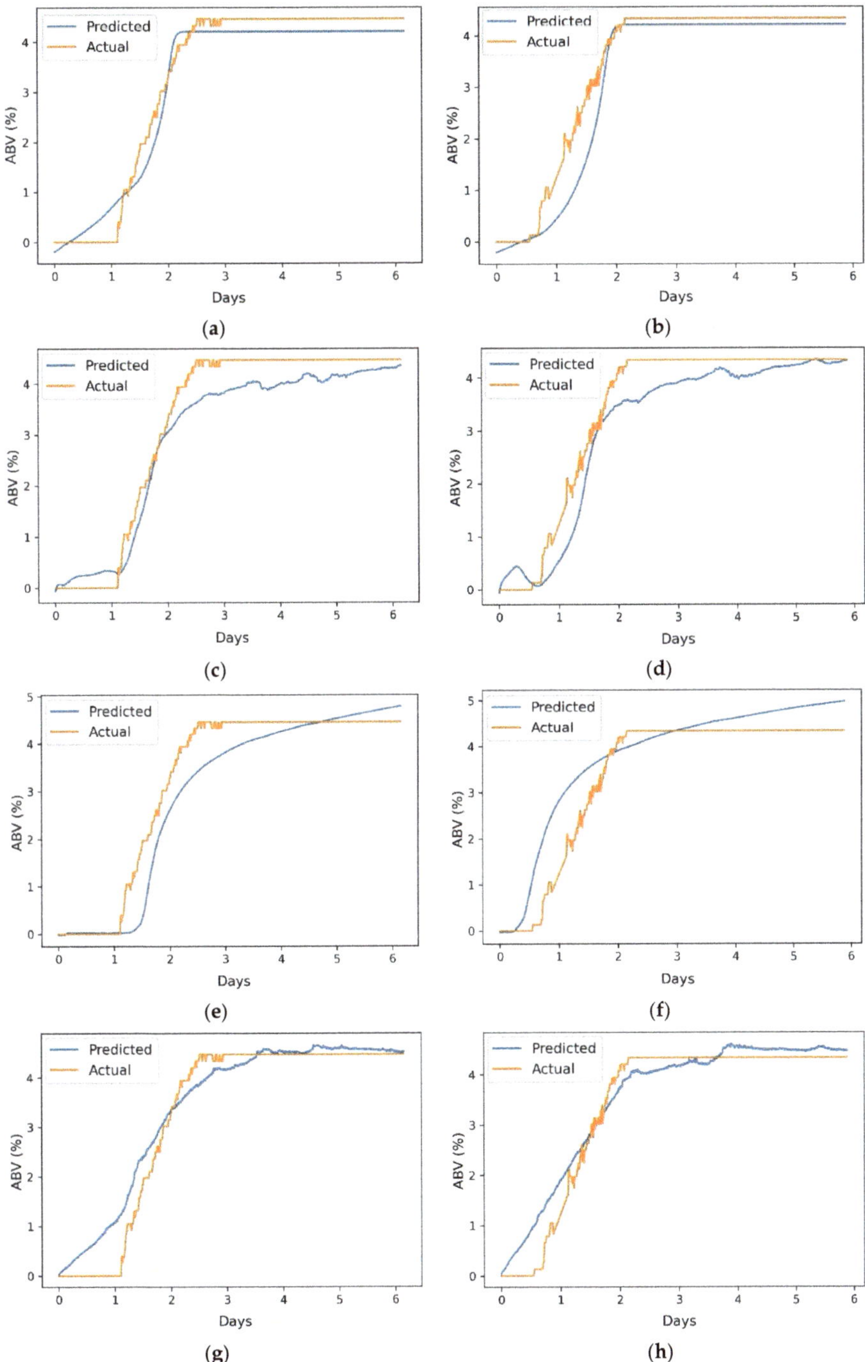

Figure 5. Predicted alcohol by volume percentage from the final LSTM models for the two test set batches (batches 12 and 13): (**a**) Model 1, Batch 12; (**b**) Model 1, Batch 13; (**c**) Model 2, Batch 12; (**d**) Model 2, Batch 13; (**e**) Model 3, Batch 12; (**f**) Model 3, Batch 13; (**g**) Model 4, Batch 12; (**h**) Model 4, Batch 13.

4. Discussion

The most accurate model (Model 1) uses features from both the 1st and 2nd reflections and the process temperature. This shows the potential of US sensors to predict the endpoint of fermentation and, as demonstrated in Figure 5a,b, accurately predict the alcohol concentration throughout the fermentation process. However, industrial implementation of this model would require the use of an invasive probe in order to obtain the 2nd reflection. In addition, an invasive temperature probe would be required to monitor the changing temperature of the fermentation media. Interestingly, the second most accurate model (Model 4) only used features from the 1st reflection and excluded the process temperature. The use of only the 1st reflection indicates that accurate results could be obtained using a non-invasive, no-transmission US sensor, similar to the techniques used in previous works by our group [23–26]. This is advantageous as it allows the alcohol volume to be accurately predicted by easily mounting a US sensor externally to an existing vessel. Therefore, it can be easily implemented into existing industrial settings at low effort and cost. The use of Model 4 would also remove the requirement for an invasive process temperature measurement. Furthermore, the performance metrics for Model 4 (R^2 = 0.948, MAE = 0.283, MSE = 0.146) are similar to those of Model 1 (R^2 = 0.952, MAE = 0.265, MSE = 0.136) indicating that no prediction accuracy would be lost through using a non-invasive and no-transmission sensor approach. This US sensing technique would also not require a hole to be bored into the vessel side, as used in this work. Instead, the US wave could be transmitted through the vessel wall.

In Model 3, the features from the 1st reflection combined with the process temperature produces a reduced accuracy. This is likely because the additional features required in Model 4 (the peak-to-peak amplitude, maximum amplitude, and minimum amplitude of the 1st reflection) contained more pertinent information about the temperature at the probe-wort interface than the non-local temperature sensor. The suggestion that the temperature sensor measured the temperature of the bulk wort instead of the region through which the 1st reflection passes is supported by the results from Model 2. When the temperature was removed as a feature, Model 2 produced a reduced accuracy compared with Model 1. This indicates that for accurate prediction using the 2nd reflection, the bulk wort temperature measurement is required as the sound wave travels through this region. The reduced accuracy obtained when combining the temperature data with the 1st reflection for Model 3 is most likely caused by the temperature at the probe-wort interface not closely following the trend of the bulk wort temperature. Therefore, using the temperature measurement as a feature increases the model complexity with little benefit, meaning it is more difficult for the network to find an optimal solution. This further supports the aforementioned point that accurate, invasive temperature measurement would not be required with a non-invasive, no-transmission US sensing technique.

Figure 5 displays the predicted ABV percentage from the trained LSTM models for the two batches used for the test set (batches 12 and 13). Model 1 (Figure 5a and b) accurately determines the fermentation endpoint. However, the final ABV prediction is not as accurate as Model 4, indicating that it may not be sensitive enough to determine differences in final ABV between batches. Whilst Model 3 appears to have no utility, Models 1, 2, and 4 all accurately followed the ABV trajectory. Owed to the real-time data acquisition of US sensors, these models suggest that the obtained data could be used to train additional anomaly detection models to provide early warning of undesired process trajectories within a batch.

Several locations in the prediction require improvement; for example, the detection of ABV plateau for Model 2 (Figure 5c,d around the 2nd day), the settling at a final ABV for Model 3 (Figure 5e,f), and the detection of the initial ABV rise for Model 4 (Figure 5g,h around the first day). This is likely due to the varying temperature throughout the fermentation having a large effect on the US properties of the wort compared with the changing density. There are also locations of decreasing ABV prediction (Figure 5d during the first day) or sudden increases in ABV prediction (Figure 5h at the end of the fourth day). This is likely due to the temperature variations being different for each batch and the particular temperature variations during the test set causing these effects. These problems would likely be reduced through obtaining more training data.

In this work, ML models were trained to predict the ABV throughout the fermentation. However, in industrial settings this may not be the most appropriate output value with which to fit a model. For example, ML models could be trained to predict the final ABV of each batch, the time remaining until the ABV plateaus, classify the end of fermentation, or provide early detection of anomalous batches. In each of these cases, the models would be trained for a more specific purpose, as such the models may perform better than indicated by Figure 5. This work is therefore demonstrative of the efficacy of real-time fermentation monitoring using US sensors and ML, and increased accuracy may be achieved through predictions of more specific outputs.

If only the 1st reflection was used in an industrial monitoring system, the sound wave could be transferred through the vessel wall. Alternatively, if the 2nd reflection was also to be used, the probe could be fitted through existing ports common to industrial fermenters. This work monitored a laboratory scale fermentation process. At industrial scale, agitation methods are uncommon in beer fermentation to prevent damage to the yeast [36]. Therefore, radial variations in alcohol concentration exist and there would be a difference in the alcohol concentration at the sensor measurement area and the bulk wort [36]. However, previous work from our group showed that a non-local probe could accurately monitor a mixing process [23]. This is because, through machine learning, the sensor data is correlated to the location of the ground truth data, rather than the sensor. In this case, a sensor would be trained to predict the alcohol concentration at the location of the hydrometer measurements or sample collection.

Future Research Directions

The largest barrier to industrial implementation of sensors and ML combined technologies is the burden of obtaining labelled data. Labelled data is used as the targets for training supervised ML models. To obtain the ground truth to label data requires another analysis method. In this work, the hydrometer readings were used due to the sample density measurements producing insufficient data points and disturbing the fermentation. In an industrial setting, the hydrometer may only be able to be used for a small number of batches for ML model development. In this case, semi-supervised learning may be required to train high accuracy models. Semi-supervised learning uses both labelled and unlabelled samples to train a model [37–39]. Firstly, unsupervised learning techniques, such as principal component analysis or autoencoders, can be used on the total dataset to learn relationships between features across the labelled and unlabelled samples. Then traditional supervised learning can be used on the new features using just the labelled samples. Secondly, a self-training (or pseudo-labelling) approach may be used to predict the labels of the unlabelled data from the trained model. These pseudo-labels may then be added to the labelled data set and the procedure repeated to improve the label predictions or to train a final model.

Alternatively, conventional sample extraction and density measurement may be used to obtain the labelled data. Either a curve may be fitted to these sparse density measurements to produce interpolated data points, or a similar semi-supervised learning procedure can be implemented. Active learning may also be used to identify data points for labelling that may be the most useful to the model [40,41]. These datapoints may be during a sparsely sampled time in the fermentation or be in a particular temperature and composition range. Operators could then analyse these samples to provide the most benefit to the ML model at the lowest investment in effort.

5. Conclusions

The transition to Industry 4.0 promises increased manufacturing efficiency, sustainability, and productivity. By implementing digital technologies such as the Internet of

Things, Cloud Computing and ML, not only can entire processes be integrated, but supply chains as well. Sensors are a key technology in this revolution by providing the real-time data to inform automatic, intelligent decision-making. Currently, beer fermentation is monitored through periodic manual sampling and off-line wort density measurements. This work has presented an in-line, low-cost US sensing technique combined with ML, which would remove the need of operator sampling. This work has shown that US sensor data combined with LSTM models are able to accurately predict the volume of alcohol during beer fermentation. The highest accuracy model (R^2 = 0.952) used a transmission-based ultrasonic sensing technique along with the process temperature. Importantly, the second most accurate model (R^2 = 0.948) only used a reflection-based technique without measurement of the temperature. This demonstrates the potential for a non-invasive, no-transmission US technique, which doesn't require invasive measurement of the process temperature. This sensing technique could be easily and inexpensively retrofitted onto existing fermentation vessels.

Author Contributions: Conceptualization, A.B., J.E. and N.W.; methodology, A.B., J.E., N.W. and M.P.; software, A.B. and J.E.; validation, A.B.; formal analysis, A.B.; investigation, A.B. and J.E.; resources, A.B.; data curation, A.B. and J.E.; writing—original draft preparation, A.B.; writing—review and editing, A.B., N.W. and M.P.; visualization, A.B.; supervision, N.W.; project administration, N.W.; funding acquisition, N.W. All authors have read and agreed to the published version of the manuscript.

Funding: This work was supported by the Engineering and Physical Sciences Research Council (EPSRC) standard research studentship (EP/R513283/1) and EPSRC network+ Connected Everything (EP/P001246/1).

Data Availability Statement: Researchers at the University of Nottingham can be contacted for access to data.

Conflicts of Interest: The authors declare no conflict of interest.

References

1. Schock, T.; Becker, T. Sensor array for the combined analysis of water–sugar–ethanol mixtures in yeast fermentations by ultrasound. *Food Control* **2010**, *21*, 362–369. [CrossRef]
2. Resa, P.; Elvira, L.; de Espinosa, F.M.; Gómez-Ullate, Y. Ultrasonic velocity in water–ethanol–sucrose mixtures during alcoholic fermentation. *Ultrasonics* **2005**, *43*, 247–252. [CrossRef]
3. Jan, M.V.S.; Guarini, M.; Guesalaga, A.; Pérez-Correa, J.R.; Vargas, Y.; Perez-Correa, J. Ultrasound based measurements of sugar and ethanol concentrations in hydroalcoholic solutions. *Food Control* **2008**, *19*, 31–35. [CrossRef]
4. Kucharczyk, K.; Tuszyński, T. The effect of wort aeration on fermentation, maturation and volatile components of beer produced on an industrial scale. *J. Inst. Brew.* **2017**, *123*, 31–38. [CrossRef]
5. De Beer, T.; Burggraeve, A.; Fonteyne, M.; Saerens, L.; Remon, J.; Vervaet, C. Near infrared and Raman spectroscopy for the in-process monitoring of pharmaceutical production processes. *Int. J. Pharm.* **2011**, *417*, 32–47. [CrossRef] [PubMed]
6. Oztemel, E.; Gursev, S. Literature review of Industry 4.0 and related technologies. *J. Intell. Manuf.* **2020**, *31*, 127–182. [CrossRef]
7. Vann, L.; Layfield, J.B.; Sheppard, J.D. The application of near-infrared spectroscopy in beer fermentation for online monitoring of critical process parameters and their integration into a novel feedforward control strategy. *J. Inst. Brew.* **2017**, *123*, 347–360. [CrossRef]
8. Corro-Herrera, V.A.; Gómez-Rodríguez, J.; Hayward-Jones, P.M.; Barradas-Dermitz, D.M.; Gschaedler-Mathis, A.C.; Aguilar-Uscanga, M.G. Real-time monitoring of ethanol production during Pichia stipitis NRRL Y-7124 alcoholic fermentation using transflection near infrared spectroscopy. *Eng. Life Sci.* **2018**, *18*, 643–653. [CrossRef] [PubMed]
9. Wang, Q.; Li, Z.; Ma, Z.; Liang, L. Real time monitoring of multiple components in wine fermentation using an on-line auto-calibration Raman spectroscopy. *Sens. Actuators B Chem.* **2014**, *202*, 426–432. [CrossRef]
10. Mazarevica, G.; Diewok, J.; Baena, J.R.; Rosenberg, E.; Lendl, B. On-Line Fermentation Monitoring by Mid-Infrared Spectroscopy. *Appl. Spectrosc.* **2004**, *58*, 804–810. [CrossRef]
11. Veale, E.; Irudayaraj, J.; Demirci, A. An On-Line Approach to Monitor Ethanol Fermentation Using FTIR Spectroscopy. *Biotechnol. Prog.* **2007**, *23*, 494–500. [CrossRef]
12. Toledo, J.; Ruiz-Díez, V.; Pfusterschmied, G.; Schmid, U.; Sánchez-Rojas, J. Flow-through sensor based on piezoelectric MEMS resonator for the in-line monitoring of wine fermentation. *Sens. Actuators B Chem.* **2018**, *254*, 291–298. [CrossRef]
13. Schalk, R.; Frank, R.; Rädle, M.; Methner, F.-J.; Beuermann, T.; Braun, F.; Gretz, N. Non-contact Raman spectroscopy for in-line monitoring of glucose and ethanol during yeast fermentations. *Bioprocess Biosyst. Eng.* **2017**, *40*, 1519–1527. [CrossRef] [PubMed]

14. Ete-Carmona, E.C.; Gallego-Martinez, J.-J.; Martin, C.; Brox, M.; Luna-Rodriguez, J.-J.; Moreno, J. A Low-Cost IoT Device to Monitor in Real-Time Wine Alcoholic Fermentation Evolution through CO_2 Emissions. *IEEE Sens. J.* **2020**, *20*, 6692–6700. [CrossRef]
15. Hussein, W.B.; Hussein, M.A.; Becker, T. Robust spectral estimation for speed of sound with phase shift correction applied online in yeast fermentation processes. *Eng. Life Sci.* **2012**, *12*, 603–614. [CrossRef]
16. Hoche, S.; Krause, D.; Hussein, M.A.; Becker, T. Ultrasound-based, in-line monitoring of anaerobe yeast fermentation: Model, sensor design and process application. *Int. J. Food Sci. Technol.* **2016**, *51*, 710–719. [CrossRef]
17. Resa, P.; Elvira, L.; De Espinosa, F.M. Concentration control in alcoholic fermentation processes from ultrasonic velocity measurements. *Food Res. Int.* **2004**, *37*, 587–594. [CrossRef]
18. Resa, P.; Elvira, L.; De Espinosa, F.M.; González, R.; Barcenilla, J. On-line ultrasonic velocity monitoring of alcoholic fermentation kinetics. *Bioprocess Biosyst. Eng.* **2008**, *32*, 321–331. [CrossRef]
19. Ojha, K.S.; Mason, T.J.; O'Donnell, C.P.; Kerry, J.P.; Tiwari, B.K. Ultrasound technology for food fermentation applications. *Ultrason. Sonochem.* **2017**, *34*, 410–417. [CrossRef]
20. Henning, B.; Rautenberg, J. Process monitoring using ultrasonic sensor systems. *Ultrasonics* **2006**, *44*, e1395–e1399. [CrossRef]
21. Schock, T.; Hussein, M.; Hitzmann, B.; Becker, T. Influence of dissolved carbon dioxide on the sound velocity and adiabatic compressibility in aqueous solutions with saccharose and ethanol. *J. Mol. Liq.* **2012**, *175*, 111–120. [CrossRef]
22. Hoche, S.; Hussein, W.B.; Hussein, M.A.; Becker, T. Time-of-flight prediction for fermentation process monitoring. *Eng. Life Sci.* **2011**, *11*, 417–428. [CrossRef]
23. Bowler, A.L.; Bakalis, S.; Watson, N.J. Monitoring Mixing Processes Using Ultrasonic Sensors and Machine Learning. *Sensors* **2020**, *20*, 1813. [CrossRef] [PubMed]
24. Escrig, J.; Woolley, E.; Simeone, A.; Watson, N. Monitoring the cleaning of food fouling in pipes using ultrasonic measurements and machine learning. *Food Control* **2020**, *116*, 107309. [CrossRef]
25. Escrig, J.E.; Simeone, A.; Woolley, E.; Rangappa, S.; Rady, A.; Watson, N. Ultrasonic measurements and machine learning for monitoring the removal of surface fouling during clean-in-place processes. *Food Bioprod. Process.* **2020**, *123*, 1–13. [CrossRef]
26. Escrig, J.; Woolley, E.; Rangappa, S.; Simeone, A.; Watson, N. Clean-in-place monitoring of different food fouling materials using ultrasonic measurements. *Food Control* **2019**, *104*, 358–366. [CrossRef]
27. Kitchn. Available online: https://www.thekitchn.com/how-to-check-and-control-alcohol-levels-the-kitchns-beer-school-2015-217260 (accessed on 14 January 2021).
28. BrewMoreBeer. Available online: http://www.brewmorebeer.com/calculate-percent-alcohol-in-beer/ (accessed on 24 February 2021).
29. Zhan, X.; Jiang, S.; Yang, Y.; Liang, J.; Shi, T.; Li, X. Inline Measurement of Particle Concentrations in Multicomponent Suspensions using Ultrasonic Sensor and Least Squares Support Vector Machines. *Sensors* **2015**, *15*, 24109–24124. [CrossRef]
30. McClements, D. Advances in the application of ultrasound in food analysis and processing. *Trends Food Sci. Technol.* **1995**, *6*, 293–299. [CrossRef]
31. Hochreiter, S.; Schmidhuber, J. Long Short-Term Memory. *Neural Comput.* **1997**, *9*, 1735–1780. [CrossRef]
32. LeCun, Y.; Bengio, Y.; Hinton, G. Deep learning. *Nature* **2015**, *521*, 436–444. [CrossRef]
33. Jain, A.; Mao, J.; Mohiuddin, K. Artificial neural networks: A tutorial. *Computer* **1996**, *29*, 31–44. [CrossRef]
34. Machine Learning Mastery. Available online: https://machinelearningmastery.com/handle-long-sequences-long-short-term-memory-recurrent-neural-networks/ (accessed on 9 February 2021).
35. Lamberti, N.; Ardia, L.; Albanese, D.; Di Matteo, M. An ultrasound technique for monitoring the alcoholic wine fermentation. *Ultrasonics* **2009**, *49*, 94–97. [CrossRef]
36. Nienow, A.W.; McLeod, G.; Hewitt, C.J. Studies supporting the use of mechanical mixing in large scale beer fermentations. *Biotechnol. Lett.* **2010**, *32*, 623–633. [CrossRef] [PubMed]
37. Kostopoulos, G.; Karlos, S.; Kotsiantis, S.; Ragos, O. Semi-supervised regression: A recent review. *J. Intell. Fuzzy Syst.* **2018**, *35*, 1483–1500. [CrossRef]
38. Forestier, G.; Wemmert, C. Semi-supervised learning using multiple clusterings with limited labeled data. *Inf. Sci.* **2016**, *361-362*, 48–65. [CrossRef]
39. Ge, Z.; Song, Z.; Ding, S.X.; Huang, B. Data Mining and Analytics in the Process Industry: The Role of Machine Learning. *IEEE Access* **2017**, *5*, 20590–20616. [CrossRef]
40. Burbidge, R.; Rowland, J.J.; King, R.D. Active Learning for Regression Based on Query by Committee. *Comput. Vis.* **2007**, *4881*, 209–218. [CrossRef]
41. Cai, W.; Zhang, Y.; Zhou, J. Maximizing expected model change for active learning in regression. In Proceedings of the 13th IEEE International Conference on Data Mining, Dallas, TX, USA, 7–10 December 2013; pp. 51–60.

 fermentation

Article

Determination of Foam Stability in Lager Beers Using Digital Image Analysis of Images Obtained Using RGB and 3D Cameras

Emmanuel Karlo Nyarko [1], Hrvoje Glavaš [1,*], Kristina Habschied [2,*] and Krešimir Mastanjević [2]

[1] Faculty of Electrical Engineering, Computer Science and Information Technology Osijek, Josip Juraj Strossmayer University of Osijek, Kneza Trpimira 2B, 31000 Osijek, Croatia; karlo.nyarko@ferit.hr
[2] Faculty of Food Technology, Josip Juraj Strossmayer University of Osijek, 31000 Osijek, Croatia; kmastanj@gmail.com
* Correspondence: hrvoje.glavas@ferit.hr (H.G.); kristinahabschied@gmail.com (K.H.); Tel.: +385-31-224-645 (H.G.); +385-31-224-300 (K.H.)

Abstract: Foam stability and retention is an important indicator of beer quality and freshness. A full, white head of foam with nicely distributed small bubbles of CO_2 is appealing to the consumers and the crown of the production process. However, raw materials, production process, packaging, transportation, and storage have a big impact on foam stability, which marks foam stability monitoring during all these stages, from production to consumer, as very important. Beer foam stability is expressed as a change of foam height over a certain period. This research aimed to monitor the foam stability of lager beers using image analysis methods on two different types of recordings: RGB and depth videos. Sixteen different commercially available lager beers were subjected to analysis. The automated image analysis method based only on the analysis of RGB video images proved to be inapplicable in real conditions due to problems such as reflection of light through glass, autofocus, and beer lacing/clinging, which make it impossible to accurately detect the actual height of the foam. A solution to this problem, representing a unique contribution, was found by introducing the use of a 3D camera in estimating foam stability. According to the results, automated analysis of depth images obtained from a 3D camera proved to be a suitable, objective, repeatable, reliable, and sufficiently sensitive method for measuring foam stability of lager beers. The applied model proved to be suitable for predicting changes in foam retention of lager beers.

Keywords: foam stability; image analysis; lager beer; foam retention

Citation: Nyarko, E.K.; Glavaš, H.; Habschied, K.; Mastanjević, K. Determination of Foam Stability in Lager Beers Using Digital Image Analysis of Images Obtained Using RGB and 3D Cameras. *Fermentation* **2021**, *7*, 46. https://doi.org/10.3390/fermentation7020046

Academic Editor: Sigfredo Fuentes

Received: 21 February 2021
Accepted: 24 March 2021
Published: 26 March 2021

1. Introduction

Ancient beer displayed weak or no foam. Uncontrolled fermentation, no addition of hops, and subsequent low carbonation resulted in a foam-less beverage. Research conducted by [1] on ancient Finnish beer Sahti showed that the investigated ancient beverage had no foam and showed a distinctive difference between today's beers, especially regarding flavor and aroma. Today's brewing industries are far from the ancient manufacturers, and stable and retentive foam head is one of the main indicators of beer freshness and quality. Even though a big and rich head of foam is a property of certain types of beer (lager, pilsner, and wheat beer among others), every consumer seeks freshness in a preferable label. Some people do not appreciate foam in their glass, regardless of the beer style, some love the lacy pattern at the bottom of the finished beer, but the majority of beer-lovers like the crystal-clear glass after finishing the last sip [2]. Cling can be described as the adhesion of beer foam to the side of the glass during beer consumption, commonly known as "lacing". For example, Belgium is known for beers that leave a lacy glass. According to BJCP Beer Style Guidelines [3], "Belgian Lace is a characteristic and persistent latticework pattern of foam left on the inside of the glass as a beer is consumed. The look is reminiscent of fine lacework from Brussels or Belgium, and is a desirable indicator of beer quality in Belgium." According to Bamforth [2], foam quality is described by many properties such as stability, retention, viscosity, whiteness, bubble size, and density. As described by Gonzalez Viejo et al. [4], the most presentable indicators of foam behavior are foamability (capacity of foam formation) and foam stability. Many factors affect foam stability and the physics behind the foam is extensive, but in short, foams are colloids comprising gas bubbles dispersed in liquid [2]. A detailed description of the physics of foam formation and stability is well described by Bamforth [2] and Hackbarth [5]. Many other authors also did an excellent job in reporting and describing the physics of foam [6–9]. A more recent contribution was provided by Gonzalez Viejo et al. [4] in an extensive review. According to [5], a decrease in surface tension reduces foam stability. This can be influenced by high fermentation temperatures resulting in fusel oils accumulation, yeast autolysis (re-leases lipids and protease A), an unpasteurized beer that allows protease A, or unclean or improperly rinsed serving glasses. Agents that aid foam stability are high molecular weight (MW > 5000) malt proteins (Z4, LTP1, lipid binding indolines, and hordeins). The choice of raw material, malting process conditions, kilning temperatures, mashing-in temperatures, excessive boiling, and excessive use of chillproofing filter-aids can influence the foam stability. A reaction between malt proteins and isomerized hop alpha acids also helps in foam stability.

Due to its importance for the brewing industry, many scholars and professionals came up with several methods for the determination and measurement of foam stability and quality [10–21]. Many of these properties, such as determination of foam stability or head retention, lacing, bubble size, whiteness, foam density, foam viscosity, foam strength, etc., are well described in a review paper by Bamfort [2]. However, modern methodologies and approaches tend to be less or even non-invasive and are therefore suitable for application in foam measurements [22–25]. Among the most popular methods are currently image analysis methods. These methods, however, have some drawbacks, especially in scenarios with automated procedures. The first problem is the reflection of light on the surface and within the beer glass, as reported by Lukinac et al. [26]. Another problem that may arise is focusing; the measuring set-up must be made under controlled directional light conditions that entail lower illuminance values and automated focusing. The third and biggest problem is foam lacing or clinging, which makes it impossible to detect the actual level of foam.

Electrode-involving method is excellent for measuring foam height (its decrease over time, and therefore its stability), but it demands regular cleaning and maintenance and is more time-consuming. The use of a camera in foam assessment demands no cleaning and is a low-maintenance, cheap, and easy method that requires minimal input from the employees, applicable in every laboratory. The implementation of a simple, easy, affordable, and non-invasive method that could solve or, at least, derail all of the above-mentioned problems would greatly help the brewing industry in an objective assessment of beer foam stability.

The aim of this paper was to analyze the applicability of an automated non-invasive, objective, and cheap image analysis method under real conditions, to follow up and measure the foam stability of lager beers produced and available on the Croatian market.

2. Materials and Methods

Sixteen samples of commercially available light lager beers packaged in brown or green glass bottles (0.5 L) were set to be analyzed using the methods described below. Ten were domestic beers and six others were foreign (Germany, Czech Republic, Denmark, and Holland). Beer samples were held at room temperature (20 °C) for two days, in order to reduce the influence of temperature fluctuations. Glasses (0.5 L; generic brand, model Lilith, h = 185 mm, Ø = 0.75 mm) were bought, washed, and degreased then rinsed in demineralized water and left to dry two days prior to the analysis. All glasses were identical and held at room temperature for two days prior to analysis. Beer was hand-poured into a degreased glass, according to the standard MEBAK (Middle European Brewing Analysis

Commission) procedure (method 2.23.1) [27], at an angle of 135°. The pourer took extra care to pour every sample evenly and uniformly into the glass (Figure 1). When the pouring was done, the sample was placed on a designated spot, and cameras were set to measure the foam stability. According to MEBAK, foam in lager beers should be stable for 3–5 min.

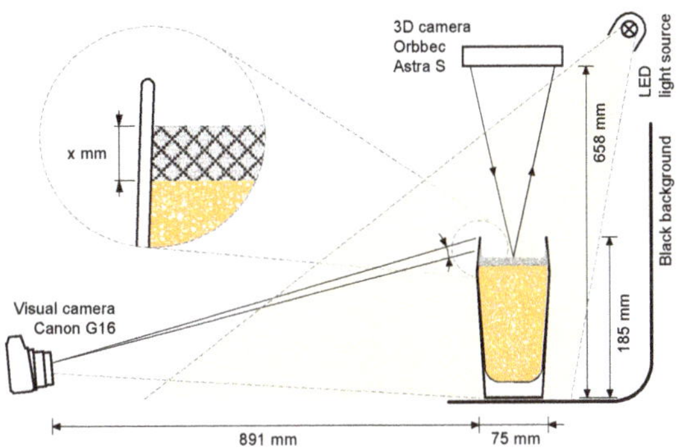

Figure 1. Experimental setup.

Two vision-based approaches to measuring foam stability were implemented: image analysis of RGB video and depth measurement using a 3D camera. Over a period of 5 min, a visual RGB camera (Canon PowerShot G16; Ota City, Tokyo, Japan) was used to take a video recording of the beer. Simultaneously, a 3D camera (Orbbec Astra S; Orbbec 3D Technology International, Inc., Troy, MI, USA) was also used to take depth measurements. Figure 1 shows the experimental setup used in data collection.

2.1. Image Analysis of Video

A visual RGB camera (Canon G16) was used to take a video recording of each sample over a period of 5 min. The recorded video had a frame rate of 30 fps. Image analysis was performed on the recorded video. The image analysis procedure to determine the height of beer foam is explained in the following seven steps below:

Step 1: For a given frame of the recorded video, define a region of interest (ROI) of known width (w) and height (l) in the image that contains beer and foam, as shown in Figure 2a. It is important that the ROI covers the whole height of the foam and part of the beer. All the following image processing steps are performed on this ROI. This ensures that all the image processing steps are directed or focused on segmenting the foam head from the rest of the image within the ROI.

Step 2: Perform color segmentation by filtering (thresholding) the ROI in HSV color space by defining the lower and upper values of the color of the foam (Figure 2b). HSV color space separates color information (chroma) from intensity (luma). Since the value is separated, thresholding can theoretically be performed using only saturation and hue. More robust color thresholding over simpler parameters can be performed in HSV color space than in RGB color space. For the purposes of the results presented in this paper, the lower HSV boundary of (0,0,230)was used, while the upper boundary was defined as (43,18,255).

Step 3: Generate a binary image of the thresholded ROI in HSV color space (Figures 2c and 3a), i.e., all pixels that have values within the defined boundaries are marked as white pixels (values are set to 255), while the remaining pixels are marked as black (values are set to 0).

Step 4: Perform morphological operations of erosion followed by dilation on the binary image (Figures 2d and 3b). These operations are needed in order to eliminate small white noises or white artifacts that appear in the binary image.

Step 5: Determine the largest contour from the list of all contours on the binary image (Figure 2e). A contour is a curve joining all the continuous points or connected components (along the boundary) having the same color or intensity. This step basically segments or marks the boundary of the foam/head.

Step 6: Determine the area (A) of the region defined by the largest contour.

Step 7: The average height (h) of the beer foam in pixels can be determined using Equation (1):

$$h = A/w. \tag{1}$$

Using the notation where h_t represents the height of foam (in pixels) at a given point in time t (in seconds), the maximum height, h_{max}, is defined as:

$$h_{max} = \max\{h_t\colon t = 0, 1, \ldots, 300\}, \tag{2}$$

and the minimum height, h_{min}, is defined as

$$h_{min} = \min\{h_t\colon t = 0, 1, \ldots, 300\}. \tag{3}$$

Based on these definitions, we define the normalized foam height at time t, h_{t_norm}, as:

$$h_{t\_norm} = (h_t - h_{min})/(h_{max} - h_{min}). \tag{4}$$

Since the beer glass is always located in the same position, the seven steps provided above can be implemented as a program to automate the procedure. One advantage of this procedure is that it can be run in both offline and online mode. For the purposes of this paper, a script written in the Python programming language [28] using the OpenCV library [29] was implemented in order to automate the process of determining the beer foam height from the recorded videos. Every 10 s, five consecutive frames were taken and the height of foam determined for each frame. The average height in pixels for these five measurements was taken to represent the height of foam every 10 s.

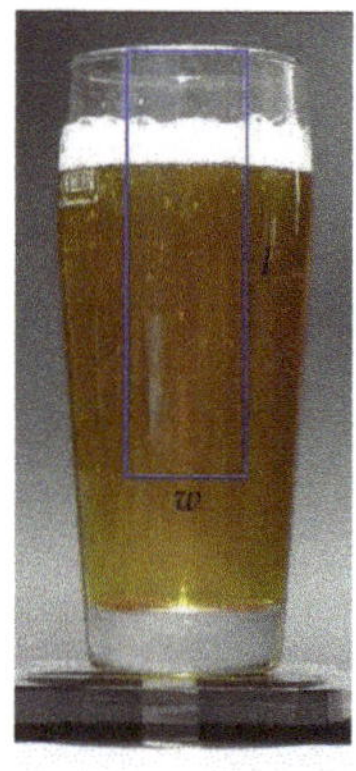

(a)

(b)

(c)

(d)

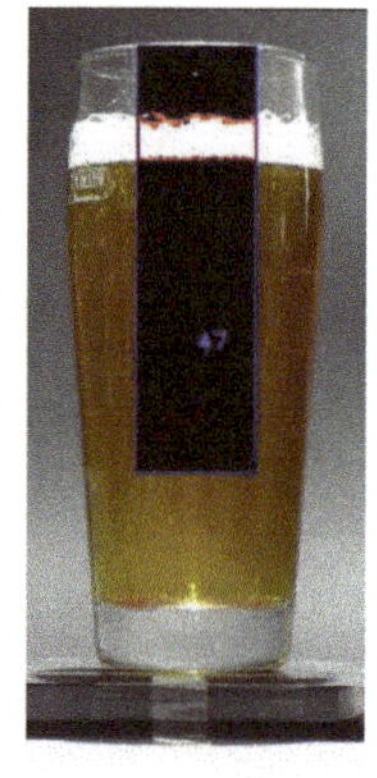

(e)

Figure 2. Estimating foam height from an image. (**a**) Region of interest (ROI) of known width (w) and height (l) defined for the image. The ROI is marked in blue; (**b**) thresholding of ROI performed in HSV color space. The pixels satisfying the threshold are marked in red; (**c**) binary image of ROI thresholded in HSV color space (a magnified image is provided in Figure 3a); (**d**) morphological operations of erosion followed by dilation performed on binary image to remove artifacts (a magnified image is provided in Figure 3b); (**e**) largest contour found marked in red. The estimated height of foam in pixels is determined using Equation (1).

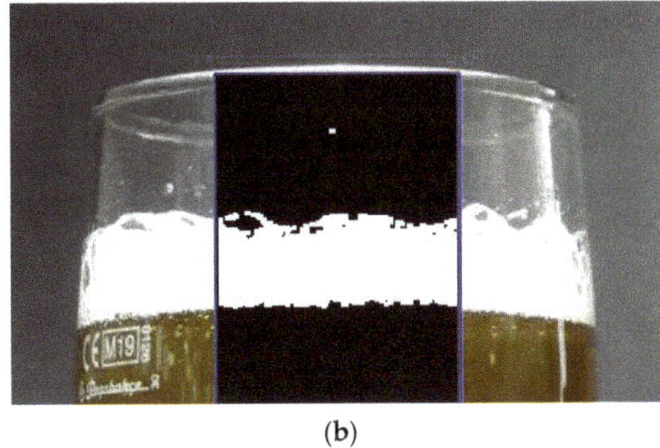

(a) (b)

Figure 3. Magnified images of Figure 2a,b: (**a**) Initial binary image of ROI thresholded in HSV color space with visible artifacts; (**b**) Final binary image of ROI thresholded in HSV color space with artifacts removed after performing morphological operations.

2.2. Depth Measurement Using a 3D Camera

A 3D camera provides a depth map or a depth image where each pixel in the image relates to the distance between the surface of the object being viewed and the camera or image plane. The Orbbec Astra S 3D camera used in this paper is based on the Structured-Light technology. The 3D camera consists of an infrared laser projector and a proprietary Infra-Red (IR) depth sensor. The depth sensor interprets 3D scene information based on continuously projected infrared structured light. It should also be mentioned that Orbbec Astra S 3D camera also consists of an RGB camera. However, this RGB camera was not used in the experiments for the purpose of this paper. An example of a depth image generated by the 3D camera is shown in Figure 4.

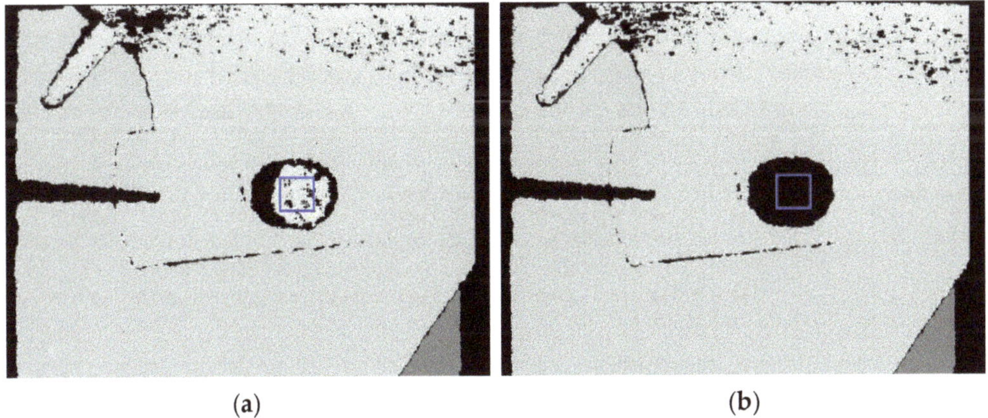

(a) (b)

Figure 4. Examples of depth images obtained using Orbbec Astra S 3D camera. Black pixels indicate that no depth information exists: (**a**) ROI defined in the middle of the glass containing foam; (**b**) the same view after a few minutes when the foam disappears.

Orbbec Astra S depth sensor has a camera resolution of 640 $\times$ 480 and a maximum frame rate of 30 Hz. Its measurement range is from 0.4 to 2 m and has a field of view of 60° horizontally and 49.5° vertically. It also has an accuracy of $+/-1$–3 mm at 1 m.

One drawback of the sensor is that it cannot detect glass nor liquids, so for example, in Figure 4, it can be seen that the edge of the beer glass cannot be detected (pixels displayed in black), and when the foam disappears no depth information can be obtained.

As displayed in Figure 1, the 3D camera was placed above the beer glass. Hence, all measurements obtained from the 3D camera actually provided the distance of the top of the foam head from the 3D camera. Thus, measurements of the distance of the foam for a given sample increased with time, as the foam in the glass decreased.

Similar to the measurements performed when using the video camera in the previous section, since the beer glass was always located in the same position, a fixed ROI was defined (Figure 4a), and a Python script was used in order to automate the process of determining the distance of the beer foam height from the camera. Every 10 s, five consecutive frames were taken, and the average distance of foam from the camera was determined for each frame. This average distance was determined using only the pixels within the defined ROI having a depth value. Pixels without depth values were excluded. The mean distance of these average distances for the five measurements was taken to represent the distance of foam from the camera every 10 s.

If d_t represents the distance of foam (in mm) from the camera at a given point in time (t = 0, 1, . . . ,300 s), and d_{table} represents the distance of the table (in mm) from the camera (d_{table} = 654 mm. Figure 1.), the height of the foam from the top of the table at time t, $diff_t$, is given by

$$diff_t = d_{table} - d_t. \quad (5)$$

The maximum height of the foam from the top of the table, $diff_{max}$, is defined as:

$$diff_{max} = \max\{diff_t\text{: } t = 0, 1, \ldots ,300\}, \quad (6)$$

and the minimum height, $diff_{min}$, is defined as

$$diff_{min} = \min\{diff_t\text{: } t = 0, 1, \ldots ,300\}. \quad (7)$$

Based on these definitions, the normalized foam height at time t, h_{t_norm}, is given by:

$$h_{t\_norm} = (diff_t - diff_{min})/(diff_{max} - diff_{min}). \quad (8)$$

It is important to emphasize that the images displayed in Figure 4 represent depth images. Areas in the images having shades of gray have depth information, while those marked in black do not. The edge of the beer glass cannot be detected in Figure 4a due to the fact that (a) the sensor cannot detect glass, since the transmitted light is not reflected, and (b) the non-defined area near the beer glass is also extended as a result of parallax, since the emitter and the sensor on the 3D camera are separated by about 7.5 cm. In Figure 4b, after the foam disappears, the transmitted beam of the 3D sensor cannot be reflected by the beer surface, and therefore it is impossible to detect the depth of the beer surface. This criterion was used for ending measurements in situations when the foam disappeared before 5 min.

3. Results and Discussion

RGB and depth video recordings were obtained for 16 samples (denoted by s01 . . . s16), each lasting 5 min. Using the procedures described in Sections 2.1 and 2.2, the estimated height (in pixels) and distance of foam from the camera (in mm) were obtained from the RGB video and depth video, respectively, every 10 s. The results of the measurements are displayed in Tables 1 and 2. Figures 5 and 6 display the corresponding normalized foam heights (%) obtained by performing image analysis on RGB images and from depth images, respectively. The actual normalized values obtained are provided in Tables A1 and A2.

Table 1. Height of foam (pixels) determined by performing image analysis on RGB videos of 16 beer samples (the actual value in mm can be obtained using the conversion 1 mm = 6.6 px).

Time (s)	s01	s02	s03	s04	s05	s06	s07	s08	s09	s10	s11	s12	s13	s14	s15	s16
0	-	46	91	176	229	207	308	264	320	308	280	187	32	169	124	318
10	-	40	32	67	137	103	231	172	269	236	183	60	8	138	108	285
20	-	37	12	28	80	45	193	120	212	155	96	24	-	110	85	280
30	-	43	3	10	49	25	163	77	181	111	62	-	-	91	79	246
40	-	44	-	-	37	26	147	58	152	72	35	-	-	80	52	219
50	-	47	-	-	-	-	135	59	126	63	33	-	-	72	41	198
60	-	51	-	-	-	-	123	58	148	55	-	-	-	64	34	179
70	-	48	-	-	-	-	111	57	124	46	-	-	-	58	23	158
80	-	49	-	-	-	-	101	-	69	41	-	-	-	56	-	143
90	-	57	-	-	-	-	93	-	51	35	-	-	-	50	-	129
100	-	56	-	-	-	-	86	-	44	29	-	-	-	46	-	115
110	-	56	-	-	-	-	84	-	45	-	-	-	-	43	-	103
120	-	59	-	-	-	-	82	-	53	-	-	-	-	34	-	91
130	-	59	-	-	-	-	80	-	52	-	-	-	-	-	-	82
140	-	55	-	-	-	-	78	-	-	-	-	-	-	-	-	81
150	-	62	-	-	-	-	77	-	-	-	-	-	-	-	-	73
160	-	58	-	-	-	-	75	-	-	-	-	-	-	-	-	69
170	-	56	-	-	-	-	74	-	-	-	-	-	-	-	-	64
180	-	53	-	-	-	-	72	-	-	-	-	-	-	-	-	73
190	-	51	-	-	-	-	72	-	-	-	-	-	-	-	-	68
200	-	50	-	-	-	-	73	-	-	-	-	-	-	-	-	66
210	-	51	-	-	-	-	73	-	-	-	-	-	-	-	-	64
220	-	50	-	-	-	-	72	-	-	-	-	-	-	-	-	63
230	-	48	-	-	-	-	71	-	-	-	-	-	-	-	-	62
240	-	43	-	-	-	-	72	-	-	-	-	-	-	-	-	60
250	-	37					69	-	-	-	-	-	-	-	-	58
260	-	41	-	-	-	-	70	-	-	-	-	-	-	-	-	58
270	-	37	-	-	-	-	68	-	-	-	-	-	-	-	-	56
280	-	35	-	-	-	-	-	-	-	-	-	-	-	-	-	54
290	-	35	-	-	-	-	-	-	-	-	-	-	-	-	-	54
300	-	34	-	-	-	-	-	-	-	-	-	-	-	-	-	53

"-" indicates that no measurements were made since there was no foam head. All values have been rounded up.

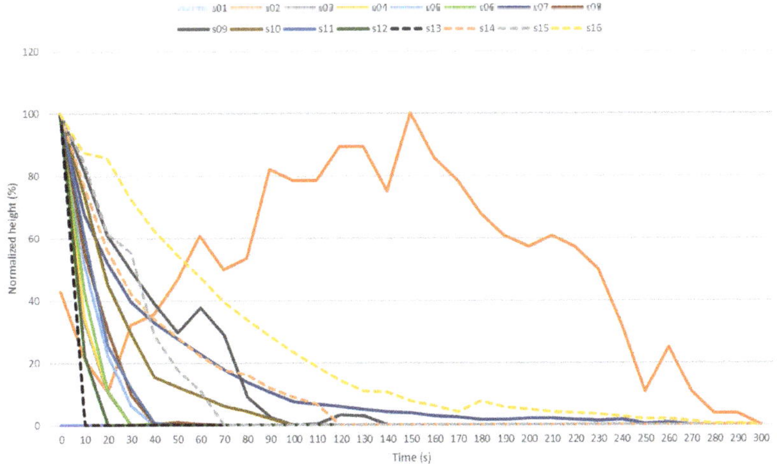

Figure 5. Normalized foam height (%) determined by performing image analysis on RGB videos of 16 beer samples.

Table 2. The distance of foam head surface (mm) from the top of the table for 16 beer samples.

Time (s)	s01	s02	s03	s04	s05	s06	s07	s08	s09	s10	s11	s12	s13	s14	s15	s16
0	158	159	161	172	171	177	181	176	178	167	177	170	134	168	165	191
10	41	158	160	169	167	169	177	171	174	166	175	162	36	166	163	189
20	-	158	152	157	156	157	173	163	169	154	165	156	-	163	160	184
30	-	158	-	153	148	152	169	156	165	145	155	-	-	161	158	181
40	-	159	-	-	146	-	166	152	161	137	149	-	-	159	157	177
50	-	159	-	-	145	-	163	150	157	134	147	-	-	157	156	174
60	-	159	-	-	-	-	161	150	153	133	145	-	-	156	155	171
70	-	159	-	-	-	-	158	149	150	132	-	-	-	156	155	169
80	-	159	-	-	-	-	156	-	146	131	-	-	-	155	-	166
90	-	159	-	-	-	-	154	-	145	131	-	-	-	155	-	164
100	-	159	-	-	-	-	152	-	144	130	-	-	-	155	-	162
110	-	159	-	-	-	-	150	-	144	-	-	-	-	155	-	160
120	-	159	-	-	-	-	149	-	144	-	-	-	-	155	-	159
130	-	160	-	-	-	-	149	-	-	-	-	-	-	-	-	157
140	-	159	-	-	-	-	148	-	-	-	-	-	-	-	-	156
150	-	159	-	-	-	-	148	-	-	-	-	-	-	-	-	155
160	-	159	-	-	-	-	147	-	-	-	-	-	-	-	-	154
170	-	159	-	-	-	-	147	-	-	-	-	-	-	-	-	154
180	-	159	-	-	-	-	147	-	-	-	-	-	-	-	-	154
190	-	159	-	-	-	-	147	-	-	-	-	-	-	-	-	153
200	-	159	-	-	-	-	147	-	-	-	-	-	-	-	-	153
210	-	159	-	-	-	-	147	-	-	-	-	-	-	-	-	153
220	-	159	-	-	-	-	147	-	-	-	-	-	-	-	-	153
230	-	159	-	-	-	-	147	-	-	-	-	-	-	-	-	153
240	-	158	-	-	-	-	147	-	-	-	-	-	-	-	-	153
250	-	158	-	-	-	-	147	-	-	-	-	-	-	-	-	153
260	-	158	-	-	-	-	147	-	-	-	-	-	-	-	-	153
270	-	158	-	-	-	-	-	-	-	-	-	-	-	-	-	153
280	-	157	-	-	-	-	-	-	-	-	-	-	-	-	-	153
290	-	157	-	-	-	-	-	-	-	-	-	-	-	-	-	153
300	-	157	-	-	-	-	-	-	-	-	-	-	-	-	-	153

"-" indicates that no measurements were made since there was no foam head. All values have been rounded up.

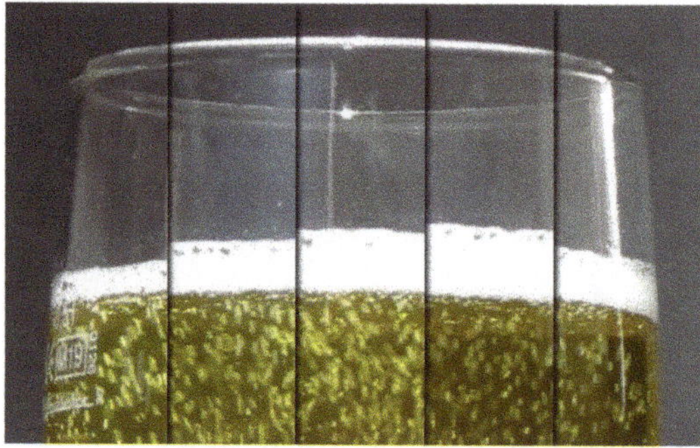

Figure 6. Changes of foam head of sample s02 over time due to the erratic behavior of CO_2 bubbles.

The processed results of the measurements obtained using the visual RGB camera are displayed in Table 1. Even though the results are displayed in pixels, the corresponding height of beer foam can be obtained by using the conversion of 1 mm = 6.6 px. The sign "-" in the table indicates that measurements were stopped since there was no foam.

A graphical representation of the normalized measurement results obtained by performing image analysis on RGB videos can be seen in Figure 5 (actual data is provided in Table A1). The behavior of the s02 sample is due to the increase in foam levels during the measurement as a result of erratic CO_2 bubbles that formed unstable foam, as can be seen in Figure 6. According to Bamforth [2], low surface tension is an important factor for foam formation. Constant surface tension withholds a pressure within a bubble that is inversely proportional to its diameter, so the gas makes an effort to pass from a smaller to larger bubble (disproportionation), so the more gas there is in foam, the greater the disproportionation, which was the case for most samples, but sample s02 was particularly erratic. If the gas fraction in liquid (beer) foams is high, the bubbles cannot form exclusively spheres, but they take forms of polyhedra separated by thin layers of the liquid phase called lamellae. Another important phenomenon is that hydrophobic particles adsorbed at the gas–liquid interface tend to compress together as bubbles contract to form barriers that prevent the continuation of disproportionation. At constant pressure, the size of bubbles is directly proportional to the surface tension. Thus, materials with lower surface tension also give smaller bubbles [2]. Coalescence or merging of two bubbles occurs upon rupture of the membrane that divides them. This leads to coarsening of foam with visible larger bubbles–fish eyes in the foam body [4]. The Young–Laplace equation describes the disproportionation as the differential pressures between the inside and outside of a bubble due to surface tension. This pressure is inversely proportional to the bubble radius, causing CO_2 gas to diffuse from smaller bubbles where pressure is higher into larger bubbles. According to Hackbath [4], "as the foam structure coarsens and larger bubbles continue to expand, their membranes thin until they reach a critical thickness. Film ruptures can be spontaneous or can be caused by fats that interfere with the film's external surface. Collapse occurs at the crown surface by rupture or by diffusion of dipolar CO_2 directly to the atmosphere through the CO_2 permeable bubble film". Comprehension of all stated data could explain the behavior of foam in sample s02. It can be presumed that this is due to the storage in unsuitable conditions in the supermarket storage space. All samples were purchased in January, when it was cold in the storage rooms of the market place, and all analyses were done in January. The temperature fluctuations in the storage room, where it is cooler, then sudden transfer to higher temperatures at the market place could cause this kind of foaming properties loss in most of the samples. As for the sample s02, it could be some type of production error in this particular batch. Apart from sample s02, samples s16 and s07 seemed to show a more stable foam in comparison to all the other samples. It appears that this foam showed significantly more stable properties, even though all samples were kept at the same temperature. This hypothesis has yet to be confirmed by detailed laboratory testing, although preliminary analysis of new samples obtained in March (which show normal behavior) lead us to this conclusion.

Depth measurements were also being taken simultaneously using a 3D camera. The distance of the beer foam surface (in mm) from the top of the table measured over time, $diff_t$, for the 16 beer samples is displayed in Table 2. The sign "-" indicates that measurements were not possible since there was no foam head.

A graphical representation of the normalized distance of beer foam (%) from 3D camera can be seen in Figure 7 (actual data is provided in Table A2). Comparing Figures 5 and 6, similar conclusions about the foam stability can be made.

One major drawback of the non-invasive automated image analysis of RGB images is that it is sensitive to foam lacing or clinging. For example, s06 has a foam height of 26 px (or about 4 mm) after 40 s (see Table 1). Figure 8 shows the RGB video frame after 40 s. On the other hand, depth measurements of the foam surface by the 3D camera were not possible after 30 s, since there was no foam on the liquid surface (scenario similar to Figure 4b). This feedback (lack of depth information) from the 3D sensor was then used to stop further measurement. It should also be noted that the measurements after 10 s (see Table 2) for samples s01 and s13 should basically be ignored, since this were unreliable measurements provided by the 3D sensor in situations where there was basically no foam.

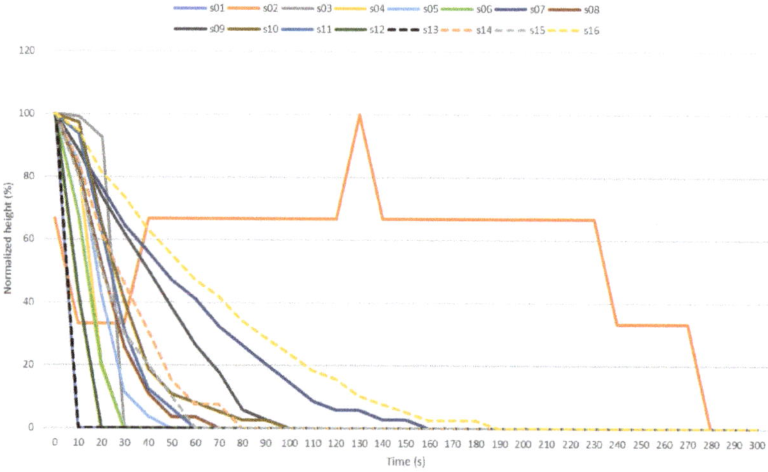

Figure 7. A normalized distance of beer foam (%) from 3D camera for 16 beer samples.

(**a**) (**b**)

Figure 8. Sample s06 after 40 s. (**a**) Foam is detected by automated image analysis of RGB image, even though this is just foam clinging to the beer glass; (**b**) 3D camera records this as lack of foam.

4. Conclusions

Beer foam stability is an important beer quality indicator. Stable beer foam after production does not have to correlate with beer foam after a certain period of storage and transport. In this research, we presented an algorithm for an automated non-invasive procedure for measuring foam height by applying image analysis of RGB images or videos. The procedure showed off as relatively robust and applicable in online and offline mode. One major drawback of this method appeared to be its sensitivity to foam lacing or cling due to poor CO_2 distribution or low foam active/stabilizing compounds concentrations (proteins or hop compounds) in beer where the camera, placed laterally in regards to the sample, could not distinguish the lacing from foam height. However, this problem was resolved by using a 3D camera, which generates depth videos or images. A 3D camera mounted directly above the glass containing the beer sample was able to measure the distance of the foam surface from the camera. By measuring the change in the distance of the foam surface from the 3D camera over time, information about the foam stability was obtained. Due to the technology used in measuring the distance of objects from the 3D camera, the camera was able to detect and recognize the foam surface, but was not able to detect the liquid surface, and therefore the distance of the liquid surface from the camera could not be measured. This usual drawback of this camera is actually an advantage in this scenario, since it provides information about the disappearance of the beer foam. Information about the lack of foam is triggered by the lack of depth information within the ROI. In any case, this could be a novel, quick, robust, precise, and accurate method for foam stability measurement.

Author Contributions: Conceptualization: K.H. and E.K.N.; methodology: E.K.N. and H.G.; software: E.K.N.; formal analysis: H.G.; investigation: K.M.; writing—original draft preparation: K.H.; writing—review and editing: E.K.N. and K.M. All authors have read and agreed to the published version of the manuscript.

Funding: This research received no external funding.

Institutional Review Board Statement: Not applicable.

Informed Consent Statement: Not applicable.

Data Availability Statement: The data presented in this study are available on request from the corresponding author.

Conflicts of Interest: The authors declare no conflict of interest.

Appendix A

Table A1. Normalized foam height (%) determined by performing image analysis on RGB videos of 16 beer samples.

Time (s)	s01	s02	s03	s04	s05	s06	s07	s08	s09	s10	s11	s12	s13	s14	s15	s16
0	0	43	100	100	100	100	100	100	100	100	100	100	100	100	100	100
10	-	21	33	34	52	43	68	56	82	74	61	22	0	77	84	88
20	-	11	10	11	22	11	52	30	61	45	26	0	-	56	61	86
30	-	32	0	0	6	0	40	10	50	29	12	-	-	42	55	73
40	-	36	-	-	0	1	33	0	39	15	1	-	-	34	29	63
50	-	46	-	-	-	0	28	1	30	12	0	-	-	28	18	55
60	-	61	-	-	-	-	23	0	38	9	-	-	-	22	11	48
70	-	50	-	-	-	-	18	-	29	6	-	-	-	18	0	40
80	-	54	-	-	-	-	14	-	9	4	-	-	-	16	-	34
90	-	82	-	-	-	-	10	-	3	2	-	-	-	12	-	29
100	-	79	-	-	-	-	8	-	0	0	-	-	-	9	-	23
110	-	79	-	-	-	-	7	-	0	-	-	-	-	7	-	19
120	-	89	-	-	-	-	6	-	3	-	-	-	-	0	-	14
130	-	89	-	-	-	-	5	-	3	-	-	-	-	-	-	11
140	-	75	-	-	-	-	4	-	0	-	-	-	-	-	-	11
150	-	100	-	-	-	-	4	-	-	-	-	-	-	-	-	8
160	-	86	-	-	-	-	3	-	-	-	-	-	-	-	-	6
170	-	79	-	-	-	-	3	-	-	-	-	-	-	-	-	4
180	-	68	-	-	-	-	2	-	-	-	-	-	-	-	-	8
190	-	61	-	-	-	-	2	-	-	-	-	-	-	-	-	6
200	-	57	-	-	-	-	2	-	-	-	-	-	-	-	-	5
210	-	61	-	-	-	-	2	-	-	-	-	-	-	-	-	4
220	-	57	-	-	-	-	2	-	-	-	-	-	-	-	-	4
230	-	50	-	-	-	-	1	-	-	-	-	-	-	-	-	3
240	-	32	-	-	-	-	2	-	-	-	-	-	-	-	-	3
250	-	11	-	-	-	-	0	-	-	-	-	-	-	-	-	2
260	-	25	-	-	-	-	1	-	-	-	-	-	-	-	-	2

Table A1. *Cont.*

Time (s)	s01	s02	s03	s04	s05	s06	s07	s08	s09	s10	s11	s12	s13	s14	s15	s16
270	-	11	-	-	-	-	0	-	-	-	-	-	-	-	-	1
280	-	4	-	-	-	-	-	-	-	-	-	-	-	-	-	0
290	-	4	-	-	-	-	-	-	-	-	-	-	-	-	-	0
300	-	0	-	-	-	-	-	-	-	-	-	-	-	-	-	0

"-" indicates that no measurements were made since there was no foam head. All values have been rounded up.

Table A2. Normalized distance of beer foam (%) from 3D camera for 16 beer samples.

Time (s)	s01	s02	s03	s04	s05	s06	s07	s08	s09	s10	s11	s12	s13	s14	s15	s16
0	100	67	100	100	100	100	100	100	100	100	100	100	100	100	100	100
10	0	33	89	84	85	68	88	81	88	97	94	43	0	85	80	95
20	-	33	0	21	42	20	76	52	74	65	63	0	-	62	50	82
30	-	33	-	0	12	0	65	26	62	41	31	-	-	46	30	74
40	-	67	-	-	4	-	56	11	50	19	13	-	-	31	20	63
50	-	67	-	-	0	-	47	4	38	11	6	-	-	15	10	55
60	-	67	-	-	-	-	41	4	26	8	0	-	-	8	0	47
70	-	67	-	-	-	-	32	0	18	5	-	-	-	8	0	42
80	-	67	-	-	-	-	26	-	6	3	-	-	-	0	-	34
90	-	67	-	-	-	-	21	-	3	0	-	-	-	0	-	29
100	-	67	-	-	-	-	15	-	0	0	-	-	-	0	-	24
110	-	67	-	-	-	-	9	-	0	-	-	-	-	0	-	18
120	-	67	-	-	-	-	6	-	0	-	-	-	-	0	-	16
130	-	100	-	-	-	-	6	-	-	-	-	-	-	-	-	11
140	-	67	-	-	-	-	3	-	-	-	-	-	-	-	-	8
150	-	67	-	-	-	-	3	-	-	-	-	-	-	-	-	5
160	-	67	-	-	-	-	0	-	-	-	-	-	-	-	-	3
170	-	67	-	-	-	-	0	-	-	-	-	-	-	-	-	3
180	-	67	-	-	-	-	0	-	-	-	-	-	-	-	-	3
190	-	67	-	-	-	-	0	-	-	-	-	-	-	-	-	0
200	-	67	-	-	-	-	0	-	-	-	-	-	-	-	-	0
210	-	67	-	-	-	-	0	-	-	-	-	-	-	-	-	0
220	-	67	-	-	-	-	0	-	-	-	-	-	-	-	-	0
230	-	67	-	-	-	-	0	-	-	-	-	-	-	-	-	0
240	-	33	-	-	-	-	0	-	-	-	-	-	-	-	-	0
250	-	33	-	-	-	-	0	-	-	-	-	-	-	-	-	0
260	-	33	-	-	-	-	0	-	-	-	-	-	-	-	-	0
270	-	33	-	-	-	-	-	-	-	-	-	-	-	-	-	0
280	-	0	-	-	-	-	-	-	-	-	-	-	-	-	-	0
290	-	0	-	-	-	-	-	-	-	-	-	-	-	-	-	0
300	-	0	-	-	-	-	-	-	-	-	-	-	-	-	-	0

"-" indicates that no measurements were made since there was no foam head. All values have been rounded up.

References

1. Ekberg, J.; Gibson, B.; Joensuu, J.J.; Krogerus, K.; Magalhães, F.; Mikkelson, A.; Seppänen-Laakso, T.; Wilpola, A. Physicochemical characterization of sahti, an'ancient' beer style indigenous to Finland. *J. Inst. Brew.* **2015**, *121*, 464–473. [CrossRef]
2. Bamforth, C.W. The foaming properties of beer. *J. Inst. Brew.* **1985**, *91*, 370–383. [CrossRef]
3. BJCP. Beer Style Guidelines, Edited by Gordon Strong, Kristen England. Available online: www.bjcp.org (accessed on 6 March 2021).
4. Gonzalez Viejo, C.; Torrico, D.D.; Dunshea, F.R.; Fuentes, S. Bubbles, Foam Formation, Stability and Consumer Perception of Carbonated Drinks: A Review of Current, New and Emerging Technologies for Rapid Assessment and Control. *Foods* **2019**, *8*, 596. [CrossRef] [PubMed]
5. Hackbarth, J.J. Multivariate Analyses of Beer Foam Stand. *J. Inst. Brew.* **2006**, *112*, 17–24. [CrossRef]
6. Prins, A.; van Marle, J.T. Foam formation in beer: Some physics behind it. *Monogr. Eur. Brew. Conv.* **1999**, *27*, 26–36.
7. Ronteltap, A.; Hollemans, M.; Bisperink, C.G.J.; Prims, A.R. Beer foam physics. *Tech. Q. Master Brew. Assoc. Am.* **1991**, *28*, 25–32.
8. Fisher, S.; Hauser, G.; Sommer, K. Influence of dissolved gases on foam. *Monogr. Eur. Brew. Conv.* **1999**, *27*, 37–46.

9. Evans, D.E.; Sheehan, M.C. Don't Be Fobbed Off: The Substance of Beer Foam—A Review. *J. Am. Soc. Brew. Chem.* **2002**, *60*, 47–57. [CrossRef]
10. ASBC. *Methods of Analysis. Method Beer-22. Foam Collapse Rate. Approved 1962, Rev. 1975*; American Society of Brewing Chemists: St. Paul, MN, USA, 2018. [CrossRef]
11. Klopper, W.J. Foam stability and foam cling. In Proceedings of the Eur. Brew. Conv. Congr., Salzburg, Austria; Elsevier Scientific: Amsterdam, The Netherlands, 1973; pp. 363–371. Available online: https://www.kruss-scientific.com/en/explore/research-and-development/research-of-foam (accessed on 21 January 2021).
12. Rudin, A. Measurement of the foam stability of beers. *J. Inst. Brew.* **1957**, *63*, 506–509. [CrossRef]
13. Rasmussen, J.N. Automated analysis of foam stability. *Carlsberg Res. Commun.* **1981**, *46*, 25–36. [CrossRef]
14. Jackson, G.; Bamforth, C. The measurement of foam-lacing. *J. Inst. Brew.* **1982**, *88*, 378–381. [CrossRef]
15. Constant, M. A practical method for characterizing poured beer foam quality. *J. Am. Soc. Brew. Chem.* **1992**, *0*, 37–47. [CrossRef]
16. Vundla, W.; Torline, P. Steps toward the formulation of a model foam standard. *J. Am. Soc. Brew. Chem.* **2007**, *65*, 21–25. [CrossRef]
17. Amerine, M.A.; Martini, L.; Mattei, W.D. Foaming Properties of Wine. *Ind. Eng. Chem.* **1942**, *34*, 152–157. [CrossRef]
18. Wilson, P.; Mundy, A. An improved method for measuring beer foam collapse. *J. Inst. Brew.* **1984**, *90*, 385–388. [CrossRef]
19. Evans, D.E.; Surrel, A.; Sheehy, M.; Stewart, D.C.; Robinson, L.H. Comparison of foam quality and the influence of hop α-acids and proteins using five foam analysis methods. *J. Am. Soc. Brew. Chem.* **2008**, *66*, 1–10. [CrossRef]
20. Evans, D.E.; Oberdieck, M.; Red, K.S.; Newman, R. Comparison of the Rudin and NIBEM Methods for Measuring Foam Stability with a Manual Pour Method to Identify Beer Characteristics That Deliver Consumers Stable Beer Foam. *J. Am. Soc. Brew. Chem.* **2012**, *70*, 70–78. [CrossRef]
21. Smith, R.J.; Davidson, D.; Wilson, R.J. Natural foam stabilizing and bittering compounds derived from hops. *J. Am. Soc. Brew. Chem.* **1998**, *56*, 52–57. [CrossRef]
22. Cimini, A.; Pallottino, F.; Menesatti, P.; Moresi, M. A low-cost image analysis system to upgrade the rudin beer foam head retention meter. *Food Bioprocess Technol.* **2016**, *9*, 1587–1597. [CrossRef]
23. Gonzalez Viejo, C.; Fuentes, S.; Li, G.; Collmann, R.; Condé, B.; Torrico, D. Development of a robotic pourer constructed with ubiquitous materials, open hardware and sensors to assess beer foam quality using computer vision and pattern recognition algorithms: RoboBEER. *Food Res. Int.* **2016**, *89*, 504–513. [CrossRef]
24. Gonzalez Viejo, C.; Fuentes, S.; Torrico, D.D.; Howell, K.; Dunshea, F.R. Assessment of Beer Quality Based on a Robotic Pourer, Computer Vision, and Machine Learning Algorithms Using Commercial Beers. *J. Food Sci.* **2018**, *83*, 1381–1388. [CrossRef] [PubMed]
25. Gonzalez Viejo, C.; Fuentes, S.; Howell, K.; Torrico, D.D.; Dunshea, F.R. Integration of non-invasive biometrics with sensory analysis techniques to assess acceptability of beer by consumers. *Physiol. Behav.* **2019**, *200*, 139–147. [CrossRef] [PubMed]
26. Lukinac, J.; Mastanjević, K.; Mastanjević, K.; Nakov, G.; Jukić, M. Computer Vision Method in Beer Quality Evaluation—A Review. *Beverages* **2019**, *5*, 38. [CrossRef]
27. Middle European Brewing Analysis Commission (MEBAK); Band II.n Brautechnische Middle European Brewing Analysis Commission (MEBAK). *Band II.n Brautechnische Analysenmethoden*, 3th ed.; Selbstverlag der MEBAK: Freising-Weihenstephan, Germany, 1997.
28. Python Programming Language. Available online: https://www.python.org/ (accessed on 21 January 2021).
29. OpenCV. Available online: https://opencv.org/ (accessed on 21 January 2021).

 *fermentation*

Article

Yeast Morphology Assessment through Automated Image Analysis during Fermentation

Mario Guadalupe-Daqui, Mandi Chen, Katherine A. Thompson-Witrick and Andrew J. MacIntosh *

Food Science and Human Nutrition Department, University of Florida, Gainesville, FL 32611, USA; mguadalupe@ufl.edu (M.G.-D.); mandi.chen@ufl.edu (M.C.); kthompsonwitrick@ufl.edu (K.A.T.-W.)
* Correspondence: andrewmacintosh@ufl.edu; Tel.: +1-352-294-3594

Abstract: The kinetics and success of an industrial fermentation are dependent upon the health of the microorganism(s) responsible. *Saccharomyces* sp. are the most commonly used organisms in food and beverage production; consequently, many metrics of yeast health and stress have been previously correlated with morphological changes to fermentations kinetics. Many researchers and industries use machine vision to count yeast and assess health through dyes and image analysis. This study assessed known physical differences through automated image analysis taken throughout ongoing high stress fermentations at various temperatures (30 °C and 35 °C). Measured parameters included sugar consumption rate, number of yeast cells in suspension, yeast cross-sectional area, and vacuole cross-sectional area. The cell morphological properties were analyzed automatically using ImageJ software and validated using manual assessment. It was found that there were significant changes in cell area and ratio of vacuole to cell area over the fermentation. These changes were temperature dependent. The changes in morphology have implications for rates of cellular reactions and efficiency within industrial fermentation processes. The use of automated image analysis to quantify these parameters is possible using currently available systems and will provide additional tools to enhance our understanding of the fermentation process.

Keywords: yeast morphology; automated image analysis; heat stress; vacuoles; cell size; computer vision

Citation: Guadalupe-Daqui, M.; Chen, M.; Thompson-Witrick, K.A.; MacIntosh, A.J. Yeast Morphology Assessment through Automated Image Analysis during Fermentation. *Fermentation* **2021**, *7*, 44. https://doi.org/10.3390/fermentation7020044

Academic Editor: Claudia Gonzalez Viejo

Received: 5 March 2021
Accepted: 20 March 2021
Published: 24 March 2021

1. Introduction

The fermentative properties of yeast have been utilized for thousands of years in a spectrum of applications including brewing, baking, biofuels, etc. For an optimal performance in most applications, the health of the yeast should be maintained throughout the fermentation [1]. Yeast health is typically measured through the determination of viability and vitality via metabolic dyes, however, there are other known morphological characteristics that have been shown to be indicators of yeast health [2,3] including size, number and shape of organelles (i.e., vacuoles). To optimize the fermentation process and quickly respond to deviations, data collection concerning yeast condition should be rapid and include automation as to be completed in a timely manner. If implemented, automated assessment can also provide supplemental information including details of fermentation rate per suspended cell which may assist both researchers and industry to better understand the fermentative process. This study used an open-source image processing to track changes in yeast morphology through an automated analysis of various characteristics during high-stress fermentations.

Commonly measured morphological characteristics of yeast include cell size (commonly expressed as cross-sectional area or volume) and the number of buds. These characteristics are used to evaluate yeast cells under a wide variety of environmental stress sources, for example, yeast cell size measurements have been used by various researchers [2,4] as an indicator of the effects of stressors such as temperature and ethanol. Additionally, cell size and the number of budding vs. single cells were successfully used to assess the effects of hyperbaric stress and various gas compositions on yeast physiological state [5]. The vacuolar structure (size and shape) and number are other yeast morphological characteristics commonly assessed. Yeast vacuoles are dynamic organelles with a changing morphology during fermentation as a response to different stress sources [6–8]. These researchers studied yeast vacuole formation under osmotic and ethanol stress environments during fermentation at low temperatures. Both studies concluded that under high osmotic environments, each cell contained at least two or more yeast vacuoles which presented a fragmented (small diameter) structure. These studies also reported that yeast vacuoles formed under high ethanol concentrations were less fragmented and were described as single swollen vacuoles that occupied almost the entire cytoplasm. Furthermore, deficiencies of this organelle in yeast (observable under the microscope) have been linked to cell death [9] and the cessation of fermentation. A better understanding of yeast performance during fermentation can be attained by combining measurements of fermentation kinetics with morphological characteristics of the cell. Tibayrenc et al. [2] studied the evolution of *Saccharomyces cerevisiae* morphology under different stressors to infer cell viability. In summary, there have been numerous studies that effectively relate yeast morphological characteristics such as cell size, length, or state, to yeast properties during fermentation [2,10,11].

The automated analysis of images facilitates the examination of microorganisms at specific times throughout a process, providing instantaneous information concerning the parameters assessed. Typically, automated image analysis is divided into three stages: image processing, variable acquisition and statistical analysis [12]. Automated analysis of yeast cell images is not a novel concept and is most commonly used to assess the number of cells suspended within media. There are existing software and devices such as the Cellometer developed by Nexcelom Bioscience LLC (Lawrence, MA, USA) that can process yeast images and assess the number of cells and viability. CalMorph v1.3 is software created to analyze yeast cells delivering information regarding cell morphology, phase, etc. However, this software is limited to microscopic images that have been fluorescence stained [13]. There are numerous examples of researchers independently using various software programs to assess images of yeast for specific properties. For example, the software Matlab v.6.1 was used by Coelho et al. [5] to fully automate the image processing procedure from an image obtained using an optical microscope. These researchers studied the effects of hyperbaric stress on single and budding yeast cells using cell area size, major and minor axis length measured through Matlab. The open-source software ImageJ has been used as an image processing software by numerous researchers [14,15] to assess properties of various microorganisms. Stolze et al. [16] developed an automated counting method for Petrifilm plates using ImageJ, reducing the time of analysis significantly.

This study aimed to combine the use of automated image analysis with a modern understanding of how yeast morphology changes in response to adverse environmental conditions. This helped to correlate how changes in the morphological state of yeast influenced the process attributes throughout the fermentation. Specifically, fermentations were conducted at extreme temperature conditions (30 °C and 35 °C) where morphological changes were expected based upon previous work. Throughout the fermentation, hundreds of images were collected and attributes such as cell cross-sectional area and vacuole size were quantified through automated analysis. These were then correlated with fermentation kinetics to enhance understanding of the changes undergone by the yeast, and the effect upon the entire fermentation. The techniques used in this paper can easily be applied in the fermentation industries where automated data collection of this nature is often already taking place allowing rapid assessment of yeast adaptions to different stress sources.

2. Materials and Methods

2.1. Media and Yeast

The media used in this research was based upon a rum fermentation and was composed of 75% *w/w* distilled water, 18.5% *w/w* sugar cane, and 6.5% *w/w* blackstrap molasses.

This broth was supplemented with 0.8%*w/w* yeast extract to increase the amount of free amino nitrogen to 320 mg/L (measured via Ninhydrin method as described in the American Society of Brewing Chemists (ASBC) Wort-12 method [17]). All the ingredients were dissolved at 50 °C and sterilized at 121 °C for 15 min. This media had a sugar concentration of 24% *w/w* and was used for both the propagation and fermentation processes. However, in the first propagation, the media was diluted to 15% *w/w* using deionized water. A *Saccharomyces cerevisiae* strain provided by Lallemand (DISTILAMAX® SR—Montreal, QC, Canada) was utilized throughout the experiments based upon recommendations from the supplier concerning survivability under adverse conditions and the ability to ferment sugars found in the raw materials used. The active dry yeast (ADY) was stored between 1 °C and 4 °C prior to its rehydration and propagation.

2.2. Rehydration and Propagation

For each experiment, 10 g of ADY were rehydrated using 100 mL of sterile tap water at 35 °C for 15 min. Rehydrated yeast was propagated in a two-step process to achieve a consistent inoculum and sufficient biomass to pitch the 4 L fermentors. The first propagation was performed using 150 mL of media (diluted to 15% *w/w*), while the second propagation was performed using 350 mL of media at (24% *w/w*). Both propagation steps were pitched at 30×10^6 viable yeast cells/mL and run for 24 h at 35 °C each as per recommendations from the supplier.

2.3. Fermentation

Each fermentation run started under identical conditions, using the same media, and pitched with the same yeast to assess the differences throughout the process by changing only the temperature. Fermentors (total volume of ~5 L) were filled with 4 L of media (24% *w/w*), oxygenated for two minutes, and pitched with 30×10^6 viable yeast cells/mL. Two temperatures were selected to complete the fermentation processes, 30 °C (low temperature) and 35 °C (high temperature). Each experiment was performed by duplicate. Samples were taken at intervals between 6 to 12 h in an attempt to monitor the reductions in the sugar concentration every 3–4% *w/w*, approximately, during peak fermentation. The volume of each sample was 125 mL and was divided into smaller aliquots to assess the fermentation attributes and the yeast morphology.

2.4. Measurement of Sugar Concentration and Viable Cells in Suspension

To measure the sugar concentration throughout the fermentation, 30 mL of sample were degassed, centrifuged and filtered. These samples were analyzed using an Anton Paar Alex 500 (Graz, Austria), which measured the density of the sample through an oscillating U-tube glass. The density measured was then used to calculate the real sugar concentration and expressed in % *w/w*, or °P (analogous to percent sugar by weight). Cells in suspension were measured during the fermentation process using a hemocytometer and a microscope with 400× magnification, as described in the ASBC Yeast-4 method [18]. Viable cells were identified and counted using the methylene blue technique described by Painting and Kirsop [19]. The number of viable cells in suspension at each sample point was calculated by multiplying the percent of viable cells by the total number of cells in suspension.

2.5. Models

The sugar concentration was modeled using the logistic model described in Yeast-14 [20] method and shown in Equation (1).

$$P_{(t)} = P_e + \frac{P_i - P_e}{1 + exp^{(-B(t-M))}} \tag{1}$$

where $P_{(t)}$ is a function that represents the sugar concentration at a specific time t (% *w/w*), P_i is the upper asymptote of the curve (%*w/w*), P_e is the lower asymptote of the curve (%*w/w*), M is the time of the inflection point of the curve (h), and B is the slope of the curve

(%w/(w*h)). To determine the consumption rate of the sugar during the fermentation, the first derivative of the logistic model (Equation (1)) was taken with respect to time (Equation (2)) as follows:

$$\frac{\partial}{\partial t}\left(P_e + \frac{P_i - P_e}{1 + exp^{(-B*(t-M))}}\right) = \left(\frac{B\,(P_i - P_e)exp^{-B(x-M)}}{\left(exp^{-B(x-M)} + 1\right)^2}\right) \quad (2)$$

The cells in suspension data were modeled using the step model (Equation (3)) as described by Rudolph et al. [21]. This model was utilized due the amplitude and the Heaviside step function that describe the maximum number of cells in suspension during the fermentation and the final number of cells at the end of the process, respectively.

$$Y_{(t)} = A\left(\frac{exp^{-\frac{1}{2}\left(\frac{\frac{t-\mu}{\sigma}}{1+(H_t*S)}\right)^2}}{1 + (H_t * S)}\right) + \left(A - \frac{A}{(1+S)}\right) * H_{(t)} \quad (3)$$

$Y_{(t)}$ is a function that represents the number of cells in suspension at a specific time t (cells/mL), A is the amplitude (cells/mL), m is the midpoint (h), σ is the width factor (h), S modifies the height of the step in this function (unitless), and $H_{(t)}$ is the Heaviside step function.

2.6. Image Collection

Prior to the image collection, yeast cells were centrifuged and washed three times using deionized water and stained with a 0.01% *w/v* methylene blue solution to differentiate living and death cells at each data point. Various images were taken from each experiment such that approximately 80 to 360 cells were captured at each sampling point. Images were captured using a Nikon Eclipse Ci-L microscope at 1000× magnification, combining a 100× oil immersion objective and a 10× eyepiece.

2.7. Manual Analysis of Vacuole Count

The manual analysis of the cell images was completed using the Nikon NIS-Elements software V5.30.01 compatible with the microscope. The cross-sectional area (CSA) of yeast and vacuoles were measured individually using an auto selection tool in conjunction with a five-point ellipse and a freehand selection tool. Once the areas were selected, the software converted the pixels into a measurement of surface area based upon previous calibrations. Budding yeast cells were assessed as two separate cells, dividing them at the cleavage. Vacuoles were clearly identifiable due to the difference in shade compared to the rest of the cell (Figure 1). The light settings on the microscope were adjusted at the beginning of the experiment to maximize this contrast. The data collected were exported and saved as an excel file for further calculations to determine the average yeast CSA and the ratio between vacuole and yeast CSA. Figure 1b shows an image analyzed manually and includes all the measurements taken using this software. There were 80–360 yeast cells assessed from images at the beginning, middle, and end of each fermentation process using this method to validate the automated image analysis method. Yeast cells that were partially in the image were not selected for data collection using this method.

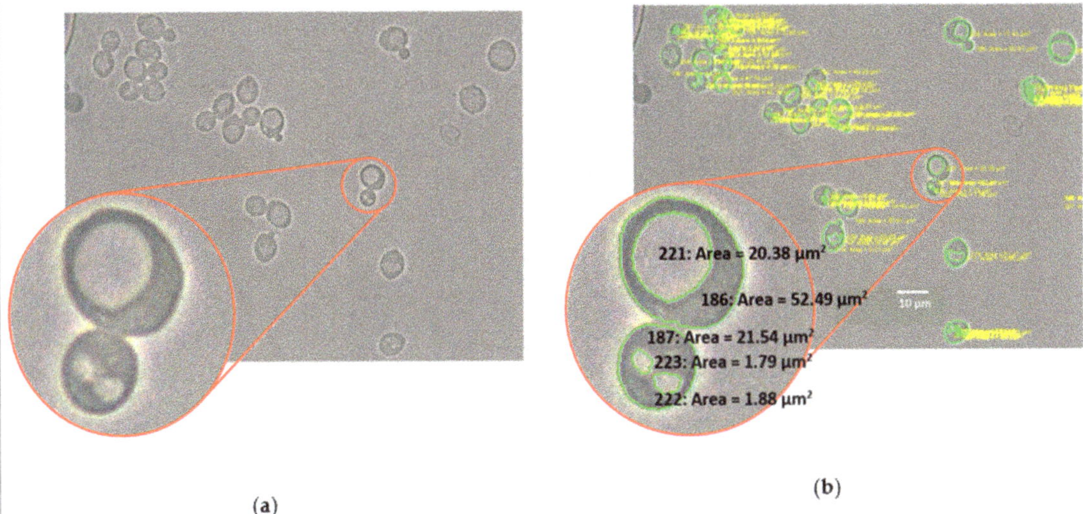

Figure 1. (**a**) Microscope image of yeast during fermentation (1000×) with enlarged budding cell (counted as two individual yeast cells) to show shading difference between vacuole and yeast cell in a raw image; (**b**) The same image after manual analysis of cell and vacuole cross-sectional area (CSA).

2.8. Automated Image Analysis of Yeast and Vacuole Cross-Sectional Area

The images of yeast taken throughout each fermentation were analyzed using ImageJ v1.8.0 software. A sample image is shown in Figure 2a. This analysis included the measurement of yeast cell CSA, total yeast CSA and total vacuole CSA. The automated image analysis was calibrated using a scale that correlated the image pixels with the scientific units of micrometers. The outline of yeast cells and the vacuoles were enhanced using tools sequentially: "binary mask", "despeckle filter", "dilation", and "pixel outlier removal". The goals of these tools were to reduce image noise that could affect the results and to enhance the distinguishing features of yeast cells from the background and the vacuoles. The binary mask converted the 32-bit image to 8-bit, which results in a black and white separation of the yeast cells (black), vacuoles (white) and background (white). The despeckle filter applied was a median filter that replaced every pixel with the median value of a 3 × 3 pixel region, and it removed sparsely occurring black pixels [22]. Then, the dilation tool was used to separate any yeast cells that may be touching each other due to proximity or budding. Finally, after dilation, the pixel outlier tool was applied to remove black particles with a radius of 4 pixels (0.363 µm) or less. A sample of the resulting processed images is shown in Figure 2b. These images were then automatically assessed with an analyze particles tool that selected for analysis, objects with an area of 10–100 μm^2 and a circularity of 0.30–1.00 (ideally, this would encompass all yeast cells). This tool enumerated and collected the cross-sectional area of all individual selected objects (Figure 2c). To measure the vacuolar cross-sectional area, everything outside of the selected objects was removed, then the image was inverted to change the black background to white and white vacuoles to black, resulting in Figure 2d. Yeast cells and the vacuoles that were partially in the image were not included in the analysis by the software. The data collected were exported and saved as an excel file for subsequent calculation of the average yeast cell cross-sectional area and the ratio between vacuole and cell cross-sectional area. The automated analysis of the cell images was completed after the fermentation processes.

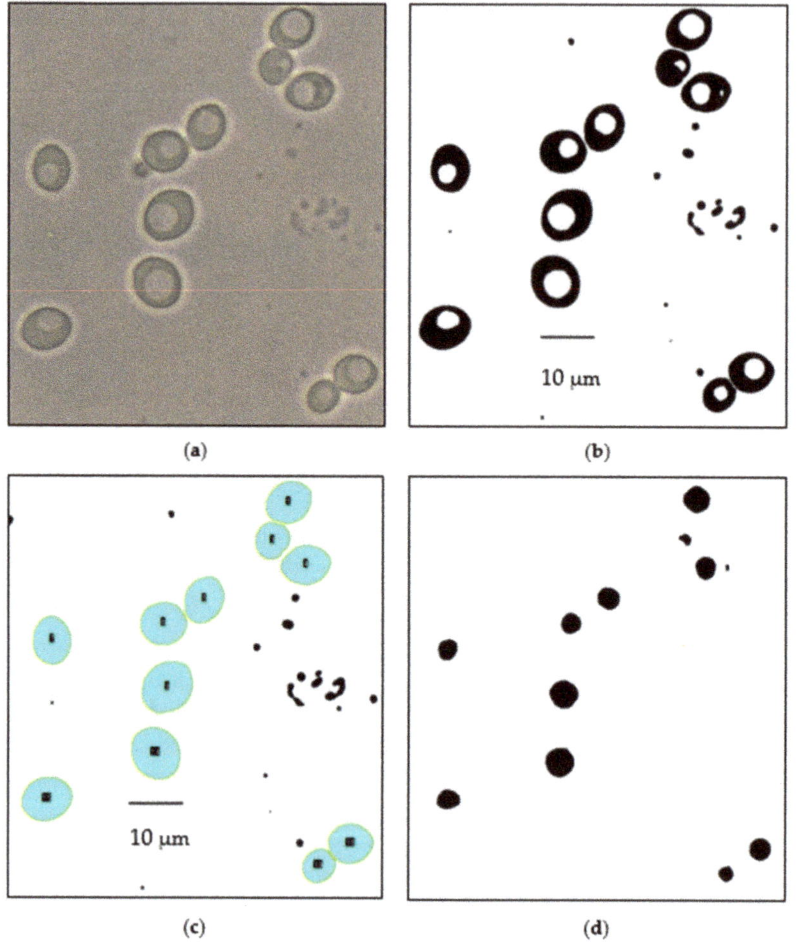

Figure 2. (**a**) Microscopic image of yeast cells (1000× magnification) at 6.5 h before processing; (**b**) After image processing; (**c**) After yeast cell area data were collected and region of interest established; (**d**) After clearing outside region of interest and inverting image and vacuolar area data was collected.

2.9. Statistical Analysis

In this study, Minitab 19 was used to determine significant differences ($p < 0.05$) among the yeast CSA and CSA ratio between vacuoles and yeast cells throughout each fermentation and between fermentations at 30 °C and 35 °C. The validation of the automated image analysis method was performed by calculating the percentage error between the results obtained manually using the Nikon software V5.30.01 and automatically using ImageJ.

3. Results

3.1. Fermentation Kinetics

The sugar consumption model (Equation (1)) and the model for number of cells in suspension (Equation (3)) were fit to the data from each fermentation. The best fit model parameters for the combined fermentation data at each temperature are presented in Table 1. The sugar concentration regression resulted in high r^2 values of 0.99, while the slightly more variable cell in suspension data resulted in values of 0.91 and 0.85 for 30 and 35 °C, respectively.

Table 1. Step and logistic model parameters.

Attribute	Experiment	Model	Parameters				
			A (Cells/mL)	μ h	σ h	S Unitless	Model Fit r^2
Cells in suspension	30 °C 35 °C	Step model Step model	168.24 79.72	20.65 13.25	13.77 11.38	0.00 0.05	0.91 0.89
			Pi %*w*/*w*	Pe %*w*/*w*	B %w/(w*h)	M h	Model Fit r^2
Sugar consumption	30 °C 35 °C	Sigmoidal 4P Sigmoidal 4P	29.92 26.93	3.14 10.00	−0.12 −0.14	11.08 11.39	0.99 0.99

Fermentations at 35 °C and 30 °C both showed a typical sigmoidal sugar consumption curve (Figure 3). The times of maximum fermentation rate (parameter M) were similar in both, at 30 °C (11.08 h) and 35 °C (11.39 h). At 30 °C, yeast metabolized more total sugars (20.8% *w/w*) compared to 35 °C (13.9% *w/w*). During fermentations at both temperatures, the number of viable cells in suspension increased (Figure 4). However, the maximum number of cells in suspension during the 35 °C fermentations was approximately 50% fewer compared to the 30 °C fermentations. These results are in accordance with the results previously reported by Torija et al. [23] and Reddy and Reddy [24] who used *Saccharomyces cerevisiae* strains to study the influence of temperature on the number of cells and the fermentation rate. Both studies found that at 35 °C, the cells in suspension were reduced approximately 50% compared to 30 °C. Similarly, they found a higher final attenuation level when fermenting at 35 °C compared to 30 °C. Torrija et al. [23] reported a drop in the ethanol yield of 11% approximately when fermenting at 35 °C. This low yield of ethanol was almost certainty linked to the reduced consumption of substrate.

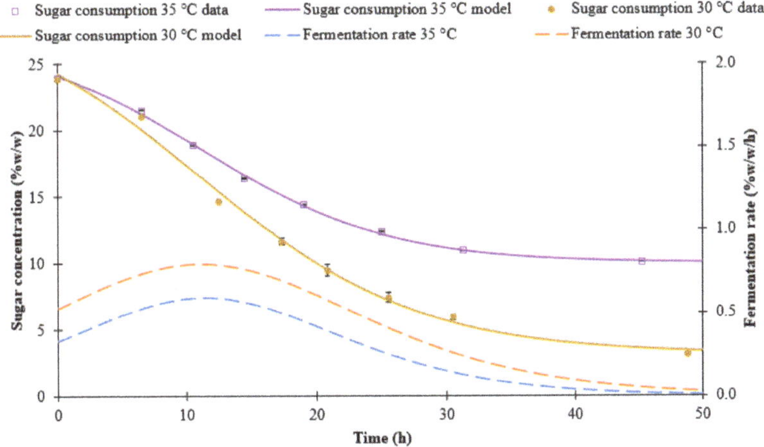

Figure 3. Sugar concentration and fermentation rate calculated by taking the 1st derivative of the sugar consumption model (Equation (2) and Equation (1), respectively). Data shown represent fermentation data at 30 °C (brown circles) and 35 °C (purple squares); sugar consumption model at 30 °C (purple line) and 35 °C (brown line); and fermentation rate model at 30 °C (orange dashed line) and 35 °C (blue dashed line). Fermentations were performed in duplicate. The error bars represent the standard deviation of the data.

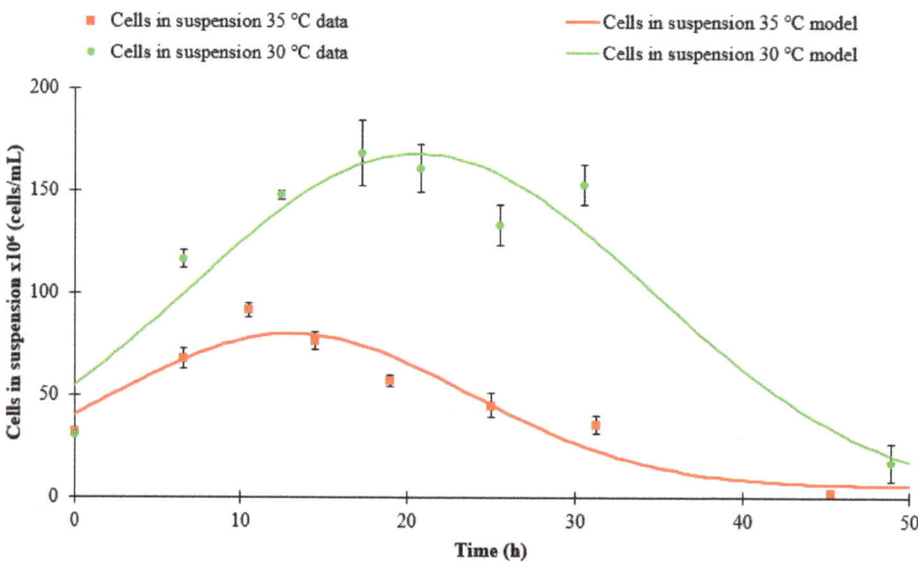

Figure 4. Viable cells in suspension throughout the fermentation. Data shown represent fermentation data at 30 °C (green circles) and 35 °C (red squares); cells in suspension model (Equation (3)) at 30 °C (green line) and 35 °C (red line). Fermentations were performed in duplicate. The error bars represent the standard deviation of the data.

3.2. *Manual Image Analysis*

During the fermentation at both temperatures, changes in yeast size, as well as in the size of the vacuoles present were observed. Figures 5 and 6 show images of yeast cells and a summary of the average of the yeast and vacuole CSA throughout the fermentation for 30 °C and 35 °C, respectively. These values were manually assessed from 80–360 cells per sampling time using the methods described above. Images at 30 °C show that the average yeast CSA decreased from 43.4 $\mu m^2 \pm 13.6\ \mu m^2$ to 37.1 $\mu m^2 \pm 11.9\ \mu m^2$, whilst at 35 °C there was a slight increase of CSA until after the fermentation peak (approximately after 15 h). Images also indicate a constant decrease in the size of vacuoles, until almost no vacuoles were present in the yeast cells towards the end of the fermentation at both temperatures. Given the large standard deviation in the size of the yeast, it was necessary to assess a large number of cells to provide statistically significant results. The manual analysis was labor-intensive, thus, substantial resources would be necessary to observe trends over the fermentation. To accomplish this goal, the automated analysis tool was developed so that many sampling times could be automatically assessed within a reasonable amount of time. Samples at the beginning, middle, and end of the fermentation processes were manually counted to validate the automated results.

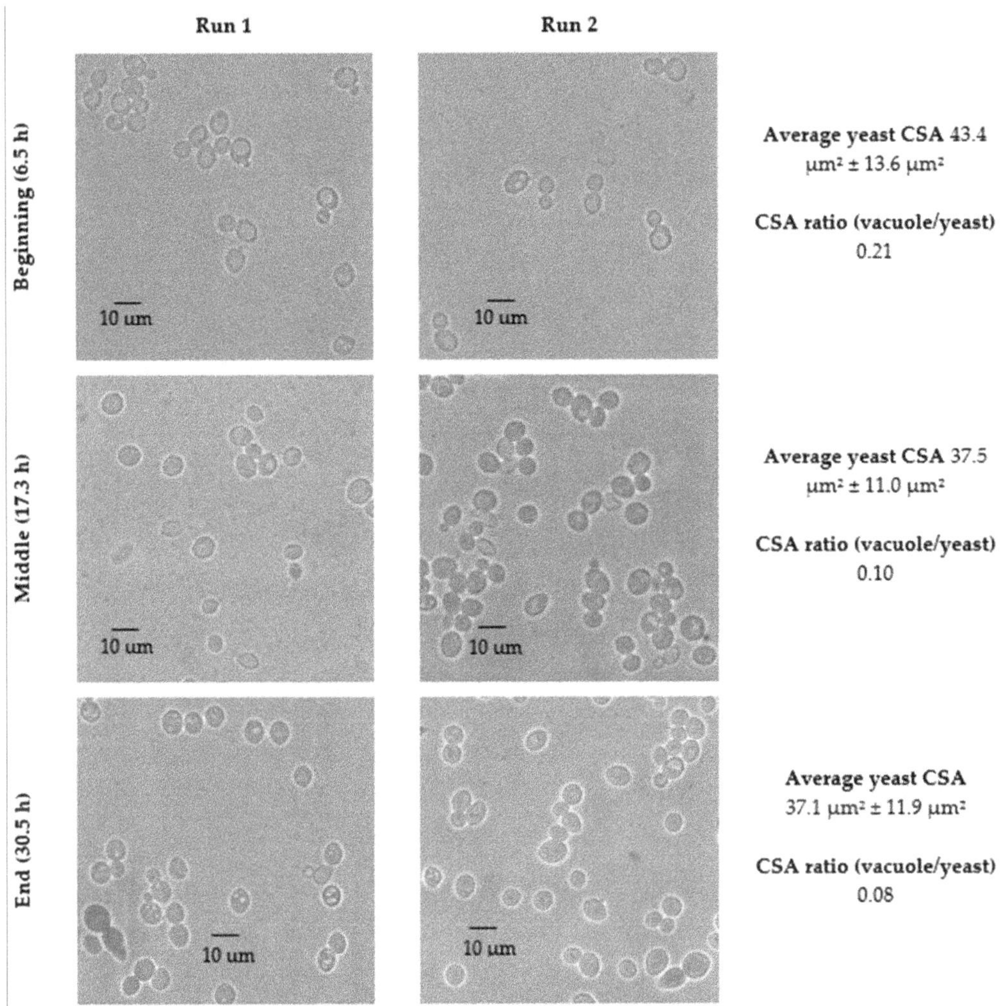

Figure 5. Bright-field microscopic image of yeast cells (1000× magnification) of the duplicate runs from beginning, middle and end of the fermentation processes at 30 °C. The media included 0.01% *w/v* methylene blue solution to differentiate non-viable cells from the analysis. The yeast CSA and CSA ratio (vacuole to yeast area) were determined manually using the Nikon software.

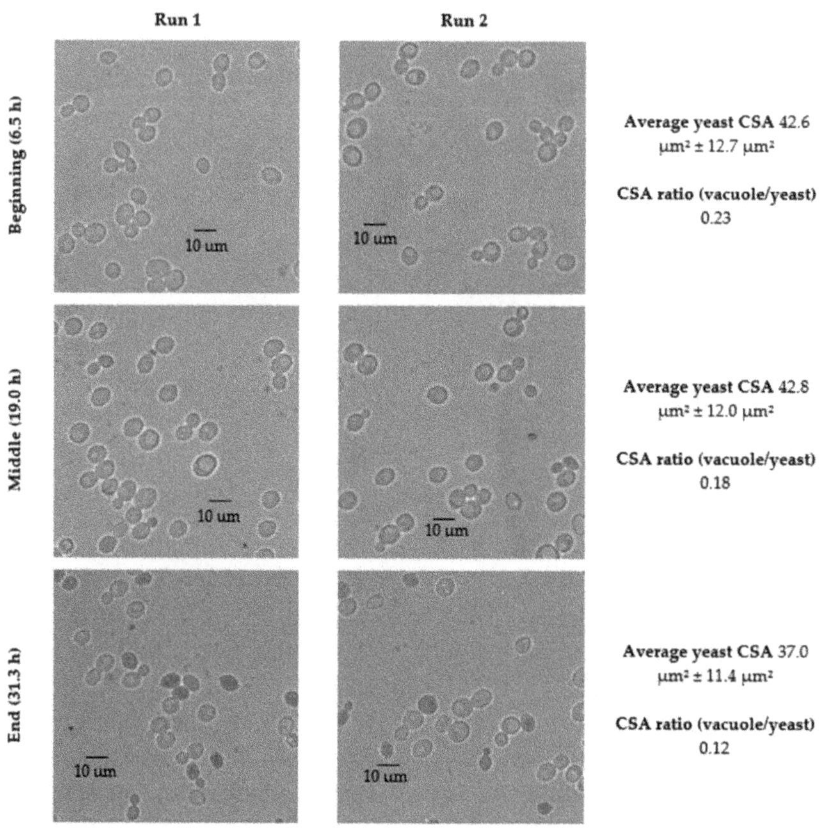

Figure 6. Bright-field microscopic image of yeast cells (1000× magnification) of the duplicate runs from beginning, middle and end of the fermentation processes at 35 °C. The media included 0.01% *w/v* methylene blue solution to differentiate non-viable cells from the analysis. The yeast CSA and CSA ratio (vacuole to yeast area) were determined manually using the Nikon software.

3.3. *Automated Image Analysis*

For this comparison, 80–360 yeast cells were automatically analyzed at each sampling period. To assess the accuracy of the automated analysis, the results for yeast CSA, total yeast CSA, and vacuoles CSA obtained through both automated and manual image analysis are compared in Table 2. The data obtained for yeast CSA and the CSA ratio vacuole to yeast, using both the automated image analysis and the manual Nikon software, resulted in an average difference of less than 4% and 2%, respectively. This shows that the data obtained in this study are reliable and comparable to the manual assessment where the Nikon software was used to measure cell morphology under the microscope.

Table 2. Comparison of automated and manual image analysis.

Temp (°C)	Time (h)	Manual	Automated			Manual			Automated				
		Average Yeast CSA (μm²)	Average Yeast CSA (μm²)	% Error	Average % Error	Total Yeast CSA (μm²)	Total Vacuole CSA (μm²)	% CSA Ratio (Vacuole/Yeast)	Total Yeast CSA (μm²)	Total Vacuole CSA (μm²)	% CSA Ratio (Vacuole/Yeast)	% Error	Average % Error
30	6.5	43.3 ± 13.6	42.1 ± 11.2	2.8%	−3.9%	6455.3	1362.6	21.1	5828.1	1225.6	22.1	−4.9%	2.0%
	17.3	37.5 ± 11.0	39.6 ± 10.1	−5.8%		9218.4	944.3	10.2	8711.3	1072.5	12.3	−19.7%	
	30.5	37.1 ± 40.4	40.4 ± 11.0	−8.8%		13,167.5	1096.6	8.3	11,450.7	661.9	5.8	30.7%	
35	6.5	42.6 ± 12.7	41.4 ± 12.0	2.7%	−2.2%	13,999.1	3158.3	22.6	13,304.5	2766.4	20.9	7.3%	1.2%
	19.0	42.8 ± 12.0	44.7 ± 11.5	−4.4%		15,583.6	2746.6	17.6	12,862.5	1828.6	15.6	17.2%	
	31.3	37.0 ± 11.4	38.8 ± 13.6	−4.8%		11,169.0	1308.1	11.7	11,133.8	1569.2	14.1	−20.7%	

3.4. Changes in Morphological State over Fermentation

At 30 °C, there was a significant difference ($p < 0.05$) between the initial and the final yeast CSA at 30, whilst at 35 °C, there was not a significant difference ($p > 0.05$) between the initial and the final yeast CSA. Results regarding the CSA ratio between vacuole and yeast CSA showed a significant difference ($p < 0.05$) at 30 °C but not at 35 °C. Table 3 shows the average of the results obtained from the morphological analysis of yeast at each temperature.

Table 3. Average and standard deviation of yeast CSA and Ratio (vacuole to yeast area) summary *.

Temp (°C)	Time (h)	Average Yeast CSA (μm^2)	% CSA Ratio (Vacuole/Yeast)
30	6.5	42.1 ± 11.2 a	22.1 ± 6.5 a
	12.5	42.4 ± 10.5 a	16.3 ± 2.5 b
	17.3	39.6 ± 10.1 a	12.3 ± 2.2 bc
	20.8	38.7 ± 9.8 a	918 ± 4.7 cd
	30.5	40.4 ± 11.0 a	5.8 ± 2.2 d
	48.8	34.7 ± 9.5 b	3.8 ± 2.0 d
35	0.0	41.7 ± 12.4 abc	13.9 ± 4.9 b
	6.5	41.4 ± 12.0 bc	20.9 ± 3.7 a
	10.5	44.7 ± 12.7 a	18.6 ± 2.8 ab
	14.5	44.9 ± 11.3 a	15.6 ± 4.0 b
	19.0	44.7 ± 11.5 ab	14.6 ± 4.2 b
	25.0	42.0 ± 12.0 abc	16.5 ±3.0 ab
	31.3	38.8 ± 13.6 c	14.1 ± 3.2 b

* The letters presented next to the standard deviation at each temperature were obtained through grouping averages based on Tukey's analysis. Different letters among each column represent significant difference.

The trend of yeast CSA for both temperatures did not show a significant change in size until the end of the fermentation (Figure 7). However, yeast at 35 °C had on average larger CSAs compared to yeast at 30 °C. These results are consistent with the finding of Gervais et al. [25] who found similar results of an increased cellular volume with increased temperature. They proposed that the increase of the cellular volume is due to a cellular response of yeast to heat shock.

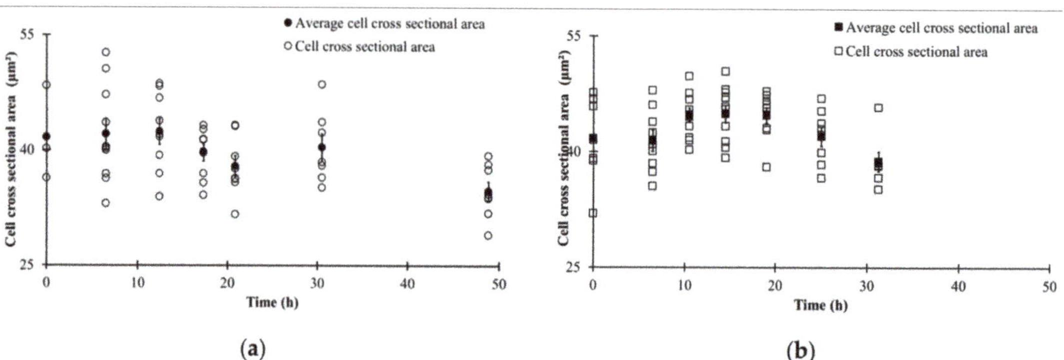

Figure 7. (**a**) Yeast CSA with respect to time at 30 °C; (**b**) Yeast CSA with respect to time at 35 °C. Cross-sectional data are represented as follows: data at 30 °C (empty circles) and 35 °C (empty squares); average data at 30 °C (filled circles) and 35 °C (filled squares). Fermentations and automated image analysis were performed in duplicate. The error bars correspond to the standard error of the data, used to represent the high confidence in the mean associated with a large number of samples.

The ratio between the total CSA of the vacuoles and the total CSA of the cell (Figure 8) had a decreasing trend for 30 °C, but not for 35 °C (Table 3). This indicates that the vacuolar space within the yeast cells was temperature-dependent. The changes undergone by the yeast show morphological changes (vacuole size) that were reported by Pratt et al. [6,7] and Izawa et al. [8]. However, the trends did not match exactly what was described in these studies. This is likely influenced by the highly stressful nature of rum fermentations. By quantifying the differences in yeast morphological state, it becomes possible to determine the impact of stress factors, and the quality of yeast for subsequent fermentations.

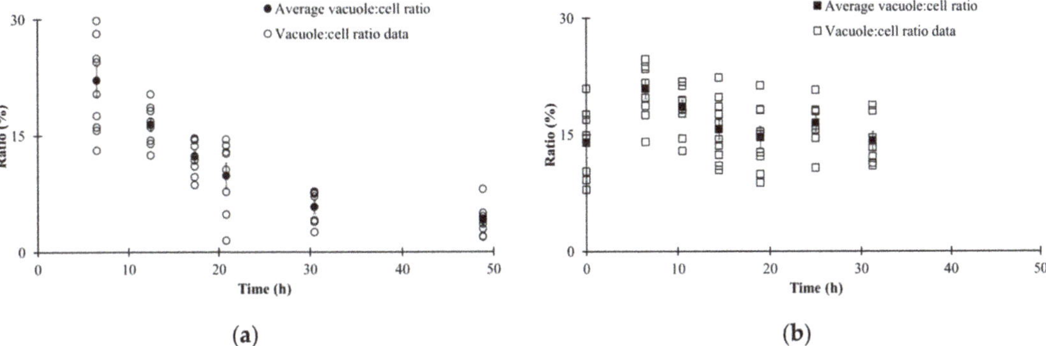

Figure 8. Vacuole and yeast CSA ratio with respect time at 30 °C (**a**) and 35 °C (**b**). Ratio data are represented as follows: data at 30 °C (empty circles) and 35 °C (empty squares); average data at 30 °C (filled circles) and 35 °C (filled squares). Fermentations and automated image analysis were performed in duplicate. The error bars represent the standard error of the data (used to represent the high confidence in the mean associated with a large number of samples).

The change in the yeast CSA, as well as the ratio between the size of vacuoles and the cell throughout the fermentation process, was obtained using an automated yeast analysis. Figures 7 and 8 clearly indicate that yeast has the capacity to undergo large morphological changes over a single fermentation, as has been observed by many researchers. Therefore, characterizing the rate of fermentation with respect to temperature, based upon the rate of sugar consumption per cell, is an incomplete assessment and it could lead to incorrect conclusions. A more detailed analysis was possible through the combination of the yeast CSAs obtained from the automated tool developed in this study, and the enumeration of viable yeast cells within suspension. Using the modeled number of cells in suspension and combining with the average surface area of yeast within suspension (estimated based on the assumption of yeast spherical shape), the total surface area of yeast in suspension over time was determined, as shown in Figure 9. This analysis has implications for reactions that are dependent upon cell wall area, such as nutrient uptake and stress responses.

Building upon this analysis, it became apparent that the rate of fermentation per cell changed over the duration of the fermentation and was dependent upon temperature. There are many factors that influence the rate of fermentation per cell, including temperature, sugar concentrations within the media, stress levels, etc. However, one parameter that could be accounted for was the total volume of yeast cells present. To determine the sugar consumption rate per volume of yeast, the authors used the CSA to determine the average volume of cells at each time period (assuming spherical cells). As the vacuole volume is not expected to be involved in cellular metabolism (and makes up a significant portion of the volume in early fermentation cells (Figure 8)), the vacuolar volume was subtracted from the average yeast volume calculation, resulting in an estimate of non-vacuolar cell volume. This value was multiplied by the model of cells in suspension to find the non-vacuolar cell volume per milliliter. Finally, the rate of sugar consumption model was divided by this value to find the sugar consumption rate per non-vacuolar cell volume; it is presented in Figure 10. This analysis shows that the sugar consumption rate per total non-vacuolar yeast volume declined with respect to time (and sugar concentration), and was higher at 35 °C. This analysis has applications for optimizing the reaction rate for fixed-bed, encapsulated yeast and other fermentations where there is not expected to be new yeast growth. It can also help predict nutrient requirements and properties of yeast populations produced under different temperature profiles for alternative fermentation processes such as chemostats reactors.

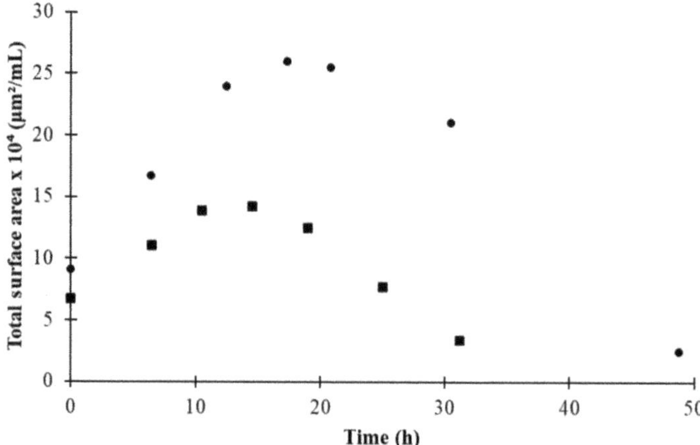

Figure 9. Total surface area of yeast in suspension per milliliter with respect to fermentation time. Data are presented as follows: average total surface area of suspended yeast per milliliter at 30 °C (circles) and 35 °C (squares).

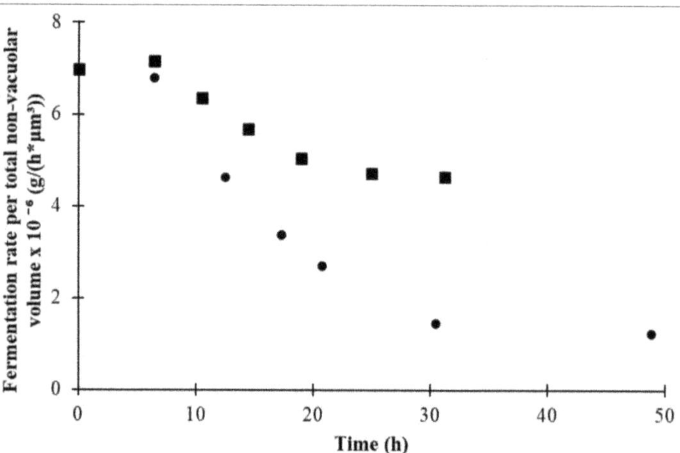

Figure 10. Fermentation rate per total non-vacuolar volume over time. Data are presented as follows: fermentation rate per total non-vacuolar volume at 30 °C (circles) and 35 °C (squares).

4. Conclusions

Fermentations were conducted in duplicate at two different temperatures resulting in differences in yeast morphology and fermentation kinetics between the experiments. Standard methods were used to evaluate the fermentation properties, while changes to morphology were assessed using manual and automated image analysis. Differences in fermentation rate, number of cells in suspension, and cell size found were consistent with previous work. An average discrepancy of only 2–4% was found between the manual and automated analysis of the images. There were high standard deviations for cellular morphology measurements (due to natural variability present within the yeast population), however, through the automated analysis of hundreds of cells, clear trends were observed. From the automated image analysis, changes to yeast CSA appeared to be temperature-dependent and progressed throughout the fermentation. Decreases in CSA ratio, defined as the ratio between the vacuole and yeast CSA, were observed at 30 °C and not at 35 °C. There were very small, or no vacuoles present in the cells at the end of the fermentation at 30 °C. Finally, the automated image analysis of cell morphology was combined with traditional measurements to provide insight into cellular surface areas and sugar consumption rates. The rate of fermentation per total non-vacuolar yeast volume decreased over the course of the fermentation and was higher at 35 °C despite a lower total rate of fermentation. This study showed that automated analysis of images can be used to track morphological changes in yeast over a fermentation, and that the combination of this analysis with traditional methods can be used to provide additional insight into yeast properties and behavior.

Author Contributions: Conceptualization, A.J.M. and M.G.-D.; methodology, M.G.-D.; software, M.C.; validation, M.G.-D., M.C.; formal analysis, A.J.M., M.G.-D.; investigation, M.G.-D.; resources, A.J.M.; data curation, M.G.-D., M.C.; writing—original draft preparation, A.J.M., M.G.-D., M.C.; writing—review and editing, K.A.T.-W.; visualization, M.G.-D., K.A.T.-W.; supervision, A.J.M.; project administration, A.J.M.; funding acquisition, A.J.M. All authors have read and agreed to the published version of the manuscript.

Funding: This research received no external funding.

Institutional Review Board Statement: Not applicable.

Informed Consent Statement: Not applicable.

Data Availability Statement: Most of the data presented in this study is available within the article itself. The (large) collection of raw image files are available upon request from the corresponding author.

Acknowledgments: The authors would like to acknowledge Lallemand Inc. for the donation of yeast used throughout this research, the Food Science and Human Nutrition department at the University of Florida for use of the facilities, and Victor Cedeño for the technical support involving microscope and software.

Conflicts of Interest: The authors declare no conflict of interest.

References

1. Wunderlich, S.; Back, W. 1—Overview of Manufacturing Beer: Ingredients, Processes, and Quality Criteria. In *Beer in Health and Disease Prevention*; Elsevier B.V.: Burlington, MA, USA, 2009; pp. 3–16. [CrossRef]
2. Tibayrenc, P.; Preziosi-Belloya, L.; Roger, J.M.; Ghommidha, C. Assessing yeast viability from cell size measurements? *J. Biotechnol.* **2010**, *149*, 74–80. [CrossRef] [PubMed]
3. Wang, J.; Ding, H.; Zheng, F.; Li, Y.; Liu, C.; Niu, C.; Li, Q. Physiological Changes of Beer Brewer's Yeast during Serial Beer Fermentation. *J. Am. Soc. Brew. Chem.* **2019**, *77*, 10–20. [CrossRef]
4. Zakhartsev, M.; Reuss, M. Cell size and morphological properties of yeast *Saccharomyces cerevisiae* in relation to growth temperature. *FEMS Yeast Res.* **2018**, *18*, foy052. [CrossRef] [PubMed]
5. Coelho, M.; Belo, I.; Pinheiro, R.; Amaral, A.; Mota, M.; Coutinho, J.; Ferreira, E. Effect of hyperbaric stress on yeast morphology: Study by automated image analysis. *Appl. Microbiol. Biotechnol.* **2004**, *66*, 318–324. [CrossRef] [PubMed]
6. Pratt, P.; Bryce, J.; Stewart, G. The Effects of Osmotic Pressure and Ethanol on Yeast Viability and Morphology. *J. Inst. Brew.* **2003**, *109*, 218–228. [CrossRef]
7. Pratt, P.; Bryce, J.; Stewart, G. The Yeast Vacuole—A Scanning Electron Microscopy Study during High Gravity Wort Fermentations. *J. Inst. Brew.* **2007**, *113*, 50–60. [CrossRef]
8. Izawa, S.; Ikeda, K.; Miki, T.; Wakai, Y.; Inoue, Y. Vacuolar morphology of *Saccharomyces cerevisiae* during the process of wine making and Japanese sake brewing. *Appl. Microbiol. Biotechnol.* **2010**, *88*, 277–282. [CrossRef] [PubMed]
9. Kida, K.; Gent, D.; Slaughter, J.C. Effect of vacuoles on viability of *Saccharomyces cerevisiae*. *J. Ferment. Bioeng.* **1993**, *76*, 284–288. [CrossRef]
10. Cahill, G.; Walsh, P.; Donnelly, D. Improved Control of Brewery Yeast Pitching Using Image Analysis. *J. Am. Soc. Brew. Chem.* **2018**, *57*, 72–78. [CrossRef]
11. O'Shea, D.; Walsh, P. The effect of culture conditions on the morphology of the dimorphic yeast Kluyveromyces marxianus var. marxianus NRRLy2415: A study incorporating image analysis. *Appl. Microbiol. Biotechnol.* **2000**, *53*, 316–322. [CrossRef] [PubMed]

12. Ginovart, M.; Carbó, R.; Blanco, M.; Portell, X. Digital Image Analysis of Yeast Single Cells Growing in Two Different Oxygen Concentrations to Analyze the Population Growth and to Assist Individual-Based Modeling. *Front. Microbiol.* **2018**, *8*, 2628. [CrossRef] [PubMed]
13. Ohtani, M.; Saka, A.; Sano, F.; Ohya, Y.; Morishita, S. Development of image processing program for yeast cell morphology. *J. Bioinform. Comput. Biol.* **2004**, *1*, 695–709. [CrossRef] [PubMed]
14. Grishagin, I. Automatic cell counting with ImageJ. *Anal. Biochem.* **2015**, *473*, 63–65. [CrossRef] [PubMed]
15. Wilson, C.; Lukowicz, R.; Merchant, S.; Valquier-Flynn, H.; Caballero, J.; Sandoval, J.; Okuom, M.; Huber, C.; Durham Brooks, T.; Wilson, E.; et al. Quantitative and Qualitative Assessment Methods for Biofilm Growth: A Mini-review. *Res. Rev. J. Eng. Technol.* **2017**, *6*. Available online: http://www.rroij.com/open-access/quantitative-and-qualitative-assessment-methods-for-biofilm-growth-a-minireview-.pdf (accessed on 5 March 2021).
16. Stolze, N.; Bader, C.; Henning, C.; Mastin, J.; Holmes, A.; Sutlief, A. Automated image analysis with ImageJ of yeast colony forming units from cannabis flowers | Elsevier Enhanced Reader. *J. Microbiol. Methods* **2019**, *164*, 105681. [CrossRef] [PubMed]
17. ASBC. Wort-12: Free Amino Nitrogen. Method, A. Ninhydrin method (international method). In *Wort Methods*; American Society of Brewing Chemists: St. Paul, MN, USA, 2011.
18. ASBC. Yeast-4: Microscopic yeast cell counting. In *Wort Methods*; American Society of Brewing Chemists: St. Paul, MN, USA, 2004.
19. Painting, K.; Kirsop, B. A quick method for estimating the percentage of viable cells in a yeast population, using methylene blue staining. *World J. Microbiol. Biotechnol.* **1990**, *6*, 346–347. [CrossRef] [PubMed]
20. ASBC. Yeast-14: Miniature fermentation assay. In *Wort Methods*; American Society of Brewing Chemists: St. Paul, MN, USA, 2011; pp. 1–3.
21. Rudolph, A.; MacIntosh, A.J.; Speers, R.A.; St. Mary, C. Modeling Yeast in Suspension during Laboratory and Commercial Fermentations to Detect Aberrant Fermentation Processes. *J. Am. Soc. Brew. Chem.* **2020**, *78*, 63–73. [CrossRef]
22. Ferreira, T.; Rasband, W. ImageJ User Guide IJ 1.46r. 2012. Available online: https://imagej.nih.gov/ij/docs/guide/user-guide-A4booklet.pdf (accessed on 5 March 2021).
23. Torija, M.; Rozès, N.; Poblet, M.; Guillamón, J.; Mas, A. Effects of fermentation temperature on the strain population of *Saccharomyces cerevisiae*. *Int. J. Food Microbiol.* **2003**, *80*, 47–53. [CrossRef]
24. Reddy, L.; Reddy, O. Effect of fermentation conditions on yeast growth and volatile composition of wine produced from mango (*Mangifer indica* L.) fruit juice. *Food Bioprod. Process.* **2011**, *89*, 487–491. [CrossRef]
25. Gervais, P.; Martínez de Marañón, I.; Evrard, C.; Ferret, E.; Moundanga, S. Cell volume changes during rapid temperature shifts. *J. Biotechnol.* **2003**, *102*, 269–279. [CrossRef]

 fermentation

Article

LC-ESI-QTOF-MS/MS Characterisation of Phenolics in Herbal Tea Infusion and Their Antioxidant Potential

Osbert Chou [1], Akhtar Ali [1], Vigasini Subbiah [1], Colin J. Barrow [2], Frank R. Dunshea [1,3] and Hafiz A. R. Suleria [1,2,*]

1 School of Agriculture and Food, Faculty of Veterinary and Agricultural Sciences, The University of Melbourne, Parkville, VIC 3010, Australia; ochou@student.unimelb.edu.au (O.C.); akali@student.unimelb.edu.au (A.A.); vsubbiah@student.unimelb.edu.au (V.S.); fdunshea@unimelb.edu.au (F.R.D.)
2 Centre for Chemistry and Biotechnology, School of Life and Environmental Sciences, Deakin University, Waurn Ponds, VIC 3217, Australia; colin.barrow@deakin.edu.au
3 Faculty of Biological Sciences, The University of Leeds, Leeds LS2 9JT, UK
* Correspondence: hafiz.suleria@unimelb.edu.au; Tel.: +61-470-439-670

Abstract: Ginger (*Zingiber officinale* R.), lemon (*Citrus limon* L.) and mint (*Mentha sp.*) are commonly consumed medicinal plants that have been of interest due to their health benefits and purported antioxidant capacities. This study was conducted on the premise that no previous study has been performed to elucidate the antioxidant and phenolic profile of the ginger, lemon and mint herbal tea infusion (GLMT). The aim of the study was to investigate and characterise the phenolic contents of ginger, lemon, mint and GLMT, as well as determine their antioxidant potential. Mint recorded the highest total phenolic content, TPC (14.35 ± 0.19 mg gallic acid equivalent/g) and 2,2′-azino-bis(3-e-thylbenzothiazoline-6-sulfonic acid), ABTS (24.25 ± 2.18 mg ascorbic acid equivalent/g) antioxidant activity. GLMT recorded the highest antioxidant activity in the reducing power assay, RPA (1.01 ± 0.04 mg ascorbic acid equivalent/g) and hydroxyl radical scavenging assay, •OH-RSA (0.77 ± 0.08 mg ascorbic acid equivalent/g). Correlation analysis showed that phenolic content positively correlated with the antioxidant activity. Venn diagram analysis revealed that mint contained a high proportion of exclusive phenolic compounds. Liquid chromatography coupled with electrospray ionisation and quadrupole time of flight tandem mass spectrometry (LC-ESI-QTOF-MS/MS) characterised a total of 73 phenolic compounds, out of which 11, 31 and 49 were found in ginger, lemon and mint respectively. These characterised phenolic compounds include phenolic acids (24), flavonoids (35), other phenolic compounds (9), lignans (4) and stilbene (1). High-performance liquid chromatography photometric diode array (HPLC-PDA) quantification showed that GLMT does contain a relatively high concentration of phenolic compounds. This study presented the phenolic profile and antioxidant potential of GLMT and its ingredients, which may increase the confidence in developing GLMT into functional food products or nutraceuticals.

Keywords: polyphenols; LC-ESI-QTOF-MS/MS; HPLC; medicinal plants; ginger; lemon; mint; herbal tea infusion; antioxidants

Citation: Chou, O.; Ali, A.; Subbiah, V.; Barrow, C.J.; Dunshea, F.R.; Suleria, H.A.R. LC-ESI-QTOF-MS/MS Characterisation of Phenolics in Herbal Tea Infusion and Their Antioxidant Potential. *Fermentation* **2021**, 7, 73. https://doi.org/10.3390/fermentation7020073

Academic Editor: Daniel Cozzolino

Received: 4 April 2021
Accepted: 6 May 2021
Published: 10 May 2021

1. Introduction

Herbal teas made from various medicinal plants are rich sources of phenolic compounds and are consumed by many cultures all around the world. Medicinal plants prepared for tea infusions have been studied for their antioxidant, anti-inflammatory and other properties [1]. Herbal teas have also been reported to exhibit synergistic antioxidant effects, which increases their value as beverages for potential health benefits [2,3]. Common medicinal plants such as ginger, lemon and mint can be fused together as ginger lemon mint tea (GLMT) and be consumed as herbal tea to improve health being.

Gingers (*Zingiber officinale* Roscoe) from the family *Zingiberaceae* [4–6], are native to Southeast Asia and have been incorporated into a diverse array of cuisines. In China and

India, gingers are regarded as medicine [5]. Ginger is reported to have various health benefits as it demonstrates antioxidant, antimicrobial and anti-inflammatory activities [7]. A gingerol is a group of phenols commonly found in the rhizomes of *Zingiberaceae* plants, which have been extensively studied for their cytotoxic potential [8]. Lemon (*Citrus limon* L.) is from the citrus family *Rutaceae* [9], and has been demonstrated to contain phenolic compounds which are beneficial to health [10]. Eriocitrin, a phenolic compound in lemons was found to increase antioxidant activities in plasma after ingestion [11]. Lemon has potent antioxidant activities in its peel and zest and its protective properties against DNA and cell damage have prompted the research on lemon as a cancer preventative [12]. Other studies have also shown that lemon's essential oil has antioxidant and antimicrobial effects [13]. Mint (*Mentha sp.*) is from the *Lamiaceae* plant family [14,15]. The essential oils of plants within the *Mentha* genus exhibit antioxidant and antimicrobial activities [16,17]. Notable antioxidants in mints were found to be rosmarinic acid, caffeic acid, α-terpinene, luteolin and eriocitrin and these compounds are believed to be the major antioxidants particularly found in peppermint [15].

Phenolic compounds are a large group of secondary plant metabolites commonly found in fruits, vegetables and spices, which are thought to play prominent roles in human nutrition and health [18]. Some of these metabolites produced by plants are classified as antioxidants [19]. Antioxidants are molecules that can reduce the damaging effects of free radicals, which are chemicals with one or more unpaired electrons produced in an organism via natural metabolism or environmental factors [20]. Free radicals can turn non-free radical molecules in the body into radicals, cascading a chain reaction that causes destructive effects within cells and tissues [21]. Thus, medicinal plants rich in antioxidants have been lauded and advised to be part of the human diet to promote health and wellbeing [22,23].

Various assays and equipment have been developed that have been useful for the estimation and analysis of phenolic and antioxidant content. For the estimation of phenolic compounds, common assays include Total Phenolic Content (TPC) Assay with Folin-Ciocalteu reagent [24], Total Flavonoid Content (TFC) Assay by the aluminium chloride method [25] and Total Tannins Content (TTC) using vanillin method [26]. 2,2′-diphenyl-1-picrylhydrazyl (DPPH) assay [27], Ferric reducing antioxidant power (FRAP) assay [28], 2,2′-azino-bis(3e-thylbenzothiazoline-6-sulfonic acid) (ABTS) radical scavenging assay [29], reducing power assay (RPA) [30], hydroxyl radical scavenging activity (•OH-RSA) [31] and ferrous ion chelating activity (FICA) [32] assays are common assays utilised to determine the antioxidant activities of samples. In addition, the phosphomolybdate method has been used in the past for the estimation of total antioxidant content (TAC). Statistical analysis was performed on the phenolic content and antioxidant activity of the samples to investigate whether the phenolic content correlated with the antioxidant activity. Liquid chromatographic-coupled mass spectrometry was useful as a separation, identification and quantification technique, enabling the characterisation of phenolic compounds. This experiment deployed HPLC-PDA and LC-ESI-QTOF-MS/MS, which enabled the quantification and characterisation of phenolic compounds from ginger, lemon, mint and GLMT. Previous experiments have also successfully utilised both HPLC and LC-MS/MS techniques to quantify and characterise phenolic compounds in various herbal teas and medicines [33–35]. The comprehensive understanding of the phenolic profile of ginger, lemon and mint, GLMT has not been extensively studied and there is a research gap that requires investigation into GLMT's antioxidant effect and its phenolic profile. Thus, the aims of this study were to: (1) investigate and compare antioxidant activities of ginger, lemon, mint and GLMT herbal tea infusion; (2) characterise and quantify their phenolic compounds through LC-ESI-QTOF-MS/MS and HPLC-PDA respectively. Results indicating high phenolic content and potent antioxidant activity from the samples could be used to inform the benefits of dietary uptake of ginger, lemon, mint and GLMT, which may also encourage commercialisation of GLMT infusions.

2. Materials and Methods

2.1. Chemicals and Reagents

Most of the chemicals utilised were analytical grade and purchased from Sigma-Aldrich (St. Louis, MO, USA). Reagents for the antioxidant assays include 2,2′-diphenyl-1-picrylhydrazyl (DPPH), 2,2′-azino-bis(3-ethylbenzothiazoline-6-sulfonic acid) (ABTS), 2,4,6-tripyridyl-s-triazine (TPTZ), Folin and Ciocalteu's phenol, hexahydrate aluminium chloride, potassium persulfate and vanillin were purchased from Sigma-Aldrich (St. Louis, MO, USA). The phenolic acid and flavonoid reference standards used such as caffeic acid, caftaric acid, catechin, chlorogenic acid, coumaric acid, epicatechin gallate, gallic acid, kaempferol, kaempferol-3-*O*-glucoside, L-ascorbic acid, *p*-hydroxybenzoic acid, protocatechuic acid, quercetin, quercetin-3-galactose, quercetin-3-*O*-glucuronide and syringic acid were also bought from Sigma-Aldrich (St Louis, MO, USA). Other chemicals including acetic acid, ethanol, ferric (III) chloride (Fe [III] $Cl_3 \cdot 6H_2O$), sodium acetate and sodium carbonate were purchased from Sigma-Aldrich (St. Louis, MO, USA). The HPLC grade acetic acid, methanol and acetonitrile used were bought from Sigma-Aldrich (St. Louis, MO, USA). The 98% sulphuric acid used in this study was bought from RCI Labscan (Rongmuang, Thailand). Anhydrous sodium carbonate, glacial acetic acid, hydrochloric acid and methanol were purchased from Thermo Fisher Scientific Inc. (Waltham, MA, USA).

2.2. Sample Preparation

Ginger rhizomes, lemon and mint were bought from local supermarkets. Each sample was blended into slurries separately using a 1.5 L blender (Russell Hobbs Classic, model DZ-1613, Melbourne, VIC, Australia). Samples were stored at −20 °C for further analysis.

The herbal infusion ratio sampled the available commercial ginger, lemon, mint green tea from Tetley (Tata Global Beverages Pty. Ltd., Richmond VIC, Australia) in which its flavouring had a ratio of 50% lemon flavouring, 33% ginger flavouring and 17% mint flavouring. Ginger rhizome, lemon and mint were sliced thin and steeped in 100 mL of boiling Milli-Q water. After 10 min, debris in the herbal infusion was filtered out. The herbal tea infusion was then stored at −20 °C until required for further analysis.

2.3. Extraction of Phenolic Compounds

Sample extracts of 1 mL (herbal infusions and herbal tea infusion) were homogenised with the Ultra-Turrax T25 Homogenizer IKA, Staufen, Germany) in 10 mL of 70% ethanol at 10,000 rpm for 30 s. The samples were then incubated in a shaking incubator (ZWYR-240, Labwit, Ashwood, VIC, Australia) at 4 °C, 120 rpm for 12 h. Samples were then centrifuged for 15 min at 5000 rpm (ROTINA 380 R centrifuge, Hettich, Victoria, Australia). The supernatant was collected and then stored at −20 °C [36]. For HPLC and LC-MS analysis, the extracts were filtered through a 0.45 µm sterile syringe filter (hydrophilic polyvinylidene fluoride—PVDF) purchased from Millipore, Merck (KGaA, Darmstadt, Germany).

2.4. Phenolic Compound Estimation and Antioxidant Assays

The following assays were performed based on the previously published methodologies with some alterations [37–39]. The absorbance of samples was measured by the Multiskan® Go microplate spectrophotometer (Thermo Fisher Scientific, Waltham, MA, USA) in triplicates. Standard curves were generated with $R^2 > 0.995$. Phenolic compound estimation was assayed through TPC, TFC and TTC while antioxidant capacity was measured by 4 antioxidant assays, TAC, ABTS, DPPH and FRAP.

2.4.1. Total Phenolic Content (TPC) Assay

TPC was measured by following a modified version of a previous method [40]. A total of 25 µL of Folin-Ciocalteu reagent (1:3 diluted with water) was mixed with 25 µL sample and 200 µL of Milli-Q water and incubated in the dark for 5 min at 25 °C. A sodium carbonate solution (10%, *w*/*w*) of 25 µL was then added and the mixtures were incubated at 25 °C for 60 min before absorbance was measured at 765 nm. Gallic acid was dissolved in 70% ethanol and was used to construct the standard curve with concentrations ranging from 0 to 200 μg/mL. Results were expressed as mg of gallic acid equivalents (GAE) per gram of fresh weight (mg GAE/$g_{f\cdot w}$).

2.4.2. Total Flavonoid Content (TFC) Assay

The flavonoid content of the samples was estimated by a previous aluminium chloride ($AlCl_3$) colourimetric-based method with some modification [26]. A total of 80 μL of the extract and 80 μL of 2% $AlCl_3$ was mixed (dilution aided by 70% ethanol). An amount of 120 μL of 50 g/L sodium acetate was added to a 96-well plate and incubated at 25 °C for 2.5 h. Sample absorbance was measured at 440 nm as triplicates. TFC of the samples was expressed as mg of quercetin per gram of fresh weight of the sample (mg QE/$g_{f\cdot w}$). The calibration curve of quercetin was constructed using various concentrations between 0 to 50 μg/mL.

2.4.3. Total Tannins Content (TTC) Assay

The following TTC is a modified iteration of a previous colourimetric method [26]. A total of 150 μL of 4% vanillin was mixed with 25 μL of sample extract and diluted with methanol. An amount of 25 μL of 32% sulphuric acid was also added to each well in the 96-well plate. The well plate was incubated in the dark at 25 °C for 15 min, and then the absorbance was measured at 500 nm. Catechin was dissolved in 70% ethanol to make varying concentrations from 0 to 1000 μg/mL, which was then used to construct the standard curve. Results were expressed as mg catechin equivalents per gram of fresh weight (mg CE/$g_{f\cdot w}$).

2.4.4. DPPH Radical Scavenging Assay

DPPH assay was used to measure the radical scavenging activity of each extract and GLMT mixture and the assay method was sourced from previous research with some modifications for this research [41]. In total, 40 μL of extract and 40 μL of 0.1 mM DPPH methanolic solution were mixed in a 96-well plate and incubated at 25 °C within a shaker for 30 min. Then the absorbance was measured at 517 nm as triplicates. The standard curve was generated results were expressed as mg AAE/$g_{f\cdot w}$ with ascorbic acid's concentration ranging from 0 to 50 μg/mL. This was then used to determine the sample's scavenging activity against DPPH free radicals.

2.4.5. Ferric Reducing Antioxidant Power (FRAP) Assay

FRAP assay assesses the ability of the sample to reduce iron in the Fe^{3+}-TPTZ complex (ferric-2,4,6-tripyridyl-s-Triazine) to Fe^{2+}-TPTZ complex [42]. The performed FRAP assay adapted a previous method with some modifications [41]. Fresh FRAP reagent was made for each FRAP assay. A sodium acetate buffer (pH = 3.6) of 300 mM, 20 mM $FeCl_3$ and 10 mM TPTZ were mixed together in a volume ratio of 10:1:1. The mixture of 1 mL of the FRAP reagent and 400 μL of the sample was incubated for 5 min at 37 °C. Then the absorbance of each sample was measured at 593 nm as triplicates. Varying concentrations between 0 to 50 μg/mL of ascorbic acid were prepared for the construction of the standard curve. The results were expressed as mg AAE/$g_{f\cdot w}$.

2.4.6. ABTS Radical Scavenging Assay

The free radical scavenging activity of the sample was also determined through the $ABTS^+$ radical cation decolourisation assay, with a slight modification to a previous methodology [41]. A total of 5 mL of 7 mmol/L ABTS solution was mixed with 88 μL of 140 mM potassium persulfate, which allowed for the generation of $ABTS^+$ for the scavenging assay. The mixture was left in the dark for 16 h before being retrieved and diluted with 70% ethanol. A total of 10 μL of sample and 290 μL of the ABTS solution were added to the 96-well plate, then the mixture was incubated at 25 °C for 6 min in the dark. Following that, the sample absorbance was read at 734 nm as triplicates. The calibration curve was generated from ascorbic acid with concentration ranging from 0 to 2000 μg/mL, in which the results were expressed as mg AAE/$g_{f \cdot w}$.

2.4.7. Reducing Power Assay (RPA)

The reducing power assay was a modified version of Ferreira, et al. [30]. A total of 10 μL of the sample, 25 μL of 0.2 M phosphate buffer (pH 6.6) and 25 μL of $K_3[Fe(CN)_6]$ were mixed sequentially and the solution was then incubated at 25 °C for 20 min. An amount of 25 μL of 10% TCA solution was added to stop the reaction, which was followed by the addition of 85 μL of water and 8.5 μL of $FeCl_3$. The solution was further incubated for 15 min at 25 °C, and then the absorbance was measured at 750 nm. Ascorbic acid with concentrations ranging from 0 to 500 μg/mL was used to construct a standard curve and results were expressed as mg AAE/$_{g.fw}$.

2.4.8. Hydroxyl Radical Scavenging Activity (•OH-RSA)

A modified Fenton-type reaction method of Smirnoff and Cumbes [31] was conducted to determine •OH-RSA of samples. In total, 50 μL of the sample was mixed with 50 μL of 6 mM H_2O_2 (30%) and with 50 μL of 6 mM $FeSO_4.7H_2O$, and the solution was incubated at 25 °C for 10 min. Then, 50 μL of 6 mM 3-hydroxybenzoic acid was added and the absorbance was measured at 510 nm. Ascorbic acid concentrations ranging from 0 to 300 μg/mL were used to create a standard curve and data were expressed as mg AAE/$_{g.fw}$.

2.4.9. Ferrous Ion Chelating Activity (FICA)

The ferrous ion chelating activity of samples was measured using a modified method of Dinis, et al. [32]. A total of 15 μL of the sample was mixed with 85 μL of water, 50 μL of 1:15 water-diluted 2 mM ferrous chloride and 50 μL of 1:6 water-diluted 5 mM ferrozine. The mixture was then incubated at 25 °C for 10 min and then the absorbance was measured at 562 nm. Ethylenediaminetetraacetic acid (EDTA) with concentrations ranging from 0 to 50 μg/mL was used to construct a standard curve and data was expressed as mg EDTA/$_{g.fw}$.

2.4.10. Total Antioxidant Content (TAC) Assay

TAC in the sample was determined by a modified phosphomolybdate method [43]. The phosphomolybdate reagent was a combination of 0.004 M ammonium molybdate, 0.028 M sodium phosphate and 0.6 M sulfuric acid. An amount of 260 μL phosphomolybdate reagent was mixed with 40 μL extracts and was then incubated at 95 °C for 10 min. Absorbance was then measured at 695 nm at 25 °C and ascorbic acid was quantified by linear regression plotting the absorbance against standard concentration (0–200 μg/mL). The amount of TAC was expressed in ascorbic acid equivalent per gram of fresh weight (mg AAE/$g_{f \cdot w}$).

2.5. LC-ESI-QTOF-MS/MS Analysis

Phenolic compounds were characterised following a previous protocol [44], which was performed with Agilent 1200 series HPLC (Agilent Technologies, Santa Clara, CA, USA) coupled with Agilent 6520 Accurate-Mass Q-TOF LC/MS (Agilent Technologies, Santa Clara, CA, USA). Synergi Hydro-RP 80A, with LC Column of 250 mm × 4.6 mm internal diameter and 4 μm particle size (Phenomenex, Torrance, CA, USA), was used for the separation of compounds with a column temperature of 25 °C and sample temperature of 10 °C. Mobile phase A was a mixture of water/acetic acid at a ratio of 98:2 (*v*/*v*). Mobile phase B was a mixture of acetonitrile/water/acetic acid at a ratio of 100:99:1 (*v*/*v*/*v*). Both mobile phases were degassed at 21 °C for 15 min. The injection volume for each infusion sample was 6 μL, the linear gradient elution of water contained 1% acetic acid and the flow rate was set at 0.8 mL/min. The gradient elution ran for 85 min with a specific mixture of mobile phase A and B following the gradient as described in a past method [45]. The gradient reset to the initial gradient after the end of the program and the column was equilibrated for 3 min before the next injection. Mass spectrometry conditions were set with nitrogen gas temperature of 300 °C with a flow rate of 5 L/min, sheath gas temperature of 250 °C with a flow rate of 11 L/min and nebuliser gas pressure of 45 psi. The nozzle voltage was set at 500 V and the capillary voltage was set at 3.5 kV. A complete mass scan ranging from m/z 50 to 1300 was used, MS/MS analyses were carried out in automatic mode with collision energy (10, 15 and 30 eV) for fragmentation. Peak identification was performed in both positive and negative modes while the instrument control, data acquisition and processing were performed using MassHunter workstation software (Qualitative Analysis, version B.03.01) (Agilent Technologies, Santa Clara, CA, USA).

2.6. HPLC-PDA Analysis

Quantification of the targeted phenolic compounds was performed following a previous methodology [45] using Agilent 1200 series HPLC (Agilent Technologies, Santa Clara, CA, USA), which was equipped with Waters Model 2998 photodiode array (PDA) detector. Settings for the column conditions were maintained as per the LC-ESI-QTOF-MS/MS method above, with only one difference in sample injection at 20 μL. Detection was set at wavelengths of 280 nm, 320 nm and 370 nm, which were selected for the identification of hydroxybenzoic acids, hydroxycinnamic acids and flavonols respectively. The spectral acquisition rate was set at 1.25 scan/second with a peak width of 0.2 min. Standards for each phenolic compound sample were diluted into 7 different concentrations to generate calibration standard curves for the quantification. Data acquisition and analysis were performed using Agilent LC-ESI-QTOF-MS/MS MassHunter Workstation Software Version B.03.01 (Agilent Technologies, Santa Clara, CA, USA).

2.7. Statistical Analysis

The one-way analysis of variance (ANOVA) via Minitab Version 19.0 (Minitab, LLC, State College, PA, USA) was calculated to differentiate the mean values between samples, using the setting Tukey's honestly significant differences (HSD) at $p \leq 0.05$. The result yielded were portrayed as mean ± standard deviation (SD). Pearson's pairwise correlation test was also performed on the same software to elucidate the relationship between the phenolic content assays and the antioxidant assays.

3. Results and Discussion

3.1. Phenolic Compound Estimation (TFC, TPC and TTC)

Ginger, lemon and mint are all considered to be medicinal plants with diverse phenolic contents. The phenolic content of ginger, lemon, mint and GLMT were measured by TFC, TPC and TTC assays with the results expressed as quercetin, gallic acid and catechin equivalent respectively. Table 1 shows that mint has the highest total phenolic content, followed by GLMT. Our ginger sample yielded 2.93 ± 0.07 mg GAE/g of TPC, which is higher than a previous study's fresh ginger (control group) TPC yield of 0.44 mg GAE/$g_{f\cdot w}$ [46]. However, when compared to the same study's oven-dried ginger samples, the TPC of the oven-dried ginger sample yielded 9.19 mg GAE/$g_{f\cdot w}$, nearly 4 times more yield than our sample. This suggests that sample drying may help increase phenolic compound yield. The phenolic content of lemon obtained in this study was considerably lower compared to a previous study on dried Beldi lemon flesh, which reported 105.55 mg GAE/$g_{f\cdot w}$ of TPC, 56.16 mg QE/$g_{f\cdot w}$ of TFC and 26.66 mg CE/$g_{f\cdot w}$ of condensed tannins, respectively [12]. The mint samples in this experiment extracted less TPC and TFC compared to a previous study, in which they were able to obtain 19.9 mg GAE/$g_{f\cdot w}$ of TPC and 13.1 mg QE/$g_{f\cdot w}$ of TFC respectively [47]. There was also another study of 6 different mint species in northeast Algeria, which found that *Mentha x piperita* produced 31.4 mg GAE/$g_{d\cdot w}$ for TPC assay and 6.50 mg CE/$g_{d\cdot w}$ for TTC assay, both of which are higher than the results obtained in this study [48].

Table 1. Phenolic compound content (TPC, TFC and TTC) and antioxidant activities (DPPH, FRAP, ABTS, RPA, •OH-RSA, FICA and TAC) of ginger, lemon, mint and GLMT per gram of fresh weight.

Assays	Ginger	Lemon	Mint	GLMT
TPC (mg GAE/g)	2.93 ± 0.07 [c]	0.12 ± 0.04 [d]	14.35 ± 0.19 [a]	5.85 ± 0.08 [b]
TFC (mg QE/g)	0.98 ± 0.02 [c]	0.07 ± 0.03 [d]	1.29 ± 0.07 [a]	1.25 ± 0.04 [b]
TTC (mg CE/g)	0.02 ± 0.04 [d]	0.04 ± 0.01 [c]	2.13 ± 0.08 [a]	0.45 ± 0.09 [b]
DPPH (mg AAE/g)	1.24 ± 0.09 [c]	0.09 ± 0.04 [d]	3.15 ± 0.12 [a]	2.24 ± 0.01 [b]
FRAP (mg AAE/g)	0.83 ± 0.03 [c]	0.08 ± 0.01 [d]	7.15 ± 0.14 [a]	1.91 ± 0.07 [b]
ABTS (mg AAE/g)	0.11 ± 0.01 [c]	0.07 ± 0.03 [d]	24.25 ± 2.18 [a]	5.48 ± 0.21 [b]
RPA (mg AAE/g)	0.08 ± 0.01 [a]	0.02 ± 0.03 [c]	0.04 ± 0.02 [b]	1.01 ± 0.04 [b]
•OH-RSA (mg AAE/g)	0.53 ± 0.03 [b]	0.39 ± 0.07 [a]	0.27± 0.01 [c]	0.77± 0.08 [c]
FICA (mg EDTA/g)	0.04 ± 0.03 [b]	-	0.09 ± 0.02 [a]	0.07 ± 0.04 [a]
TAC (mg AAE/g)	0.73 ± 0.08 [c]	0.03 ± 0.04 [d]	6.74 ± 0.58 [a]	1.35 ± 0.41 [b]

Data expressed as mean ± SD of three replicates; Different letters [a,b,c,d] indicate that the data is significantly different from the other data of the same row ($p \leq 0.05$), in which [a] is assigned to the largest value, then [b] assigned to second largest and so forth. The significant difference was calculated through one-way analysis of variance (ANOVA) and Tukey's HSD Test.

The variation of phenolic content difference with the previous study may be due to differences in variety or growing condition. Using lemon as an example, the variety in the previous study was identified to be Beldi, whereas the variety of the lemon used within this study could be any of the common lemon varieties in Australia such as Eureka or Lisbon [49]. Different varieties of the same fruit may exhibit different phenotypic traits and, thus, this could be a contributing factor to the observed difference in phenolic content. Growing conditions can also alter the phenolic content in plants such as a change in metabolism when dealing with environmental stress [50]. Finally, as seen in the ginger sample comparison with a previous study, sample treatment such as drying may increase phenolic content dramatically. Similarly, sample extraction and preparation will directly affect the yield of phenolic compounds in the sample.

3.2. Antioxidant Activities (DPPH, FRAP, ABTS, RPA, •OH-RSA, FICA and TAC)

Antioxidant radical scavenging activities of samples were measured through DPPH, FRAP, ABTS, RPA, •OH-RSA, FICA and TAC assays. These are common assays used for determining the antioxidant capacity of plant materials. Table 1 shows that ginger, lemon, mint and GLMT's antioxidant activities significantly differ from each other according to Tukey's HSD Test. Across most assays, the highest antioxidant activity has been recorded in mint samples, followed by GLMT sample, then ginger and lastly lemon. This pattern matches that of the phenolic compound content discussed above (Section 3.1), with mint recording the highest phenolic content. Previous studies have demonstrated that there is a significant positive correlation between antioxidant activity and phenolic content [51]. This correlation may explain why mint exhibited the highest antioxidant activity out of all samples. However, other non-phenolic phytochemicals may also contribute to the antioxidant assays and mint does contain other phytochemicals that can also act as antioxidant such as α-terpinene [15].

Contrary to past research, lemon did not produce high antioxidant activity or a high phenolic compound content. For antioxidant assays that are predicated on the antioxidants donating hydrogen to reduce the free radical, the amount of phenolic antioxidants will be correlated with the amount of antioxidant activity, thus, the lower the phenolic content, the more likely it is that the antioxidant capacity of that sample will be lower [52]. Since mint displayed the highest antioxidant activity for most assays, the antioxidant activity in GLMT may be contributed predominantly by mint's polyphenols or other phytochemicals.

DPPH radical scavenging assay measures antioxidant activity, which is indicated by the colourimetric change of the DPPH radical solution from the oxidised violet form to an antioxidant-reduced yellow form [53]. A previous study's peppermint DPPH assay yielded 147.5 mg AAE/g [54], which is drastically higher than our result. Possible reasons for the dramatic difference may be that the previous study dried all leaf samples before extracting the sample whereas this study did not dry the sample. A previous characterisation study of phenolic compounds from a variety of *Citrus* fruits found that sun dried and grounded Eureka lemon exhibited 8.26 mg TEAC/$g_{d \cdot w}$ [55], which is much higher than this study's lemon DPPH result. As discussed, dried samples may result in increased phenolic compound measured in a sample and thus increased antioxidant activity of that sample per gram of fresh weight. This study's ginger result was slightly higher compared to a previous study on phenolic compound-rich vegetables, which found that ginger yielded 0.21 ± 0.00 mg AAE/g for the DPPH assay [56]. The size and age of the sampled ginger rhizome may also be a factor in the differences.

FRAP assay is a colourimetric assay that measures the absorbance of the developed blue solution from the reduction of Fe(III)-TPTZ complex to Fe(II)-TPTZ by antioxidants in acidic conditions, in which the amount of absorption provides an estimate of the antioxidant activity [28]. In a previous phenolic compound-rich vegetable antioxidant study with similar methods, the FRAP assay yielded 0.04 mg AAE/g [56], which is very similar to the result obtained in this study. A study on fruit peels found that lemon peels exhibited 14.03 mg AAE/g from the FRAP assay [57], much higher than this study's lemon results. Many studies have consistently reported that fruit peels yield more polyphenols and thus more antioxidant activity. A study on tropical and temperate tea drinks found that mint and peppermint samples produced 31 mg GAE/g and 37 mg GAE/g, respectively, from the FRAP assay [58]. These results are much higher than the mint result from this study, which may potentially be attributed to the source of mint and also the aqueous extraction method performed.

ABTS assay determines the sample's antioxidant radical scavenging activity by measuring the colourimetric change of the sample solution from blueish green to colourless as ABTS free radicals become scavenged and reduced by the sample's antioxidants in the presence of persulfate salt [59]. In a previous study on vegetable phenolic content and antioxidants, the ginger samples yielded 1.09 mg AAE/g from ABTS assay, roughly 10 times higher than the results obtained in this study [56]. A study on spices and food condiments in South Korea found that ginger rhizome and mint recorded 7.49 mg AAE/$g_{d \cdot w}$ and 19.89 mg AAE/$g_{d \cdot w}$, respectively for the ABTS assay [60]. A previous Singaporean study found that blended lemon flesh produced approximately 0.93 mg AAE/g from ABTS assay [61], which is higher than this study's lemon ABTS result. The difference may be due to a different variety of lemon studied and regional variance of the fruit.

RPA is an assay that measures the sample's ability to reduce the ferric ion in potassium ferricyanide into ferrous ion in the form of potassium ferrocyanide [62]. GLMT recorded the highest RPA at 1.01 ± 0.04 mg AAE/g and markedly contrasts with the RPA of other samples (ranging from 0.02 to 0.08 mg AAE/g). A past study observed that ginger roots using methanolic and ethanolic extraction produced the best RPA results [63]. Additionally, they also reported that their RPA values correlated with the total tannin content of the sample. A study on spearmint recorded RPA activities in a spearmint sample [64].•OH-RSA determines the sample's ability to scavenge hydroxyl radicals. GLMT recorded the highest value of 0.77 ± 0.08 mg AAE/g, followed by 0.53 ± 0.03 mg AAE/g from ginger sample. Previous research showed that ginger displayed the ability to scavenge hydroxyl radicals [65].

FICA measures the sample's ability to compete with ferrozine to chelate ferrous ions. Our mint sample produced the highest FICA at 0.09 ± 0.02 mg EDTA/g, followed by GLMT at 0.07 ± 0.04 mg EDTA/g. In a past study, mint has demonstrated the ability to chelate ferrous ion and its ability to chelate ferrous ion was stated to be proportional to its polyphenol content [66]. Our lemon sample did not appear to have chelated ferrous ions. A previous study also did not observe ferrous ion chelation in their lemon fruit sample [67].

Phosphomolybdate assay was chosen to measure the TAC of samples. The presence of antioxidants turns the greenish solution blue upon reduction of the phosphomolybdenum complex's molybdenum ion from Mo^{6+} to Mo^{5+} or Mo^{4+} [43]. A study on culinary spices found that ginger yielded 1.5 μg AAE/mL and peppermint yielded 1 μg AAE/mL [68]. A study on lemon's total antioxidant activity using phosphomolybdenum assay reported 1.75 mg of butylated hydroxytoluene equivalent per gram of sample (BHT/g) when using 70% methanol as solvent [69].

3.3. Phenolic Content and Antioxidant Activity Correlation Analysis

Pairwise Pearson's correlation test was performed to determine whether the phenolic compound content from the samples contributed to their corresponding antioxidant activities. The correlation test results are as shown in Table 2. The *r* values for the pairwise correlations between TTC and the antioxidant assays were the most significant, with TTC significantly correlated with the FRAP, ABTS and TAC assays. TTC's high correlation with the antioxidant assays suggests that the tannins are likely the compounds that are primarily responsible for the antioxidant activities measured. TFC did not correlate well with the antioxidant assays except for DPPH and FICA. TPC was also significantly correlated with DPPH, FRAP, ABTS, FICA and TAC assays, which indicates that the observed antioxidant activities were likely caused primarily by the phenolic compounds extracted and not by other non-phenolic compounds.

Table 2. Pearson's correlation coefficient (*r*) for the pairwise correlation between the phenolic content assays (TPC, TFC, TTC) and the antioxidant assays (DPPH, FRAP, ABTS, RPA, •OH-RSA, FICA and TAC).

Assays	TPC	TFC	TTC	DPPH	FRAP	ABTS	RAP	OH-RSA	FICA
TFC	0.747								
TTC	0.971 *	0.565							
DPPH	0.942 *	0.915 *	0.839						
FRAP	0.988 *	0.639	0.994 **	0.879					
ABTS	0.976 *	0.583	1.000 **	0.852	0.996 **				
RAP	0.004	0.440	−0.150	0.293	−0.126	−0.126			
•OH-RSA	−0.357	0.269	−0.537	−0.030	−0.492	−0.515	0.886 *		
FICA	0.907 *	0.948 *	0.783	0.995 **	0.831 *	0.798	0.353	0.055	
TAC	0.977 *	0.604	0.994 **	0.850	0.998 **	0.993 **	−0.192	−0.546	0.799

* indicates significant correlation with $p \leq 0.1$, while ** indicates highly significant correlation with $p \leq 0.01$.

FRAP was significantly correlated with ABTS and TAC assays and TAC was significantly correlated with ABTS assay. The correlation between DPPH with TPC and TTC was low compared to other antioxidant assays. This could mean that the phenolic compounds found in the samples were not as capable of scavenging DPPH radicals. Although DPPH assays are used widely to determine the antioxidant capacity of bioactive compounds, there have been reports of its limitations. The sensitivity of DPPH radicals to some compounds including oxygen and their reactivity with other radicals in the solution are among some of the drawbacks of the DPPH assay [53]. Interestingly, both RPA and •OH-RSA did not significantly correlate with any other assays, except for their correlation with each other at $r = 0.886$.

Mint is known to contain various phenolic compounds such as rosmarinic acid, ferulic acid, caffeic acid, apigenin and more [70]. Mint's phenolic content has been shown to be correlated with its antioxidant activity [48]. A previous study reported that ginger's phenolic content in the leaves and rhizome correlated with its antioxidant activity [71]. A study on Sudanese wasted *Citrus* fruits (including wasted lemons) showed that these wasted peels had higher phenolic content and was significantly correlated with antioxidant activity [72]. These findings in the literature further increase our confidence that the recorded antioxidant activities were due to the samples' phenolic contents.

3.4. LC-MS/MS Analysis

LC-MS/MS is a commonly utilised technique to characterise phenolic compounds from a diverse source of organic samples. In tandem mass spectrometry, product ions are generated, in which the product ion pattern is characteristic of a particular molecule, and this feature allows for the characterisation of compounds of a particular sample [73]. Since its conception, LC-MS/MS has been widely used for sample analysis in food, biomedical and pharmaceutical industries. For polyphenol characterisation, LC-MS/MS has been regarded as a powerful and accurate technique to identify and quantify polyphenols due to its versatility [74]. Out of all the screened phenolic compounds, 73 were characterised in this study.

3.4.1. Phenolic Distribution—Venn Diagram Analysis

A total of 292 phenolic compounds were screened from ginger (51), lemon (167) and mint (247) using LC-MS analysis. It was found that 23 phenolic compounds were common in all three samples as shown in Figure 1. The majority of phenolic acids were identified in mint, and mint shared a relatively large proportion of its phenolic acid profile with ginger and lemon. However, the phenolic acid profiles of ginger and lemon were relatively dissimilar when compared to the similarity of their phenolic acid profiles with mint. For flavonoids, ginger shared a large portion of its flavonoids profile with both lemon and mint. Despite sharing a significant portion of its flavonoids profile with mint, lemon did contain a relatively large portion of flavonoids that were found only in lemon. Mint had the highest number of other polyphenols identified. Ginger and lemon both share a large portion of their other polyphenol profile with mint. Despite ginger's seemingly low phenolic compound profile, ginger is still considered to be a medicinal plant that offers beneficial phytochemicals with potent antioxidant and anti-inflammatory activities such as gingerol and shogaol [75]. Thus, it is still considered worthwhile to infuse ginger into herbal teas to incorporate ginger's other potentially healthy and useful phytochemicals.

3.4.2. LC-ESI-QTOF-MS/MS Characterization

The untargeted shotgun qualitative analysis of ginger, lemon and mint was performed via LC-ESI-QTOF-MS/MS in both positive and negative ionisation modes ($[M-H]^-$/$[M+H]^+$). The phenolic compounds from the samples were identified based on their *m/z* value and MS spectra in both ionisation modes using Agilent's LC-MS Qualitative Software and Personal Compound Database and Library (PCDL). Mass error less than ± 5 ppm and a PCDL library score of more than 80 were used as a benchmark to select further MS/MS identification, *m/z* characterisation and to verify the compounds (Supplementary Materials, Figures S1 and S2). In this research, through LC-MS/MS, 73 compounds were identified and characterised out of ginger (11 compounds), lemon (31 compounds) and mint (49 compounds), which include phenolic acids (24), flavonoids (35), other phenolic compounds (9), lignans (4) and stilbene (1) as listed in Table 3. It is possible that the relatively high diversity of phenolic compounds found in mint samples may explain its relatively higher antioxidant activities compared with ginger and lemon samples.

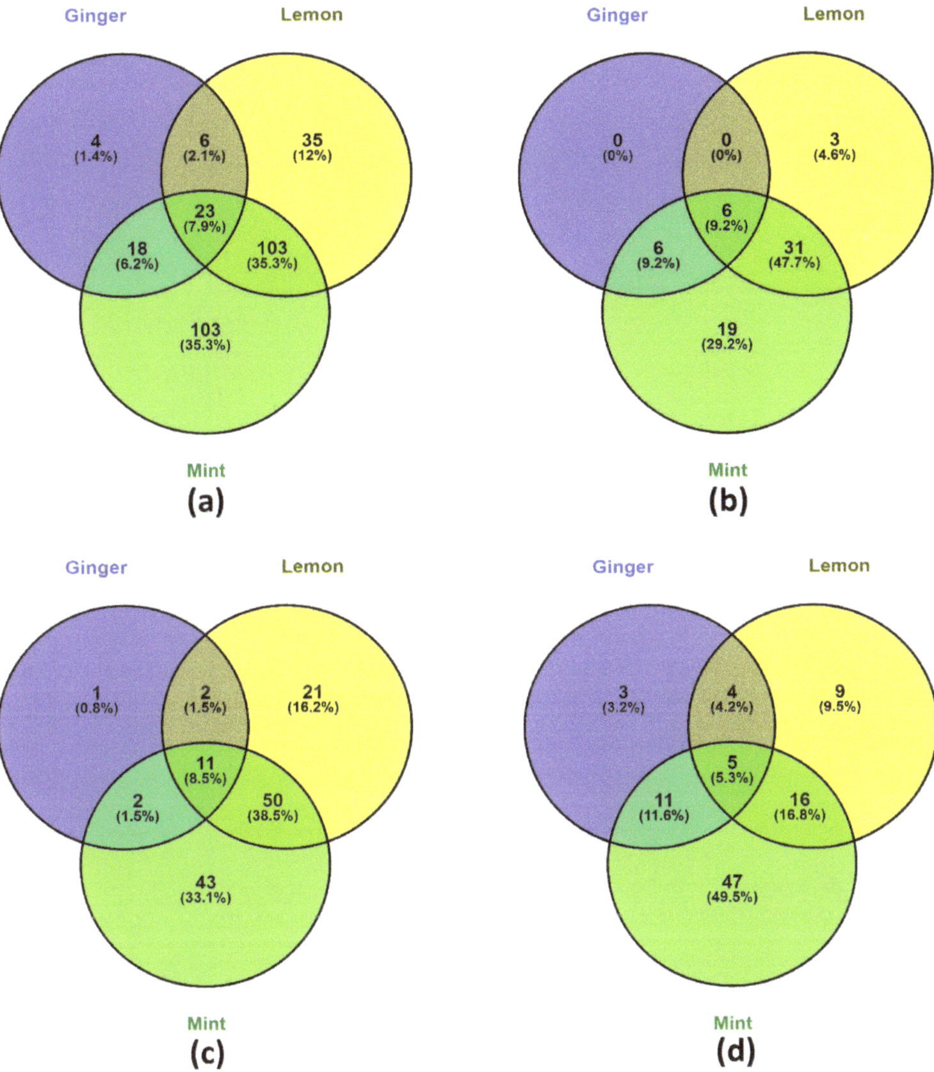

Figure 1. Venn Diagram of the distribution of screened phenolic compounds. (**a**) Total phenolic compounds found in ginger, lemon and mint; (**b**) Phenolic acids in ginger, lemon and mint; (**c**) Flavonoids in ginger, lemon and mint; (**d**) Other polyphenols (including lignans and stilbenes) in ginger, lemon and mint.

Table 3. LC-ESI-QTOF-MS/MS characterisation of phenolic compounds in ginger, lemon and mint samples.

No.	Proposed Compounds	Molecular Formula	RT (min)	Ionisation (ESI^+/ESI^-)	Molecular Weight	Theoretical (*m/z*)	Observed (*m/z*)	Error (ppm)	MS^2 Productions	Samples
					Phenolic acids					
	Hydroxybenzoic Acids									
1	Gallic acid 4-*O*-glucoside	$C_{13}H_{16}O_{10}$	6.580	$[M-H]^-$	332.0743	331.067	331.0684	4.2	169, 125	M
2	4-Hydroxybenzoic acid 4-*O*-glucoside	$C_{13}H_{16}O_8$	11.898	$[M-H]^-$	300.0845	299.0772	299.077	−0.7	255, 137	M
3	2,3-Dihydroxybenzoic acid	$C_7H_6O_4$	12.378	$[M-H]^-$	154.0266	153.0193	153.0194	0.7	109	M
4	Protocatechuic acid 4-*O*-glucoside	$C_{13}H_{16}O_9$	13.256	$[M-H]^-$	316.0794	315.0721	315.0723	0.6	153	M
5	2-Hydroxybenzoic acid	$C_7H_6O_3$	19.932	** $[M-H]^-$	138.0317	137.0244	137.0245	0.7	93	M
6	Paeoniflorin	$C_{23}H_{28}O_{11}$	58.033	$[M-H]^-$	480.1632	479.1559	479.1577	3.8	449, 357, 327	L
	Hydroxycinnamic Acids									
7	*p*-Coumaroyl tartaric acid	$C_{13}H_{12}O_8$	8.232	** $[M-H]^-$	296.0532	295.0459	295.0466	2.4	115	*G, L
8	Cinnamic acid	$C_9H_8O_2$	9.166	$[M-H]^-$	148.0524	147.0451	147.0448	−2.0	103	L
9	Caffeoyl tartaric acid	$C_{13}H_{12}O_9$	13.438	$[M-H]^-$	312.0481	311.0408	311.0413	1.6	161	M
10	Ferulic acid	$C_{10}H_{10}O_4$	15.708	** $[M-H]^-$	194.0579	193.0506	193.0515	4.7	178, 149,134	L, *M
11	Caffeic acid 3-*O*-glucuronide	$C_{15}H_{16}O_{10}$	16.354	** $[M-H]^-$	356.0743	355.067	355.0685	4.2	179	M
12	3-*p*-Coumaroylquinic acid	$C_{16}H_{18}O_8$	17.695	** $[M-H]^-$	338.1002	337.0929	337.0943	4.2	265, 173, 162	M
13	*p*-Coumaric acid 4-*O*-glucoside	$C_{15}H_{18}O_8$	19.137	$[M-H]^-$	326.1002	325.0929	325.093	0.3	163	M
14	*m*-Coumaric acid	$C_9H_8O_3$	19.153	$[M-H]^-$	164.0473	163.04	163.0395	−3.1	119	M
15	Caffeic acid 4-sulfate	$C_9H_8O_7S$	19.248	$[M+H]^+$	259.9991	261.0064	261.0057	−2.7	179, 135	M
16	3-Caffeoylquinic acid	$C_{16}H_{18}O_9$	19.766	** $[M-H]^-$	354.0951	353.0878	353.0878	0.0	253, 190, 144	L, *M
17	Feruloyl tartaric acid	$C_{14}H_{14}O_9$	22.185	$[M-H]^-$	326.0638	325.0565	325.0573	2.5	193, 149	M
18	Ferulic acid 4-*O*-glucoside	$C_{16}H_{20}O_9$	25.779	** $[M-H]^-$	356.1107	355.1034	355.1019	−4.2	193, 178, 149, 134	M
19	Chicoric acid	$C_{22}H_{18}O_{12}$	35.138	$[M-H]^-$	474.0798	473.0725	473.0753	5.0	293, 311	M
20	1-Sinapoyl-2-feruloylgentiobiose	$C_{33}H_{40}O_{18}$	35.768	$[M-H]^-$	724.2215	723.2142	723.2184	4.1	529, 499	M
21	1,5-Dicaffeoylquinic acid	$C_{25}H_{24}O_{12}$	48.341	** $[M-H]^-$	516.1268	515.1195	515.122	4.9	353, 335, 191, 179	M
	Hydroxyphenylacetic Acids									
22	3,4-Dihydroxyphenylacetic acid	$C_8H_8O_4$	10.059	** $[M-H]^-$	168.0423	167.035	167.0354	2.4	149, 123	G, *M
23	2-Hydroxy-2-phenylacetic acid	$C_8H_8O_3$	15.310	** $[M-H]^-$	152.0473	151.04	151.0408	3.8	136, 92	M
	Hydroxyphenylpropanoic Acids									
24	Dihydroferulic acid 4-*O*-glucuronide	$C_{16}H_{20}O_{10}$	6.978	** $[M-H]^-$	372.1056	371.0983	371.0999	4.3	195	M

Table 3. *Cont.*

No.	Proposed Compounds	Molecular Formula	RT (min)	Ionisation (ESI$^+$/ESI$^-$)	Molecular Weight	Theoretical (*m*/*z*)	Observed (*m*/*z*)	Error (ppm)	MS^2 Productions	Samples
					Flavonoids					
	Flavanols									
25	4′-*O*-Methylepigallocatechin	$C_{16}H_{16}O_7$	33.560	[M + H]$^+$	320.0896	321.0969	321.0958	−3.4	302	L, *M
26	4′-*O*-Methyl-(-)-epigallocatechin 7-*O*-glucuronide	$C_{22}H_{24}O_{13}$	48.622	[M − H]$^-$	496.1217	495.1144	495.115	1.2	451, 313	M
27	(+)-Catechin 3-*O*-gallate	$C_{22}H_{18}O_{10}$	50.445	** [M − H]$^-$	442.09	441.0827	441.0833	1.4	289, 169, 125	M
28	(+)-Gallocatechin 3-*O*-gallate	$C_{22}H_{18}O_{11}$	63.448	** [M − H]$^-$	458.0849	457.0776	457.0794	3.9	305, 169	M
	Flavones									
29	Apigenin 6,8-di-C-glucoside	$C_{27}H_{30}O_{15}$	26.791	** [M − H]$^-$	594.1585	593.1512	593.1556	4.7	503, 473	G, *L, M-T
30	Apigenin 7-*O*-apiosyl-glucoside	$C_{26}H_{28}O_{14}$	31.067	[M + H]$^+$	564.1479	565.1552	565.1545	−1.2	296	*G, L, M-T
31	7,4′-Dihydroxyflavone	$C_{15}H_{10}O_4$	37.337	[M + H]$^+$	254.0579	255.0652	255.0659	2.7	227, 199, 171	M
32	6-Hydroxyluteolin 7-*O*-rhamnoside	$C_{21}H_{20}O_{11}$	39.131	** [M − H]$^-$	448.1006	447.0933	447.095	3.8	301	L, *M
33	Rhoifolin	$C_{27}H_{30}O_{14}$	43.471	** [M − H]$^-$	578.1636	577.1563	577.1582	3.3	413, 269	L, *M
34	Cirsilineol	$C_{18}H_{16}O_7$	45.338	[M + H]$^+$	344.0896	345.0969	345.0966	−0.9	330, 312, 297, 284	L, *M
35	Apigenin 7-*O*-glucuronide	$C_{21}H_{18}O_{11}$	47.673	[M + H]$^+$	446.0849	447.0922	447.0901	−4.7	271, 253	M
36	Chrysoeriol 7-*O*-glucoside	$C_{22}H_{22}O_{11}$	54.565	[M + H]$^+$	462.1162	463.1235	463.1234	−0.2	445, 427, 409, 381	L, *M
37	Diosmin	$C_{28}H_{32}O_{15}$	59.17	** [M + H]$^+$	608.1741	609.1814	609.1819	0.8	301, 286	L, *M
	Flavanones									
38	Naringin 4′-*O*-glucoside	$C_{33}H_{42}O_{19}$	25.233	[M − H]-	742.232	741.2247	741.2279	4.3	433, 271	M
39	Neoeriocitrin	$C_{27}H_{32}O_{15}$	36.946	** [M − H]$^-$	596.1741	595.1668	595.1658	−1.7	431, 287	*L, M
40	Hesperidin	$C_{28}H_{34}O_{15}$	42.745	** [M + H]$^+$	610.1898	611.1971	611.1956	−2.5	593, 465, 449, 303	*L, M
	Flavonols									
41	Quercetin 3′-*O*-glucuronide	$C_{21}H_{18}O_{13}$	12.512	** [M − H]$^-$	478.0747	477.0674	477.067	−0.8	301	*L, M
42	Quercetin 3-*O*-glucosyl-xyloside	$C_{26}H_{28}O_{16}$	15.395	[M − H]$^-$	596.1377	595.1304	595.1299	−0.8	265, 138, 116	L
43	Kaempferol 3,7-*O*-diglucoside	$C_{27}H_{30}O_{16}$	23.162	** [M − H]$^-$	610.1534	609.1461	609.1486	4.1	447, 285	L, *M
44	Kaempferol 3-*O*-(2″-rhamnosyl-galactoside) 7-*O*-rhamnoside	$C_{33}H_{40}O_{19}$	34.660	** [M − H]$^-$	740.2164	739.2091	739.2114	3.1	593, 447, 285	G, *L, M-T
45	Kaempferol 3-*O*-glucosyl-rhamnosyl-galactoside	$C_{33}H_{40}O_{20}$	37.254	** [M − H]$^-$	756.2113	755.204	755.2037	−0.4	285	*G, L, M-T
46	Myricetin 3-*O*-rhamnoside	$C_{21}H_{20}O_{12}$	39.479	** [M − H]$^-$	464.0955	463.0882	463.0874	−1.7	317	L, *M
47	3-Methoxysinensetin	$C_{21}H_{22}O_8$	61.671	[M + H]$^+$	402.1315	403.1388	403.1388	0.0	388, 373, 355, 327	M
48	Myricetin 3-*O*-rutinoside	$C_{27}H_{30}O_{17}$	81.239	** [M − H]$^-$	626.1483	625.141	625.1404	−1.0	301	L, *M

Table 3. *Cont.*

No.	Proposed Compounds	Molecular Formula	RT (min)	Ionisation (ESI^+/ESI^-)	Molecular Weight	Theoretical (*m/z*)	Observed (*m/z*)	Error (ppm)	MS^2 Productions	Samples
	Dihydroflavonols									
49	Dihydromyricetin 3-*O*-rhamnoside	$C_{21}H_{22}O_{12}$	39.926	$[M-H]^-$	466.1111	465.1038	465.1051	2.8	301	M
	Anthocyanins									
50	Isopeonidin 3-*O*-arabinoside	$C_{21}H_{21}O_{10}$	29.965	$[M+H]^+$	433.1135	434.1208	434.1213	1.2	271, 253, 243	M
	Isoflavonoids									
51	3′-*O*-Methylviolanone	$C_{18}H_{18}O_6$	12.494	** $[M-H]^-$	330.1103	329.103	329.1041	3.3	314, 299, 284, 256	G, *M
52	Sativanone	$C_{17}H_{16}O_5$	14.051	$[M-H]^-$	300.0998	299.0925	299.0919	−2.0	284, 269, 225	M
53	2′-Hydroxyformononetin	$C_{16}H_{12}O_5$	28.896	$[M+H]^+$	284.0685	285.0758	285.076	0.7	270, 229	*L, M
54	5,6,7,3′,4′-Pentahydroxyisoflavone	$C_{15}H_{10}O_7$	31.563	** $[M+H]^+$	302.0427	303.05	303.0501	0.3	285, 257	*L, M
55	3′-Hydroxygenistein	$C_{15}H_{10}O_6$	39.466	** $[M+H]^+$	286.0477	287.055	287.0543	−2.4	269, 259	*G, L, M-T
56	2′,7-Dihydroxy-4′,5′-dimethoxyisoflavone	$C_{17}H_{14}O_6$	41.246	$[M+H]^+$	314.079	315.0863	315.085	−4.1	300, 282	M
57	6″-*O*-Acetylglycitin	$C_{24}H_{24}O_{11}$	45.345	** $[M+H]^+$	488.1319	489.1392	489.1378	−2.9	285, 270	*L, M
58	3′-Hydroxydaidzein	$C_{15}H_{10}O_5$	46.895	$[M+H]^+$	270.0528	271.0601	271.0603	0.7	253, 241, 225	L, *M
59	Glycitin	$C_{22}H_{22}O_{10}$	70.633	$[M+H]^+$	446.1213	447.1286	447.1276	−2.2	285	M
					Other Phenolic Compounds					
	Hydroxycoumarins									
60	Coumarin	$C_9H_6O_2$	60.230	$[M+H]^+$	146.0368	147.0441	147.0436	−3.4	103, 91	M
	Hydroxybenzaldehydes									
61	4-Hydroxybenzaldehyde	$C_7H_6O_2$	19.932	$[M-H]^-$	122.0368	121.0295	121.0299	3.3	77	M
	Hydroxybenzoketones									
62	2,3-Dihydroxy-1-guaiacylpropanone	$C_{10}H_{12}O_5$	13.157	** $[M-H]^-$	212.0685	211.0612	211.062	3.8	167, 123, 105, 93	M
	Phenolic Terpenes									
63	Carnosic acid	$C_{20}H_{28}O_4$	80.86	$[M-H]^-$	332.1988	331.1915	331.1922	2.1	287, 269	L
	Tyrosols									
64	Hydroxytyrosol 4-*O*-glucoside	$C_{14}H_{20}O_8$	10.49	$[M-H]^-$	316.1158	315.1085	315.109	1.6	153, 123	M
65	Oleoside 11-methylester	$C_{17}H_{24}O_{11}$	14.217	$[M-H]^-$	404.1319	403.1246	403.1262	4.0	223, 165	M
66	3,4-DHPEA-AC	$C_{10}H_{12}O_4$	33.080	** $[M-H]^-$	196.0736	195.0663	195.0671	4.1	135	*G, L, M-T
	Other Phenolic Compounds									
67	Lithospermic acid	$C_{27}H_{22}O_{12}$	49.119	** $[M-H]^-$	538.1111	537.1038	537.1054	3.0	493, 339, 295	M
68	Salvianolic acid B	$C_{36}H_{30}O_{16}$	76.568	** $[M-H]^-$	718.1534	717.1461	717.1491	4.2	519, 339, 321, 295	M

Table 3. *Cont.*

No.	Proposed Compounds	Molecular Formula	RT (min)	Ionisation (ESI$^+$/ESI$^-$)	Molecular Weight	Theoretical (m/z)	Observed (m/z)	Error (ppm)	MS^2 Productions	Samples
					Lignans					
69	7-Oxomatairesinol	$C_{20}H_{20}O_7$	30.089	$[M + H]^+$	372.1209	373.1282	373.1297	4.0	358, 343, 328, 325	L
70	Conidendrin	$C_{20}H_{20}O_6$	45.653	$[M + H]^+$	356.126	357.1333	357.1325	−2.2	339, 221, 206	M
71	Pinoresinol	$C_{20}H_{22}O_6$	50.544	$[M - H]^-$	358.1416	357.1343	357.1364	1.3	342, 327, 313, 221	M
72	Schisandrin C	$C_{22}H_{24}O_6$	80.132	** $[M + H]^+$	384.1573	385.1646	385.163	−4.2	370, 315, 300	G, *L
					Stilbenes					
73	4-Hydroxy-3,5,4′-trimethoxystilbene	$C_{17}H_{18}O_4$	76.456	$[M + H]^+$	286.1205	287.1278	287.1266	−4.2	271, 241, 225	G

RT is short for "retention time". * Signals the sample in which the displayed data was obtained from. ** Indicates that the compound was detected in both negative $[M - H]^-$ and positive $[M + H]^+$ mode of ionisation, with only one ionisation mode presented in the table. Samples G, L and M are abbreviations for ginger, lemon and mint respectively while—T represents that the compound was also characterised in the herbal tea infusion.

Phenolic Acids

Phenolic acids are a subclass of phenolic compounds with a carboxyl group. Phenolic acids mainly comprise hydroxybenzoic acids and hydroxycinnamic acids and they have been extensively studied for their antioxidant, antimicrobial and anti-inflammatory effects [76]. Phenolic acid species can be found in ginger, lemon and mint samples [77–79]. In this study, five subclasses of phenolic acid were identified, which include hydroxybenzoic acids (6), hydroxycinnamic acids (15), hydroxyphenylacetic acids (2) and hydroxyphenylpropanoic acids (1).

Compound **8** was identified to be cinnamic acid based on the product ion formed at m/z 103, which represents the loss of carbon dioxide (44 Da) from the precursor ion. Cinnamic acids are found in a variety of dietary plant materials such as *Citrus* fruits, tea, *Brassica* vegetables, cereals and more [80]. Cinnamic acids and their derivatives possess antioxidant properties, especially cinnamic acid derivatives with cinnamoyl and hydroxyl moieties in which these moieties reportedly increase cinnamic acid derivatives' antioxidant activities [81]. Compound **10** found in lemon and mint samples, had a precursor ion at $[M - H]^-$ m/z 193.0515 and was assigned to be ferulic acid based on the MS/MS fragmentation product ions produced. The product ions of ferulic acid were at m/z 178, m/z 149 and m/z 134, which coincide with the loss of methyl group (15 Da), carbon dioxide (44 Da) and both methyl group and carbon dioxide (59 Da total) from the precursor ion respectively. A previous study on phenolic compounds found in *Phoenix dactylifera*'s male flowers also reported product ions at m/z 178 and m/z 134 for ferulic acid [82]. Previously, as ferulic acid has been reported in mint, it is likely that ferulic acid contributed to the antioxidant activity observed in this study's mint sample.

Two hydroxyphenylacetic acids were identified. Compound **22** ($[M - H]^-$ m/z at 167.0354) was found in both ginger and mint samples and was identified as 3,4-dihydroxyphenylacetic acid (DOPAC). The result was two daughter ions, which were m/z 149 and m/z 123, corresponding to the loss of hydroxide group and carbon dioxide, respectively. To the best of our knowledge, this is the first time that DOPAC has been found in both ginger and mint. A previous study on the bioactive compounds of buckwheat found that DOPAC exhibits radical scavenging antioxidant activity [83]. DOPAC at micromolar concentrations was shown to display antioxidative activity against lipid peroxidation in vitro rat plasma model [84]. Although some other polyphenols such as quercetin may be more potent antioxidants, DOPAC has been suggested as a suitable additive or supplement alternative due to its relatively lowered cytotoxicity as it is a naturally occurring metabolite of dopamine within the human body [85].

Flavonoids

Flavonoids are a diverse group of phenolic compounds that are present in many dietary plant foods. Flavonoids have attracted interest due to their antioxidant, anti-inflammatory effects and their ability to modulate certain enzymatic functions [86]. Flavonoids are divided into further subgroups based on the composition and structure of their B and C rings [86]. In this experiment, compounds from seven flavonoid subclasses were identified, which include flavanols (4), flavones (9), flavanones (3), flavonols (8), dihydroflavonols (1), anthocyanins (1) and isoflavonoids (9).

Compound **39** was found in both lemon and mint samples. And had a precursor ion at $[M - H]^-$ m/z 595.1658 and was identified to be neoeriocitrin as the precursor ion generated product ions at m/z 431 (loss of rhamnose, 164 Da) and m/z 287 (loss of rhamnose and glucose moieties, total 308 Da) in MS/MS fragmentation. A similar fragmentation pattern of neoeriocitrin was also reported in a study on Exocarpium Citri grandis flavonoid metabolites in human urine after oral administration [87]. Neoeriocitrin was previously shown to have high antioxidant activity through the superoxide radical scavenging assay and low-density lipoprotein oxidation assay [88]. Compound **40** was present in both ionisation modes and had a precursor ion at $[M + H]^+$ m/z 611.1956. Compound **40** was identified to be hesperidin based on the MS/MS fragment ion peaks at m/z 593, m/z 465, m/z 449 and m/z 303. Hesperidin has been reported in both peppermint and lemon previously [89,90]. Studies have demonstrated hesperidin as a potent radical scavenger, with beneficial in vitro effects such as antimicrobial and anticancer effects and has been suggested for the management of cutaneous functions [89]. Hesperidin is thought to be one of the main contributors to lemon peel's antioxidant properties [90]. Hesperidin was highly likely to have contributed to the observed antioxidant activity of this study's mint sample. The presence of neoeriocitrin and hesperidin is consistent with the literature, as these are two of the main flavanones found in *Citrus* fruits such as lemon [60].

Compound **41** was found in lemon and mint samples in both negative and positive modes. The precursor ion at $[M-H]^-$ m/z 477.067 generated a product ion at m/z 301 through MS/MS fragmentation, which was the loss of glucuronide (176 Da) from the precursor ion. This confirmed the identity of Compound **41** as quercetin 3′-*O*-glucuronide and, to the best of our knowledge, this is the first time quercetin 3′-*O*-glucuronide was characterised in lemon and mint. Quercetin 3′-*O*-glucuronide has been demonstrated to exhibit antioxidant activities [91,92]. Additionally, it has been suggested that quercetin 3′-*O*-glucuronide can protect cell membranes from lipid peroxidation through its catechol structure [93]. As one of the metabolites of dietary quercetin with health benefits, this further reaffirms the value of adopting a polyphenol-rich diet.

Other Phenolic Compounds

Compounds identified as other phenolic compound were further divided into hydroxycoumarins (1), hydroxybenzaldehydes (1), hydroxybenzoketones (1), phenolic terpenes (1), tyrosols (3) and other phenolic compounds (2). A total of 9 phenolic compounds classified as other phenolic compounds were found in the samples.

Compound **60** was the only hydroxycoumarin identified and was found in mint samples in positive mode, with a precursor ion at $[M+H]^+$ m/z 147.0436. The product ion generated from the MS/MS fragmentation peaked at m/z 103 $[M+H-CO_2]$ and m/z 91 $[M+H-2CO]$, which identified Compound **60** as coumarin. A previous study also observed the same fragmentation pattern at m/z 103 and m/z 91 [94]. Previously coumarin and its derivatives have been reported to be associated with beneficial health effects such as reducing inflammation and risk of cancer, which is thought to be due to their antioxidant properties [95]. Compound **61** is a hydroxybenzaldehyde, which was identified in mint samples in negative mode. Compound **61** was identified as 4-hydroxybenzaldehyde because the precursor ion at $[M-H]^-$ m/z 121.0299 produced a product ion at m/z 77, representing the loss of carbon dioxide (44 Da) from the parent ion. A previous vanilla extract study showed that 4-hydroxybenzaldehyde exhibited little DPPH scavenging activity [96]. Referring to the literature, it is postulated that 4-hydroxybenzaldehyde contributed little to mint's observed antioxidant activity.

Three tyrosols were identified, of which one was present in all samples and the other two were exclusively present only in mint samples. Present in the negative mode, Compound **65** had a precursor ion at $[M-H]^-$ m/z 403.1262 and was identified as oleoside 11-methylester based on the product ions produced at m/z 223 and m/z 165. A previous study on olive polyphenols also characterised oleoside 11-methylester, with product ion at m/z 223 [97]. These product ions correspond to the loss of glycoside (180 Da) and the loss of glycoside moiety and methyl ester (238 Da) from the precursor ion, respectively. To the best of our knowledge, this is the first time that oleoside 11-methylester has been characterised in mint.

Lignans and Stilbenes

Only one stilbene was identified which was only present in ginger samples. Four lignans were identified across ginger, lemon and mint samples. Compounds **71** and **72** were identified as pinoresinol and schisandrin C respectively from the product ions they produced. Compound **71** was identified in negative mode in mint samples, while Compound **72** was identified in both positive and negative modes and was present in ginger and lemon samples. Compound **71** precursor ion $[M-H]^-$ at m/z 357.1364 produced 4 product ions at m/z 342, m/z 327, m/z 313 and m/z 221. A study on fringe tree found that pinoresinol possesses considerable antioxidant activity [98]. This is the first time that pinoresinol has been characterised in mint. Compound **72** precursor ion in positive mode $[M + H]^+$ at m/z 385.1646 was identified to be schisandrin C based on the product ions produced at m/z 370, m/z 315 and m/z 300. A study on *Schisandra chinensis* fruit lignans using tandem mass spectrometry also reported schisandrin C's product ions at m/z 370, m/z 315 and m/z 300, as well as other fragments [99]. Previously, it was reported that schisandrin C has potential anti-inflammatory effects, was capable of inducing autophagy and enhanced C2C12 skeletal muscle cells' ability to deal with oxidative stress [100]. To the best of our knowledge, this is the first time schisandrin C has been characterised in both ginger rhizome and lemon fruit.

3.5. HPLC Quantification Analysis

From the characterised and identified phenolic compounds, 10 compounds were selected to be quantified using HPLC-PDA. The results in Table 4 were generated through calculation from the calibration curve. The lemon samples had a relatively lower amount of the selected phenolic compounds compared to other samples. Ginger samples recorded the greatest amount of certain molecules, such as quercetin, kaempferol and *p*-hydroxybenzoic acid. This is consistent with the literature as kaempferol and certain flavonoids comprise a significant portion of the phenolic compounds found within ginger [101]. This is unlike the TPC assay result, clearly demonstrating HPLC's ability to generate higher-quality data. GLMT consistently scored relatively well in terms of quantity of the selected phenolic compounds, and had relatively higher levels of chlorogenic acid, caffeic acid and catechin compared to other samples. Therefore, there is still an advantage to ingesting herbal tea infusions because the combination of different plant parts results in the incorporation of more of a variety of phenolic compounds and other phytochemicals. This reinforces the importance of herbal tea infusions and combinations of diverse sources of phenolic compounds with antioxidant properties for consumption. The relatively high quantity of phenolic content observed in GLMT may be viewed as a promising attribute for GLMT to be considered as a healthy herbal tea infusion, which could be exploited for commercialisation.

Table 4. Quantification of phenolic compounds in ginger, lemon, mint and GLMT.

No.	Compound Name	Ginger (mg/g)	Lemon (mg/g)	Mint (mg/g)	GLMT (mg/g)
1	Gallic acid	3.21 ± 0.15^{d}	4.42 ± 0.25^{c}	7.21 ± 0.12^{a}	6.85 ± 0.08^{b}
2	Protocatechuic acid	2.36 ± 0.14^{b}	-	4.27 ± 0.13^{a}	2.16 ± 0.11^{c}
3	*p*-Hydroxybenzoic acid	7.87 ± 0.23^{a}	3.17 ± 0.023^{c}	6.37 ± 0.31^{b}	7.84 ± 0.36^{a}
4	Chlorogenic acid	15.78 ± 1.12^{d}	21.45 ± 1.72^{b}	18.79 ± 1.05^{c}	31.47 ± 1.86^{a}
5	Caffeic acid	4.39 ± 0.18^{b}	2.16 ± 0.02^{d}	3.47 ± 0.05^{c}	8.73 ± 0.40^{a}
6	Catechin	11.95 ± 0.48^{c}	4.56 ± 0.09^{d}	17.87 ± 0.91^{b}	21.56 ± 1.42^{a}
7	Epicatechin	2.34 ± 0.03^{c}	-	5.43 ± 0.33^{a}	3.71 ± 0.02^{b}
8	Epicatechin gallate	-	1.25 ± 0.05^{b}	3.42 ± 0.14^{a}	3.32 ± 0.10^{a}
9	Quercetin	32.56 ± 1.00^{a}	6.78 ± 0.26^{d}	8.45 ± 0.40^{c}	17.76 ± 0.66^{b}
10	Kaempferol	14.37 ± 0.66^{a}	7.98 ± 0.34^{d}	11.43 ± 0.29^{b}	9.74 ± 0.32^{c}

Data expressed as mean ± SD of three replicates; Different letters ^{a,b,c,d} indicates that the data is significantly different from the other data of the same row ($p \leq 0.05$), in which ^{a} is assigned to the largest value, then ^{b} assigned to second largest and so forth. The significant difference was calculated through one-way analysis of variance (ANOVA) and Tukey's HSD Test.

A previous study on Malaysian ginger varieties found that 16-week old Halia Bentong rhizomes were found to contain 0.803 $mg/g_{d \cdot w}$ of quercetin, 0.360 $mg/g_{d \cdot w}$ of catechin and 0.045 $mg/g_{d \cdot w}$ of kaempferol [102]. Compared to that study, our study's samples produced significantly higher yields of those flavonoids from HPLC quantification. The differences in flavonoids observed may be attributed to many factors such as different variety, age, growth condition, harvesting method and more. For lemon, the content of certain phenolic acids and flavonoids demonstrated in this study was higher in comparison with previous studies on lemon phenolic quantity. A previous study on flavonoids in common plants showed that lemon yielded the equivalent of 11 μg/g of quercetin [103], and another study showed that Eureka lemons contained no quercetin nor kaempferol, but was high in eriocitrin and hesperidin [104]. In a different study on the phenolic content of five varieties of lemon, the juices contained a range of 0.38 μg/g to 7.62 μg/g of gallic acid and a range of 2.70 μg/g to 22.08 μg/g of chlorogenic acid [105]. In the same study, it was revealed that the peel contained the most phenolic acid and flavonoids compared to the pulp and juice of lemon. Thus, methods that increase the infusion of phenolic compounds from the lemon peel into the herbal tea would reap higher phenolic content and, subsequently, higher antioxidant activity. All ten compounds in Table 4 were present and quantified in mint and were all found in relatively high quantities. In a previous study, *Mentha x piperita* (mint) crude extract yielded 1.86 mg/g of caffeic acid, 0.73 mg/g of chlorogenic acid and 0.84 mg/g of quercetin but no kaempferol was observed [106]. Another study was also unable to quantify any kaempferol in mint but was able to quantify small amounts of caffeic acid and catechin (0.027 mg/g and 0.147 mg/g, respectively). However, previous studies revealed that mint and other related species do contain conjugated kaempferol and kaempferol derivatives [107,108].

4. Conclusions

GLMT is a herbal infusion composed of ingredients with marked antioxidant properties and with unique polyphenols. To the best of our knowledge, our examination of GLMT's phenolic and antioxidant properties was unprecedented. The samples do exhibit antioxidant activities and the activities were likely attributed to the variety of phenolic compounds and possibly other phytochemicals found within the samples. Correlation analysis suggested that the antioxidant activities recorded were significantly correlated with the phenolic content from the samples. Furthermore, the LC-ESI-QTOF-MS/MS characterisation aided the identification of the phenolic compounds within the samples, in which a few were identified in the samples for the first time. Most of the phenolic compounds identified through LC-MS/MS were found in mint (49). Ten compounds were selected for HPLC quantification in which some phenolic compounds were abundant in ginger and GLMT samples. As discussed, although ginger did not appear to contain a diverse array of phenolic compounds from the analysis of this study, a wealth of literature has reported other phytochemicals in ginger that are also considered to be beneficial to health. With an understanding of the phenolic content in each sample from the characterisation analysis, further research should consider the optimal ingredient ratio for the maximum antioxidant activity or in vitro studies. Follow-up research could increase the acceptance of functional foods like GLMT, which may encourage commercialisation and promote a health-conscious society.

Supplementary Materials: The following are available online at https://www.mdpi.com/article/10.3390/fermentation7020073/s1, Figure S1: LC-ESI-QTOF-MS/MS basic peak chromatograph (BPC) for characterisation of phenolic compounds of herbal tea. Figure S2: Extracted ion chromatogram and their mass spectrum of characterised compounds in herbal tea.

Author Contributions: Conceptualization, methodology, formal analysis, validation and investigation, O.C., A.A., V.S. and H.A.R.S.; resources, H.A.R.S., C.J.B. and F.R.D.; writing—original draft preparation, O.C.; writing—review and editing, O.C., V.S., C.J.B., H.A.R.S. and F.R.D.; supervision, H.A.R.S.; A.A. and F.R.D. ideas sharing, H.A.R.S.; A.A.; C.J.B. and F.R.D.; funding acquisition, H.A.R.S., F.R.D. and C.J.B. All authors have read and agreed to the published version of the manuscript.

Funding: This research was funded by the University of Melbourne under the "McKenzie Fellowship Scheme" (Grant No. UoM-18/21), the "Richard WS Nicholas Agricultural Science Scholarship" and the "Faculty Research Initiative Funds" funded by the Faculty of Veterinary and Agricultural Sciences, The University of Melbourne, Australia and "The Alfred Deakin Research Fellowship" funded by Deakin University, Australia.

Institutional Review Board Statement: Not applicable.

Informed Consent Statement: Not applicable.

Data Availability Statement: Not applicable.

Acknowledgments: We would like to thank Michael Leeming, Nicholas Williamson and Shuai Nie from the Mass Spectrometry and Proteomics Facility, Bio21 Molecular Science and Biotechnology Institute, the University of Melbourne, VIC, Australia for providing access and support for the use of HPLC-PDA and LC-ESI-QTOF-MS/MS and data analysis.

Conflicts of Interest: The authors declare no conflict of interest.

References

1. Buyukbalci, A.; El, S.N. Determination of in vitro antidiabetic effects, antioxidant activities and phenol contents of some herbal teas. *Plant. Foods Hum. Nutr.* **2008**, *63*, 27–33. [CrossRef] [PubMed]
2. Gupta, R.K.; Chawla, P.; Tripathi, M.; Shukla, A.K.; Pandey, A.J.I.J.O.P.; Sciences, P. Synergistic antioxidant activity of tea with ginger, black pepper and tulsi. *Int. J. Pharm. Pharm. Sci.* **2014**, *6*, 477–479.
3. Malongane, F.; McGaw, L.J.; Mudau, F.N. The synergistic potential of various teas, herbs and therapeutic drugs in health improvement: A review. *J. Sci. Food Agric.* **2017**, *97*, 4679–4689. [CrossRef] [PubMed]
4. Bekkouch, O.; Harnafi, M.; Touiss, I.; Khatib, S.; Harnafi, H.; Alem, C.; Amrani, S. In vitro antioxidant and in vivo lipid-lowering properties of zingiber officinale crude aqueous extract and methanolic fraction: A follow-up study. *Evid. Based Complement Altern. Med.* **2019**, *2019*, 9734390. [CrossRef]
5. Nair, K.P. *Turmeric (Curcuma Longa l.) and Ginger (Zingiber Officinale Rosc.)-World's Invaluable Medicinal Spices: The Agronomy and Economy of Turmeric and Ginger*; Springer Nature: Berlin/Heidelberg, Germany, 2019.
6. Yeh, H.-Y.; Chuang, C.-H.; Chen, H.-C.; Wan, C.-J.; Chen, T.-L.; Lin, L.-Y. Bioactive components analysis of two various gingers (zingiber officinale roscoe) and antioxidant effect of ginger extracts. *Lwt Food Sci. Technol.* **2014**, *55*, 329–334. [CrossRef]
7. Mao, Q.Q.; Xu, X.Y.; Cao, S.Y.; Gan, R.Y.; Corke, H.; Beta, T.; Li, H.B. Bioactive compounds and bioactivities of ginger (zingiber officinale roscoe). *Foods* **2019**, *8*, 185. [CrossRef]
8. Jitoe, A.; Masuda, T.; Tengah, I.G.P.; Suprapta, D.N.; Gara, I.W.; Nakatani, N. Antioxidant activity of tropical ginger extracts and analysis of the contained curcuminoids. *J. Agric. Food Chem.* **1992**, *40*, 1337–1340. [CrossRef]
9. Curk, F.; Ollitrault, F.; Garcia-Lor, A.; Luro, F.; Navarro, L.; Ollitrault, P. Phylogenetic origin of limes and lemons revealed by cytoplasmic and nuclear markers. *Ann. Bot.* **2016**, *117*, 565–583. [CrossRef]
10. Shimizu, C.; Wakita, Y.; Inoue, T.; Hiramitsu, M.; Okada, M.; Mitani, Y.; Segawa, S.; Tsuchiya, Y.; Nabeshima, T. Effects of lifelong intake of lemon polyphenols on aging and intestinal microbiome in the senescence-accelerated mouse prone 1 (samp1). *Sci. Rep.* **2019**, *9*, 3671. [CrossRef] [PubMed]
11. Miyake, Y.; Shimoi, K.; Kumazawa, S.; Yamamoto, K.; Kinae, N.; Osawa, T.J. Identification and antioxidant activity of flavonoid metabolites in plasma and urine of eriocitrin-treated rats. *J. Agric. Food Chem.* **2000**, *48*, 3217–3224. [CrossRef] [PubMed]
12. Makni, M.; Jemai, R.; Kriaa, W.; Chtourou, Y.; Fetoui, H. Citrus limon from tunisia: Phytochemical and physicochemical properties and biological activities. *Biomed. Res. Int.* **2018**, *2018*, 6251546. [CrossRef] [PubMed]
13. Moosavy, M.; Hassanzadeh, P.; Mohammadzadeh, E.; Mahmoudi, R.; Khatibi, S.; Mardani, K.J.J.o.f.q. Antioxidant and antimicrobial activities of essential oil of lemon (citrus limon) peel in vitro and in a food model. *J. Food Qual. Hazards Control* **2017**, *4*, 42–48.
14. Bunsawat, J.; Elliott, N.E.; Hertweck, K.L.; Sproles, E.; Alice, L.A.J.S.B. Phylogenetics of mentha (lamiaceae): Evidence from chloroplast DNA sequences. *Syst. Bot.* **2004**, *29*, 959–964. [CrossRef]
15. Riachi, L.G.; De Maria, C.A. Peppermint antioxidants revisited. *Food Chem.* **2015**, *176*, 72–81. [CrossRef]
16. Singh, R.; Shushni, M.A.M.; Belkheir, A. Antibacterial and antioxidant activities of mentha piperita L. *Arab. J. Chem.* **2015**, *8*, 322–328. [CrossRef]
17. Wu, Z.; Tan, B.; Liu, Y.; Dunn, J.; Martorell Guerola, P.; Tortajada, M.; Cao, Z.; Ji, P. Chemical composition and antioxidant properties of essential oils from peppermint, native spearmint and scotch spearmint. *Molecules* **2019**, *24*, 2825. [CrossRef]
18. Peng, D.; Zahid, H.F.; Ajlouni, S.; Dunshea, F.R.; Suleria, H.A.R. LC-ESI-QTOF/MS Profiling of Australian Mango Peel By-Product Polyphenols and Their Potential Antioxidant Activities. *Processes* **2019**, *7*, 764. [CrossRef]
19. Imran, M.; Butt, M.S.; Akhtar, S.; Riaz, M.; Iqbal, M.J.; Suleria, H.A.R. Mangiferin from Mango Peel. *J. Food Process. Preserv.* **2016**, *40*, 760–769. [CrossRef]
20. Halliwell, B.; Murcia, M.A.; Chirico, S.; Aruoma, O.I. Free radicals and antioxidants in food and in vivo: What they do and how they work. *Crit. Rev. Food Sci. Nutr.* **1995**, *35*, 7–20. [CrossRef] [PubMed]
21. Riley, P.A. Free radicals in biology: Oxidative stress and the effects of ionizing radiation. *Int. J. Radiat. Biol.* **1994**, *65*, 27–33. [CrossRef]
22. Briskin, D.P.J.P.p. Medicinal plants and phytomedicines. Linking plant biochemistry and physiology to human health. *Plant Physiol.* **2000**, *124*, 507–514. [CrossRef] [PubMed]

23. Awan, K.A.; Butt, M.S.; Haq, I.U.; Ashfaq, F.; Suleria, H.A.R. Storage Stability of Garlic Fortified Chicken Bites. *J. Food Chem. Nanotechnol.* **2017**, *3*, 80–85. [CrossRef]
24. Feng, Y.; Dunshea, F.R.; Suleria, H.A.R. LC-ESI-QTOF/MS characterization of bioactive compounds from black spices and their potential antioxidant activities. *J. Food Sci. Technol.* **2020**, *57*, 4671–4687. [CrossRef] [PubMed]
25. Chang, C.-C.; Yang, M.-H.; Wen, H.-M.; Chern, J.-C. Estimation of total flavonoid content in propolis by two complementary colorimetric methods. *J. Food Drug Anal.* **2002**, *10*, 178–182.
26. Stavrou, I.J.; Christou, A.; Kapnissi-Christodoulou, C.P. Polyphenols in carobs: A review on their composition, antioxidant capacity and cytotoxic effects, and health impact. *Food Chem.* **2018**, *269*, 355–374. [CrossRef]
27. Shotorbani, N.Y.; Jamei, R.; Heidari, R.J.A.J.o.P. Antioxidant activities of two sweet pepper capsicum annuum l. Varieties phenolic extracts and the effects of thermal treatment. *Avicenna J. Phytomed.* **2013**, *3*, 25.
28. Benzie, I.F.; Strain, J.J.J.A.b. The ferric reducing ability of plasma (frap) as a measure of "antioxidant power": The frap assay. *Anal. Biochem.* **1996**, *239*, 70–76. [CrossRef] [PubMed]
29. Re, R.; Pellegrini, N.; Proteggente, A.; Pannala, A.; Yang, M.; Rice-Evans, C.J.F. Antioxidant activity applying an improved abts radical cation decolorization assay. *Free Radic. Biol. Med.* **1999**, *26*, 1231–1237. [CrossRef]
30. Ferreira, I.C.F.R.; Baptista, P.; Vilas-Boas, M.; Barros, L. Free-radical scavenging capacity and reducing power of wild edible mushrooms from northeast portugal: Individual cap and stipe activity. *Food Chem.* **2007**, *100*, 1511–1516. [CrossRef]
31. Smirnoff, N.; Cumbes, Q.J. Hydroxyl radical scavenging activity of compatible solutes. *Phytochemistry* **1989**, *28*, 1057–1060. [CrossRef]
32. Dinis, T.C.P.; Madeira, V.M.C.; Almeida, L.M. Action of phenolic derivatives (acetaminophen, salicylate, and 5-aminosalicylate) as inhibitors of membrane lipid peroxidation and as peroxyl radical scavengers. *Arch. Biochem. Biophys.* **1994**, *315*, 161–169. [CrossRef]
33. Kadam, D.; Palamthodi, S.; Lele, S.S. Lc-esi-q-tof-ms/ms profiling and antioxidant activity of phenolics from l. Sativum seedcake. *J. Food Sci. Technol.* **2018**, *55*, 1154–1163. [CrossRef]
34. Pekal, A.; Drozdz, P.; Biesaga, M.; Pyrzynska, K. Screening of the antioxidant properties and polyphenol composition of aromatised green tea infusions. *J. Sci. Food Agric.* **2012**, *92*, 2244–2249. [CrossRef]
35. Seo, C.-S.; Lee, M.-Y. Hplc–pda and lc–ms/ms analysis for the simultaneous quantification of the 14 marker components in sojadodamgangki-tang. *Appl. Sci.* **2020**, *10*, 2804. [CrossRef]
36. Yang, D.; Dunshea, F.R.; Suleria, H.A.R. Lc-esi-qtof/ms characterization of australian herb and spices (garlic, ginger, and onion) and potential antioxidant activity. *J. Food Process. Preserv.* **2020**, *44*, e14497. [CrossRef]
37. Tang, J.; Dunshea, F.R.; Suleria, H.A.R. Lc-esi-qtof/ms characterization of phenolic compounds from medicinal plants (hops and juniper berries) and their antioxidant activity. *Foods* **2019**, *9*, 7. [CrossRef]
38. Subbiah, V.; Zhong, B.; Nawaz, M.A.; Barrow, C.J.; Dunshea, F.R.; Suleria, H.A. Screening of phenolic compounds in australian grown berries by lc-esi-qtof-ms/ms and determination of their antioxidant potential. *Antioxidants* **2021**, *10*, 26. [CrossRef] [PubMed]
39. Wang, Z.; Barrow, C.J.; Dunshea, F.R.; Suleria, H.A.R. A comparative investigation on phenolic composition, characterization and antioxidant potentials of five different australian grown pear varieties. *Antioxidants* **2021**, *10*, 151. [CrossRef] [PubMed]
40. Kılıç, C.; Can, Z.; Yılmaz, A.; Yıldız, S.; Turna, H. Antioxidant properties of some herbal teas (green tea, senna, corn silk, rosemary) brewed at different temperatures. *Int. J. Second. Metab.* **2017**, 148–154. [CrossRef]
41. Sogi, D.S.; Siddiq, M.; Greiby, I.; Dolan, K.D. Total phenolics, antioxidant activity, and functional properties of 'tommy atkins' mango peel and kernel as affected by drying methods. *Food Chem.* **2013**, *141*, 2649–2655. [CrossRef]
42. Alam, M.N.; Bristi, N.J.; Rafiquzzaman, M. Review on in vivo and in vitro methods evaluation of antioxidant activity. *Saudi Pharm. J. Spj Off. Publ. Saudi Pharm. Soc.* **2013**, *21*, 143–152. [CrossRef]
43. Prieto, P.; Pineda, M.; Aguilar, M.J.A.b. Spectrophotometric quantitation of antioxidant capacity through the formation of a phosphomolybdenum complex: Specific application to the determination of vitamin E. *Anal. Biochem.* **1999**, *269*, 337–341. [CrossRef] [PubMed]
44. Suleria, H.A.; Barrow, C.J.; Dunshea, F.R.J.F. Screening and characterization of phenolic compounds and their antioxidant capacity in different fruit peels. *Foods* **2020**, *9*, 1206. [CrossRef]
45. Zhong, B.; Robinson, N.A.; Warner, R.D.; Barrow, C.J.; Dunshea, F.R.; Suleria, H.A.R. Lc-esi-qtof-ms/ms characterization of seaweed phenolics and their antioxidant potential. *Mar. Drugs* **2020**, *18*, 331. [CrossRef] [PubMed]
46. Ghafoor, K.; Al Juhaimi, F.; Özcan, M.M.; Uslu, N.; Babiker, E.E.; Mohamed Ahmed, I.A. Total phenolics, total carotenoids, individual phenolics and antioxidant activity of ginger (zingiber officinale) rhizome as affected by drying methods. *Lwt* **2020**, *126*, 109354. [CrossRef]
47. Farnad, N.; Heidari, R.; Aslanipour, B. Phenolic composition and comparison of antioxidant activity of alcoholic extracts of peppermint (mentha piperita). *J. Food Meas. Charact.* **2014**, *8*, 113–121. [CrossRef]
48. Benabdallah, A.; Rahmoune, C.; Boumendjel, M.; Aissi, O.; Messaoud, C. Total phenolic content and antioxidant activity of six wild mentha species (lamiaceae) from northeast of algeria. *Asian Pac. J. Trop. Biomed.* **2016**, *6*, 760–766. [CrossRef]
49. Sarooshi, R.; Broadbent, P.J.A.J.o.E.A. Evaluation of rootstocks for eureka and lisbon lemons in replant ground in new south wales. *Aust. J. Exp. Agric.* **1992**, *32*, 205–209. [CrossRef]

50. Rahmati, M.; Vercambre, G.; Davarynejad, G.; Bannayan, M.; Azizi, M.; Genard, M. Water scarcity conditions affect peach fruit size and polyphenol contents more severely than other fruit quality traits. *J. Sci. Food Agric.* **2015**, *95*, 1055–1065. [CrossRef]
51. Velioglu, Y.; Mazza, G.; Gao, L.; Oomah, B.D. Antioxidant activity and total phenolics in selected fruits, vegetables, and grain products. *J. Agric. Food Chem.* **1998**, *46*, 4113–4117. [CrossRef]
52. Contreras-López, E.; Castañeda-Ovando, A.; Jaimez-Ordaz, J.; Cruz-Cansino, N.d.S.; González-Olivares, L.G.; Rodríguez-Martínez, J.S.; Ramírez-Godínez, J.J.A.S. Release of antioxidant compounds of zingiber officinale by ultrasound-assisted aqueous extraction and evaluation of their in vitro bioaccessibility. *Appl. Sci.* **2020**, *10*, 4987. [CrossRef]
53. Kedare, S.B.; Singh, R.P. Genesis and development of dpph method of antioxidant assay. *J. Food Sci. Technol* **2011**, *48*, 412–422. [CrossRef]
54. Chrpova, D.; Kouřimská, L.; Gordon, M.H.; Heřmanová, V.; Roubíčková, I.; Panek, J.J.C.J.o.F.S. Antioxidant activity of selected phenols and herbs used in diets for medical conditions. *Czech J. Food Sci.* **2010**, *28*, 317–325. [CrossRef]
55. Sun, Y.; Qiao, L.; Shen, Y.; Jiang, P.; Chen, J.; Ye, X. Phytochemical profile and antioxidant activity of physiological drop of citrus fruits. *J. Food Sci.* **2013**, *78*, C37–C42. [CrossRef]
56. Gu, C.; Howell, K.; Dunshea, F.R.; Suleria, H.A.R. Lc-esi-qtof/ms characterisation of phenolic acids and flavonoids in polyphenol-rich fruits and vegetables and their potential antioxidant activities. *Antioxidants* **2019**, *8*, 405. [CrossRef]
57. Mittal, N. Estimation of antioxidant levels in pomegranate, banana, orange, lemon, sweet lime. *Stud. Ethno-Med.* **2019**, *13*. [CrossRef]
58. Chan, E.W.C.; Lim, Y.Y.; Chong, K.L.; Tan, J.B.L.; Wong, S.K. Antioxidant properties of tropical and temperate herbal teas. *J. Food Compos. Anal.* **2010**, *23*, 185–189. [CrossRef]
59. Ilyasov, I.R.; Beloborodov, V.L.; Selivanova, I.A.; Terekhov, R.P. Abts/pp decolorization assay of antioxidant capacity reaction pathways. *Int J. Mol Sci.* **2020**, *21*, 1131. [CrossRef]
60. Assefa, A.D.; Keum, Y.-S.; Saini, R.K. A comprehensive study of polyphenols contents and antioxidant potential of 39 widely used spices and food condiments. *J. Food Meas. Charact.* **2018**, *12*, 1548–1555. [CrossRef]
61. Leong, L.; Shui, G.J.F.c. An investigation of antioxidant capacity of fruits in singapore markets. *Food Chem.* **2002**, *76*, 69–75. [CrossRef]
62. Bhalodia, N.R.; Nariya, P.B.; Acharya, R.N.; Shukla, V.J. In vitro antioxidant activity of hydro alcoholic extract from the fruit pulp of cassia fistula linn. *Ayu* **2013**, *34*, 209–214. [CrossRef]
63. Prakash, J.J.J.o.M.P.R. Chemical composition and antioxidant properties of ginger root (zingiber officinale). *J. Med. Plants Res.* **2010**, *4*, 2674–2679.
64. Kanatt, S.R.; Chander, R.; Sharma, A.J.F.C. Antioxidant potential of mint (*mentha spicata* L.) in radiation-processed lamb meat. *Food Chem.* **2007**, *100*, 451–458. [CrossRef]
65. Stoilova, I.; Krastanov, A.; Stoyanova, A.; Denev, P.; Gargova, S. Antioxidant activity of a ginger extract (zingiber officinale). *Food Chem.* **2007**, *102*, 764–770. [CrossRef]
66. Dorman, H.D.; Koşar, M.; Kahlos, K.; Holm, Y.; Hiltunen, R.J.J. Antioxidant properties and composition of aqueous extracts from mentha species, hybrids, varieties, and cultivars. *J. Agric. Food Chem.* **2003**, *51*, 4563–4569. [CrossRef]
67. Muthiah, P.; Umamaheswari, M.; Asokkumar, K.J.I.J.o.P. In vitro antioxidant activities of leaves, fruits and peel extracts of citrus. *Int. J. Phytopharm.* **2012**, *2*, 13–20.
68. Ramkissoon, J.S.; Mahomoodally, M.F.; Ahmed, N.; Subratty, A.H. Therapeutic potential of common culinary herbs and spices of mauritius. In *Chemistry: The Key to Our Sustainable Future*; Springer: Berlin, Germany, 2014; pp. 147–162.
69. Agarwal, M.; Kumar, A.; Gupta, R.; Upadhyaya, S.J.O.J.o.C. Extraction of polyphenol, flavonoid from emblica officinalis, citrus limon, cucumis sativus and evaluation of their antioxidant activity. *Orient. J. Chem.* **2012**, *28*, 993. [CrossRef]
70. Brown, N.; John, J.A.; Shahidi, F. Polyphenol composition and antioxidant potential of mint leaves. *Food Prod. Process. Nutr.* **2019**, *1*, 1–14. [CrossRef]
71. Ghasemzadeh, A.; Jaafar, H.Z.; Rahmat, A. Antioxidant activities, total phenolics and flavonoids content in two varieties of malaysia young ginger (zingiber officinale roscoe). *Molecules* **2010**, *15*, 4324–4333. [CrossRef]
72. Sir Elkhatim, K.A.; Elagib, R.A.A.; Hassan, A.B. Content of phenolic compounds and vitamin c and antioxidant activity in wasted parts of sudanese citrus fruits. *Food Sci. Nutr.* **2018**, *6*, 1214–1219. [CrossRef]
73. Vogeser, M.; Parhofer, K.G. Liquid chromatography tandem-mass spectrometry (lc-ms/ms)—technique and applications in endocrinology. *Exp. Clin. Endocrinol. Diabetes* **2007**, *115*, 559–570. [CrossRef] [PubMed]
74. Ali, A.; Wu, H.; Ponnampalam, E.N.; Cottrell, J.J.; Dunshea, F.R.; Suleria, H.A.R. Comprehensive Profiling of Most Widely Used Spices for Their Phenolic Compounds through LC-ESI-QTOF-MS2 and Their Antioxidant Potential. *Antioxidants* **2021**, *10*, 721. [CrossRef]
75. Dugasani, S.; Pichika, M.R.; Nadarajah, V.D.; Balijepalli, M.K.; Tandra, S.; Korlakunta, J.N. Comparative antioxidant and anti-inflammatory effects of [6]-gingerol, [8]-gingerol, [10]-gingerol and [6]-shogaol. *J. Ethnopharmacol.* **2010**, *127*, 515–520. [CrossRef]
76. Kumar, N.; Goel, N. Phenolic acids: Natural versatile molecules with promising therapeutic applications. *Biotechnol. Rep. (Amst)* **2019**, *24*, e00370. [CrossRef]

77. Klimek-Szczykutowicz, M.; Szopa, A.; Ekiert, H. Citrus limon (lemon) phenomenon-a review of the chemistry, pharmacological properties, applications in the modern pharmaceutical, food, and cosmetics industries, and biotechnological studies. *Plants* **2020**, *9*, 119. [CrossRef] [PubMed]
78. Tahira, R.; Naeemullah, M.; Akbar, F.; Masood, M.S.J.P.J.B. Major phenolic acids of local and exotic mint germplasm grown in islamabad. *Pak. J. Bot.* **2011**, *43*, 151–154.
79. Tohma, H.; Gülçin, İ.; Bursal, E.; Gören, A.C.; Alwasel, S.H.; Köksal, E. Antioxidant activity and phenolic compounds of ginger (zingiber officinale rosc.) determined by hplc-ms/ms. *J. Food Meas. Charact.* **2016**, *11*, 556–566. [CrossRef]
80. Guzman, J.D. Natural cinnamic acids, synthetic derivatives and hybrids with antimicrobial activity. *Molecules* **2014**, *19*, 19292–19349. [CrossRef]
81. Pontiki, E.; Hadjipavlou-Litina, D.; Litinas, K.; Geromichalos, G. Novel cinnamic acid derivatives as antioxidant and anticancer agents: Design, synthesis and modeling studies. *Molecules* **2014**, *19*, 9655–9674. [CrossRef]
82. Ben Said, R.; Hamed, A.I.; Mahalel, U.A.; Al-Ayed, A.S.; Kowalczyk, M.; Moldoch, J.; Oleszek, W.; Stochmal, A. Tentative characterization of polyphenolic compounds in the male flowers of phoenix dactylifera by liquid chromatography coupled with mass spectrometry and dft. *Int. J. Mol. Sci.* **2017**, *18*, 512. [CrossRef]
83. Gimenez-Bastida, J.A.; Zielinski, H.; Piskula, M.; Zielinska, D.; Szawara-Nowak, D. Buckwheat bioactive compounds, their derived phenolic metabolites and their health benefits. *Mol. Nutr. Food Res.* **2017**, *61*. [CrossRef]
84. Raneva, V.; Shimasaki, H.; Ishida, Y.; Ueta, N.; Niki, E.J.L. Antioxidative activity of 3, 4-dihydroxyphenylacetic acid and caffeic acid in rat plasma. *Lipids* **2001**, *36*, 1111. [CrossRef] [PubMed]
85. Tang, Y.; Nakashima, S.; Saiki, S.; Myoi, Y.; Abe, N.; Kuwazuru, S.; Zhu, B.; Ashida, H.; Murata, Y.; Nakamura, Y. 3,4-dihydroxyphenylacetic acid is a predominant biologically-active catabolite of quercetin glycosides. *Food Res. Int.* **2016**, *89*, 716–723. [CrossRef]
86. Panche, A.N.; Diwan, A.D.; Chandra, S.R. Flavonoids: An overview. *J. Nutr. Sci.* **2016**, *5*, e47. [CrossRef] [PubMed]
87. Zeng, X.; Su, W.; Zheng, Y.; Liu, H.; Li, P.; Zhang, W.; Liang, Y.; Bai, Y.; Peng, W.; Yao, H. Uflc-q-tof-ms/ms-based screening and identification of flavonoids and derived metabolites in human urine after oral administration of exocarpium citri grandis extract. *Molecules* **2018**, *23*, 895. [CrossRef]
88. Yu, J.; Wang, L.; Walzem, R.L.; Miller, E.G.; Pike, L.M.; Patil, B.S. Antioxidant activity of citrus limonoids, flavonoids, and coumarins. *J. Agric. Food Chem.* **2005**, *53*, 2009–2014. [CrossRef]
89. Man, M.Q.; Yang, B.; Elias, P.M. Benefits of hesperidin for cutaneous functions. *Evid Based Complement. Altern. Med.* **2019**, *2019*, 2676307. [CrossRef]
90. Mcharek, N.; Hanchi, B.J.J.A.B.F.Q. Maturational effects on phenolic constituents, antioxidant activities and lc-ms/ms profiles of lemon (citrus limon) peels. *J. Appl. Bot. Food Qual.* **2017**, *90*, 1–9.
91. Csepregi, K.; Neugart, S.; Schreiner, M.; Hideg, E. Comparative evaluation of total antioxidant capacities of plant polyphenols. *Molecules* **2016**, *21*, 208. [CrossRef] [PubMed]
92. Zhou, M.; Lin, Y.; Fang, S.; Liu, Y.; Shang, X. Phytochemical content and antioxidant activity in aqueous extracts of cyclocarya paliurus leaves collected from different populations. *PeerJ* **2019**, *7*, e6492. [CrossRef]
93. Shirai, M.; Moon, J.-H.; Tsushida, T.; Terao, J.J.J.o.A.; Chemistry, F. Inhibitory effect of a quercetin metabolite, quercetin 3-o-β-d-glucuronide, on lipid peroxidation in liposomal membranes. *J. Agric. Food Chem.* **2001**, *49*, 5602–5608. [CrossRef] [PubMed]
94. Ren, Z.; Nie, B.; Liu, T.; Yuan, F.; Feng, F.; Zhang, Y.; Zhou, W.; Xu, X.; Yao, M.; Zhang, F. Simultaneous determination of coumarin and its derivatives in tobacco products by liquid chromatography-tandem mass spectrometry. *Molecules* **2016**, *21*, 1511. [CrossRef]
95. Al-Majedy, Y.; Al-Amiery, A.; Kadhum, A.A.; BakarMohamad, A. Antioxidant activity of coumarins. *Syst. Rev. Pharm.* **2016**, *8*, 24–30. [CrossRef]
96. Shyamala, B.; Naidu, M.M.; Sulochanamma, G.; Srinivas, P. Studies on the antioxidant activities of natural vanilla extract and its constituent compounds through in vitro models. *J. Agric. Food Chem.* **2007**, *55*, 7738–7743. [CrossRef]
97. Melliou, E.; Zweigenbaum, J.A.; Mitchell, A.E. Ultrahigh-pressure liquid chromatography triple-quadrupole tandem mass spectrometry quantitation of polyphenols and secoiridoids in california-style black ripe olives and dry salt-cured olives. *J. Agric. Food Chem.* **2015**, *63*, 2400–2405. [CrossRef]
98. Gülçin, İ.; Elias, R.; Gepdiremen, A.; Boyer, L. Antioxidant activity of lignans from fringe tree (chionanthus virginicus L.). *Eur. Food Res. Technol.* **2006**, *223*, 759–767. [CrossRef]
99. Huang, X.; Song, F.; Liu, Z.; Liu, S. Studies on lignan constituents from schisandra chinensis (turcz.) baill. Fruits using high-performance liquid chromatography/electrospray ionization multiple-stage tandem mass spectrometry. *J. Mass Spectrom* **2007**, *42*, 1148–1161. [CrossRef] [PubMed]
100. Kim, J.S.; Yi, H.K. Schisandrin c enhances mitochondrial biogenesis and autophagy in c2c12 skeletal muscle cells: Potential involvement of anti-oxidative mechanisms. *Naunyn Schmiedebergs Arch. Pharm.* **2018**, *391*, 197–206. [CrossRef] [PubMed]
101. Tanweer, S.; Mehmood, T.; Zainab, S.; Ahmad, Z.; Shehzad, A. Comparison and hplc quantification of antioxidant profiling of ginger rhizome, leaves and flower extracts. *Clin. Phytosci.* **2020**, *6*. [CrossRef]
102. Ghasemzadeh, A.; Jaafar, H.Z.; Rahmat, A. Variation of the phytochemical constituents and antioxidant activities of zingiber officinale var. Rubrum theilade associated with different drying methods and polyphenol oxidase activity. *Molecules* **2016**, *21*, 780. [CrossRef] [PubMed]

103. Mattila, P.; Astola, J.; Kumpulainen, J. Determination of flavonoids in plant material by hplc with diode-array and electro-array detections. *J. Agric. Food Chem.* **2000**, *48*, 5834–5841. [CrossRef]
104. Kawaii, S.; Tomono, Y.; Katase, E.; Ogawa, K.; Yano, M.J. Quantitation of flavonoid constituents in citrus fruits. *J. Agric. Food Chem.* **1999**, *47*, 3565–3571. [CrossRef] [PubMed]
105. Xi, W.; Lu, J.; Qun, J.; Jiao, B. Characterization of phenolic profile and antioxidant capacity of different fruit part from lemon (citrus limon burm.) cultivars. *J. Food Sci. Technol.* **2017**, *54*, 1108–1118. [CrossRef]
106. Mišan, A.; Mimica-Dukić, N.; Mandić, A.; Sakač, M.; Milovanović, I.; Sedej, I. Development of a rapid resolution hplc method for the separation and determination of 17 phenolic compounds in crude plant extracts. *Open Chem.* **2011**, *9*, 133–142. [CrossRef]
107. Dolzhenko, Y.; Bertea, C.M.; Occhipinti, A.; Bossi, S.; Maffei, M.E. Uv-b modulates the interplay between terpenoids and flavonoids in peppermint (mentha x piperita L.). *J. Photochem. Photobiol. B* **2010**, *100*, 67–75. [CrossRef] [PubMed]
108. Kapp, K. *Polyphenolic and Essential Oil Composition of Mentha and Their Antimicrobial Effect*; Helsingin Yliopisto: Helsinki, Finland, 2015.

 *fermentation*

Article

Oenological Processes and Product Qualities in the Elaboration of Sparkling Wines Determine the Biogenic Amine Content

Aina Mir-Cerdà [1], Anaïs Izquierdo-Llopart [1], Javier Saurina [1,*] and Sonia Sentellas [1,2]

[1] Department of Chemical Engineering and Analytical Chemistry, University of Barcelona, 08028 Barcelona, Spain; ainamir17@gmail.com (A.M.-C.); anais.izquierdo.llopart@gmail.com (A.I.-L.); sonia.sentellas@ub.edu (S.S.)
[2] Serra Húnter Lecturer, Generalitat de Catalunya, 08007 Barcelona, Spain
* Correspondence: xavi.saurina@ub.edu

Abstract: The biogenic amine (BA) content in wines is dependent on the fermentation processes and other oenological practices, as well as on grape quality. These compounds can participate in different cellular functions in humans; however, the intake of high amounts can provoke some toxicological effects. For that reason, controlling the evolution of biogenic amines in wine production processes is of extreme importance. This work aims to assess the occurrence of biogenic amines in sparkling wines and related samples, including musts, base wines, stabilized wines, and three-month and seven-month aged sparkling wines obtained from Pinot Noir and Xarel lo grape varieties. The determination of BA content relies on liquid chromatography with fluorescence detection (HPLC–FLD) with precolumn derivatization of analytes with dansyl chloride. The analysis has shown that putrescine is the most abundant amine in these types of samples. Ethanolamine, tyramine, spermine, and histamine concentrations are also remarkable. Principal component analysis has been applied to try to extract featured information concerning overall patterns dealing with wine production steps and qualities. Interesting conclusions have been drawn on BA formation depending on different factors. BA concentrations are quite low in must but rise, especially after the first alcoholic fermentation. Moreover, BA levels are much lower in the range of products elaborated with grapes of the best qualities while they significantly increase when using grapes of lower qualities. The results obtained pointed out the analytical potential of using BAs to control the quality of wine and its production processes, thus providing valuable information for both wineries and consumers.

Keywords: sparkling wine; fermentation; biogenic amines; wine quality; liquid chromatography; principal component analysis

Citation: Mir-Cerdà, A.; Izquierdo-Llopart, A.; Saurina, J.; Sentellas, S. Oenological Processes and Product Qualities in the Elaboration of Sparkling Wines Determine the Biogenic Amine Content. *Fermentation* **2021**, *7*, 144. https://doi.org/10.3390/fermentation7030144

Academic Editors: Claudia Gonzalez Viejo and Sigfredo Fuentes

Received: 28 June 2021
Accepted: 30 July 2021
Published: 4 August 2021

1. Introduction

Biogenic amines (BAs) are low molecular nitrogenous compounds present in different types of food, but are especially abundant in wine, cheese, meat, and fish as well as in spoiled products [1–4]. In addition, the presence of high amounts of BAs in foods can be a sign of processing under poor hygiene conditions [5].

BAs are bioactive compounds that can participate in different cellular functions of humans but toxicological problems, such as migraines, headaches, hypo- or hypertension, effects on the vascular or nervous system, and even anaphylactic shocks, can occur when ingested in high concentrations [4,6]. More specifically, histamine has been extensively studied because it can be related to most biogenic amine foodborne intoxication, causing headaches, hypotension, and digestive problems. Histamine, together with other amines such as tyramine and serotonin, can also affect, directly or indirectly, the human vascular and nervous system. Aromatic amines (tyramine and phenylethylamine) can cause migraines and hypertension. Moreover, tyramine, tryptamine, and phenylethylamine show vasoconstrictor activity while others (histamine and serotonin) present a vasodilator effect. Psychoactive amines, like dopamine and serotonin, are neurotransmitters of the central nervous system, while tyramine and histamine act as hormonal mediators in humans and animals. As a complementary aspect, some BA can react with nitrite to generate carcinogenic nitrosamines [7–9]. Finally, it is worth mentioning that the presence of ethanol in alcoholic beverages may potentiate BA toxicological effects due to the inhibition of amino oxidases which are the enzymes responsible for their metabolism. As far as we know, legislation concerning BA in foods and beverages does not exist and just some recommendations regarding the upper limit for histamine are given, in general in the range from 2 mg to 8 mg L^{-1} [10]. Therefore, effective and selective methods are required for the determination of these amines in food products to assess their quality and avoid the commercialization of dangerous or spoiled foodstuffs [11–14].

Sparkling wine (cava) is generally elaborated using the traditional Champenoise method [15–17] from white or rosé base wines. When grapes arrive at the cellar, they are carefully pressed in batches separated by variety and quality to produce the must. Next, the alcoholic fermentation is developed in stainless steel tanks from 15 to 18 °C, thus resulting in the base wine. If necessary, malolactic fermentation (MLF) is applied after this step. Subsequently, different base wines are blended depending on the sparkling wine to be produced and the mix is clarified and stabilized to avoid further precipitation of tartrate salts. The key step in the elaboration of a cava is the second fermentation that takes place in the bottle, triggered by adding the tirage liquor—a mixture of wine, sugars, and yeasts—to the stabilized base wine. Cava is aged in contact with the yeast for at least a minimum of nine months before its commercialization to consolidate its organoleptic features [15].

Cava is mainly produced from the classical Catalonian varieties of white grapes Macabeu, Xarel lo, and Parellada. In recent years, however, other white (e.g., Chardonnay) and red varieties (e.g., Pinot Noir) have been introduced as well [18]. Xarel lo and Pinot Noir varieties have been selected for this study. Xarel lo is characterized by medium-sized and round white grapes with thin skin. It offers freshness, acidity, and high alcohol content, thus resulting in an excellent grape variety for long-aged wines. Pinot Noir is a typical French grape: small, round, and with an intense violet color. It gives aroma and color to the product and promotes the formation of bubbles in the sparkling wine cup.

BA concentrations may provide valuable information regarding the quality of cava as well as the fermentation processes underwent during its production. This issue has been scarcely studied [19–23]. Furthermore, interesting conclusions have been extracted, indicating that BA content is low in the must and more remarkable after fermentations. The content of amino acids and biogenic amines in wine depends on agricultural practices involved in the production of the grapes as well as on the vinification and aging processes. Nitrogen fertilization of the soil, poor health status of the grape due to molds, and the high must pH are factors that can favor the moderate occurrence of biogenic amines in the must. Subsequently, during alcoholic and malolactic fermentation, the presence of certain yeasts and bacteria significantly increases BA content in wine. BAs are generated by the decarboxylation of specific amino acids by the action of microorganisms. The activity of decarboxylases under the specific conditions of pH, temperature, etc., will determine the amount of BA produced, which can vary substantially depending on the oenological conditions and, especially, on microorganisms and strains used in the fermentation processes. In addition, it is worth highlighting the high stability of these compounds once generated. The most abundant BA in wines is, in general, putrescine, ethanolamine, tyramine, and histamine [24–26]. As commented, beyond toxicological and quality issues, the role of BAs as descriptors of different wine features should not be underestimated. For instance, Garcia-Villar et al. concluded that this family of components could be used as a source of information to classify wines according to aging into young, crianza, reserve, and grand reserve classes [27]. Conclusions on the grape varieties could be extracted as well. For characterization purposes, modern analytical platforms offer excellent possibilities for the generation of great amounts of data, thus resulting in complex tables/datasets that should be conveniently interpreted to try to extract the underlying analytical information. In this regard, multivariate statistics and chemometric methods of analysis may be fundamental to achieve a comprehensive insight into the set of samples under study [24].

In this paper, the compositional profiles of BAs in various types of samples from sparkling wine (cava) production has been evaluated to try to extract information on the influence of different oenological factors such as winemaking steps, grape/wine quality, and variety. An HPLC–FLD method with offline derivatization with dansyl chloride (Dns-Cl) has been applied to determine the BA content. Once all of the samples have been analyzed, chemometric methods have been applied to find trends and descriptors within the groups of samples. As a result, data can be more effectively interpreted, and more solid and global conclusions can be extracted than when explored more conventionally.

2. Materials and Methods

2.1. Reagents

Formic acid (>96%, Merck KGaA, Darmstadt, Germany), phosphoric acid (85% *w*/*w*, Merck), acetonitrile (UHPLC PAI-ACS SuperGradient, Panreac, Castellar del Valles, Barcelona, Spain), methanol (UHPLC-Supergradient, Panreac ApplyChem, Castellar del Valles, Barcelona, Spain), and water (Elix3, Millipore, Bedford, MA, USA) were used to prepare the solvents and mobile phases of the chromatographic method.

Biogenic amines were obtained from the following sources: 1,5-diaminopentane (cadaverine, 98%), 1,4-diaminobutane dihydrochloride (putrescine, 99%), spermidine trihydrochloride (99%), and spermine tetrahydrochloride (99%) were from Alfa Aesar (Kandel, Germany); histamine hydrochloride (≥99%), 2-phenylethylamine hydrochloride (≥99%), tryptamine hydrochloride (≥98%), tyramine hydrochloride (≥97%), octopamine hydrochloride (≥99%), and agmatine sulfate (≥99%) were from Fluka (Buchs, Switzerland); ethanolamine hydrochloride (≥98%) and hexylamine (≥98%) were from TCI (Tokyo, Japan), the latter was used as the internal standard. Lysine monohydrochloride (99%) was obtained from Fluka (Buchs, Switzerland). Pure standard stock solutions of each amine at a concentration of 1000 mg L^{-1} were prepared in Milli-Q water. These stock solutions were stored at 4 °C until use and were used to prepare mixtures of the working standard solutions within a concentration range from 0.1 to 50 mg L^{-1}.

Amines were derivatized with dansyl chloride (dansyl-Cl 98%, Thermo Fisher Scientific, Waltham, MA, USA). The reagent solution was prepared by dissolving 100 mg of dansyl-Cl in 10 mL of acetone (LichroSolv, Merck, Darmstadt, Germany). The buffer solution consisted of 0.1 mol L^{-1} sodium tetraborate (Analytical grade, Merck KGaA, Darmstadt, Germany). Chloroform (≥99.8%, Fluka, Buchs, Switzerland) was used for the extraction of derivatives.

2.2. Samples

A total of 40 samples including musts, wines, and sparkling wines were kindly provided by the Raventós Codorníu Group (Sant Sadurni d'Anoia, Spain), corresponding to white (from Xarel lo grapes) and rosé (from Pinot Noir grapes) cavas (vintage of 2020) produced in Penedès and Costers del Segre regions (both from Catalonia, Spain). For each variety, four musts, eight wines (four monovarietal and four stabilized), and eight sparkling wines (four of three months in rhyme and four of seven months) were available. In each class, four different qualities were defined according to the quality of grapes, namely: A, B, C, and D, with A referring to the best ones and D the ones of the lowest quality. The quality was ranked by the cava manufacturer and depended on multiple factors such as the type of soil, the type of grape plantation (ecological or conventional), the type of harvest and transport (manual for class A or mechanized for B, C, and D), the climatic conditions, the age of the vines, the yield of grapes obtained (ca. from 6000 for A to 10,000 kg per hectare or more for D), the type of yeasts, etc. Moreover, for grapes of C and D qualities, malolactic fermentation was required to reduce the wine acidity. More specific details on these issues were confidential and have not been provided by the company. Additional information on the codes of samples according to classes, qualities, and varieties is given in Table 1.

Table 1. List of samples under study. Sample codes are as follows: M, must; BW, base wine; SW, stabilized wine; C3, 3 months in rhyme cava wine (sparkling wine); C7, 7 months in rhyme cava wine (sparkling wine); P, Pinot Noir; X, Xarel lo; A, quality A; B, quality B; C, quality C; D, quality D.

Grape Variety	Quality	Must	Base Wine	Stabilized Wine	3 Months Sparkling Wine	7 Months Sparkling Wine
Pinot Noir	A	MPA	BWPA	SWPA	C3PA	C7PA
	B	MPB	BWPB	SWPB	C3PB	C7PB
	C	MPC	BWPC	SWPC	C3PC	C7PC
	D	MPD	BWPD	SWPD	C3PD	C7PD
Xarel lo	A	MXA	BWXA	SWXA	C3XA	C7XA
	B	MXB	BWXB	SWXB	C3XB	C7XB
	C	MXC	BWXC	SWXC	C3XC	C7XC
	D	MXD	BWXD	SWXD	C3XD	C7XD

Samples were analyzed in triplicate, from three independent derivatization processes. Furthermore, quality control (QC) samples were prepared by mixing 50 µL of each must/wine/cava sample. Specific QC for Xarel lo and Pinot Noir samples were prepared as well. QCs were used to evaluate the reproducibility of the analytical methods and the significance of the PCA models.

2.3. Amine Derivatization

A precolumn offline derivatization was performed to enhance the sensitivity of the detection of the analytes as well as to facilitate their separation. The procedure described elsewhere was followed with some modifications [28–31]. Briefly, the reaction was carried out in a glass vial mixing a 250 µL biogenic amine standard (or sample) solution, 250 µL dansyl-Cl reagent solution, and 250 µL buffer solution. The reaction took place at pH 9.2 in a thermostatic water bath (Tectron 473-100, J.P. Selecta, Barcelona, Spain) for 60 min at 40 °C. Next, 750 µL of chloroform was added to the reaction vial and the solution was shaken for 10 min (Vortex 3 IKA, Staufen, Germany with a VG 3.31 Test tube attachment accessory) to extract derivatives from the mixture. Then the organic phase was separated. This extraction process was repeated 3 times to quantitatively extract the corresponding derivatives. Subsequently, the organic fractions were pooled, evaporated to dryness under a nitrogen current, and re-dissolved in 600 µL of acetonitrile/water (50:50, *v*/*v*). The resulting solutions were ready to be analyzed chromatographically.

2.4. Chromatographic Method

An Agilent 1100 Series HPLC instrument was used which was equipped with degasser (G1379A), binary pump (G1312A), automatic injector (G1392A), diode-array UV-vis detector (G1315B), and fluorescence detector (FLD, G1321A), all of them from Agilent Technologies (Waldbronn, Germany). Data were acquired and processed with Agilent Chemstation software (Rev. A 10.02).

Compounds were separated by reversed-phase mode using a Kinetex C18 column (150 mm × 2.6 mm I.D., 2.6 µm particle size) from Phenomenex (Torrance, CA, USA) and 0.1% (*v*:*v*) formic acid aqueous solution and acetonitrile (ACN) as the components of the mobile phase. The elution gradient profile was as follows: 0 to 14 min, 55 to 75% ACN; 14 to 20 min, 75 to 95% ACN; 20 to 22 min, 95% ACN (column cleaning); 22 to 22.1, 95 to 55% ACN; 22.1 to 25 min, 55% ACN (column stabilization). The flow rate was 0.7 mL min^{-1} and the injection volume was 10 µL. Chromatograms were acquired at 254 nm in the UV absorption detection mode and selecting 320/523 nm as the excitation and emission wavelength in FLD. Samples were analyzed randomly, and the QCs were repeatedly injected every 10 samples.

The LC–MS system used for structure confirmation was the Agilent 1100 Series liquid chromatograph—described above—coupled to an AB Sciex 4000 QTrap hybrid triple quadrupole/linear ion trap mass spectrometer (AB Sciex, Framingham, MA, USA). Chromatographic conditions were as indicated above. BA derivatives were detected in positive electrospray ionization (ESI) tandem mass spectrometry (MS/MS) in the multiple reaction monitoring (MRM) mode. The ion spray voltage was set at 4500 V. The source temperature was set at 500 °C. Nitrogen was used as nebulizer and auxiliary gas and was set at 10, 50, and 50 arbitrary units for the curtain gas, ion source gas 1 and ion source gas 2, respectively. Declustering potential (DP), collision energy (CE), collision exit cell potential (CXP), and ion transitions pairs were optimized for each analyte and are given in Table S1.

2.5. *Data Analysis*

The huge amount of information resulting from the analysis of the large set of samples required the application of multivariate methods such as Principal Component Analysis (PCA) for a more efficient recovery of the underlying chemical patterns related to the overall vinification process as well as each given stage or quality. The chemometric software used was the PLS-Toolbox for MATLAB (Mathworks, Natick, MA, USA). Other preliminary and statistical studies were carried out with Microsoft Excel (Microsoft, Redmon WA, USA). The resulting data, which consisted of concentrations of the biogenic amines in the different samples, was arranged in a matrix of responses (X matrix) in which each column corresponded to a given analyte and each row referred to a sample. As a result, dimensions of the X matrix were 135 × 11, for 120 samples plus 15 QCs and 11 amines, respectively (see structures of analytes in Figure S1 in Supplementary Material).

PCA is highly powerful and versatile for exploratory studies dealing with the influence of grape qualities and vinification process on the content of BAs. PCA allows the chemical information contained originally to be concentrated into the so-called principal components (PCs). Commonly, the scatter plot of scores, for instance of PC1 vs PC2, is used to find out some sample patterns that, in this case, could be related to features such as grape qualities, varieties, and winemaking practices. Complementarily, the plot of loadings may show relationships or correlations among variables (i.e., among the levels of BAs) and, even more importantly, it may reveal the most significant descriptors or biomarkers of the different sample classes. Additional information on PCA can be found elsewhere [32].

3. Results and Discussion

3.1. *Method Development and Validation*

The HPLC–FLD method used in this paper was based on previous publications [24–27]. Slight modifications in the experimental conditions were applied to adapt the separation to the current circumstances of the analytical column, the internal standard used to control the extraction step, and the sample matrix features. In particular, the elution gradient was optimized for the core-shell Kinetex C18 column to achieve a good chromatographic separation of common BA present on different kinds of fermented foodstuffs, with the minimum running time to speed up the analysis, which is especially critical when dealing with large sets of samples. Regarding the derivatization, it was confirmed by LC–MS (MRM mode) the formation of mono derivatives of ethanolamine, tryptamine, hexylamine, and phenylethylamine, di-dansyled derivatives for cadaverine, putrescine, histamine, agmatine and tyramine, and tri- and tetra-dansyled derivatives for spermidine and spermine, respectively (Table S1). As an example, Figure S2 shows the FLD chromatogram of a dansyl derivatized standard solution of amines at a concentration of 2 mg L^{-1} each. As can be seen, a good separation resolution among all the analytes was obtained in a reasonably short time of analysis (25 min, including column cleaning and conditioning steps).

In addition, the performance of the method was evaluated by estimating some quality parameters under the optimal conditions of the HPLC–FLD method (see Table 2). Results are summarized in Table S2 in the Supplementary Material. Briefly, the relationship between peak areas and concentrations was linear at least up to 20 mg L^{-1}, with determination coefficients (R^2) higher than 0.994. Limits of detection and quantification ranged from 0.006 (cadaverine) to 0.15 (tyramine) mg L^{-1} and from 0.021 to 0.49 mg L^{-1}, respectively. In terms of repeatability, RSD values were below 0.4% for retention time and below 0.9%

for peak areas. Compared with other existing methods for BA determination, the proposed one provides high sampling throughput, quite a low detection limit (for some analytes in the order of magnitude of 10 ng L^{-1} or below), and other remarkable figures of merit, thus being high, appropriate for the analysis of a large series of samples such as those generated in the quality control of cavas.

Table 2. Quality parameters obtained for the derivatized amines using the HPLC–FLD optimized method.

Amine [a]	Sensitivity (I_f min L mg^{-1}) [b]	Determination Coefficient (R^2)	Retention Time Repeatability (RSD, %)	Peak Area Repeatability (RSD, %)	LOD (mg L^{-1})	LOQ (mg L^{-1})
Agm	0.00355	0.995	0.4	0.7	0.15	0.511
Tryp	0.0217	0.9991	0.3	0.6	0.045	0.149
Phe	0.03195	0.995	0.3	0.5	0.019	0.064
Put	0.07335	0.996	0.2	0.5	0.011	0.037
Cad	0.08465	0.9992	0.2	0.5	0.006	0.021
His	0.00215	0.998	0.2	0.9	0.038	0.128
Oct	0.00473	0.9993	0.2	0.7	0.158	0.527
Tyr	0.0098	0.999	0.1	0.4	0.147	0.488
Smd	0.0667	0.998	0.1	0.4	0.027	0.089
Spm	0.07145	0.998	0.1	0.3	0.007	0.022
Eth	0.0332	0.997	0.1	0.4	0.015	0.050

[a] Amine identification: Agm: agmantine; Tryp: tryptamine; Phe: phenylethylamine; Put: putrescine; Cad: cadaverine; His: histamine; Oct: octopamine; Tyr: tyramine; Smd: spermidine; Spm: spermine; Eth: ethanolamine [b] I_f: fluorescence intensity in arbitrary units.

3.2. Sample Analysis

The set of 40 samples, including musts, base wines, stabilized wines, and aged sparkling wines, was analyzed according to the proposed HPLC–FLD method (see details in the experimental section). As an example, Figure 1 shows various representative FLD chromatograms of dansyled derivatives of Pinot Noir samples of quality D taken throughout the different vinification steps. Some remarkable differences can be observed suggesting the occurrence of important compositional changes associated with the processes. Furthermore, apart from biogenic amines, other peaks identified as derivatives of free amino acids were also detected in the chromatograms. These compounds appeared in the time range from 2 to 10 min with lysine, ornithine, cysteine, leucine/isoleucine, and phenylalanine being the most prominent ones.

A preliminary statistical analysis was performed by calculating the mean concentration and standard deviation of each amine in the full set of samples (Table S2). In terms of abundance and variability among classes, putrescine was the most remarkable amine, with an average concentration of 7 mg L^{-1} and a standard deviation of 6 mg L^{-1}, meaning that putrescine concentrations varied dramatically depending on the oenological practices as well as on the grape quality. Ethanolamine was the second most abundant amine with an average concentration of 5 mg L^{-1} and a standard deviation of 1.5 mg L^{-1}. Ethanolamine was even present at ca. 3 mg L^{-1} in the must samples. Tyramine, spermine, and histamine, with concentrations between 1 and 3 mg L^{-1} were also important quantitatively. Tyramine and histamine displayed interesting variations in their concentration levels as a function of sample type while spermine was quite homogeneous. Phenylethylamine, cadaverine, and spermidine occurred in amounts below 1 mg L^{-1}; despite being trace amines, their potential descriptive ability should not be underestimated since variations in their profiles among sample classes were noticeable. Finally, agmatine was detected at concentrations of ca. 0.1 mg L^{-1} while tryptamine and octopamine were not detected.

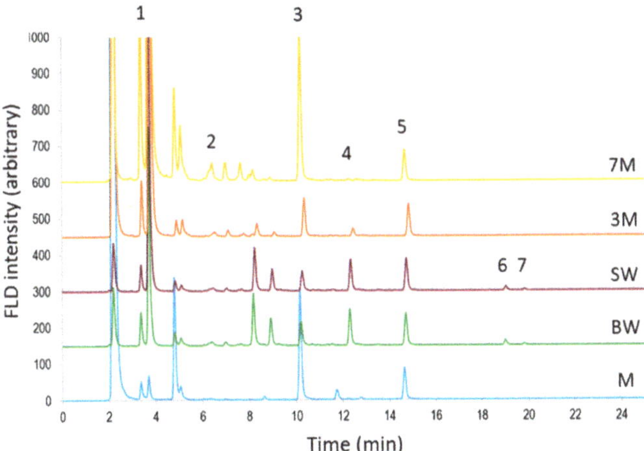

Figure 1. FLD chromatograms of the five types of sample classes from Pinot Noir variety using grapes of D quality. Sample assignment: M = must; BW = base wine; SW = stabilized wine; 3M = sparkling wine of 3 months aging; 7M = sparkling wine of 7 months aging. Peak assignment: 1 = Ethanolamine; 2 = Agmatine; 3 = Lysine; 4 = Cadaverine; 5 = Internal standard; 6 = Tyramine; 7 = Spermidine.

A preliminary comparison of varieties showed that, in general, BA levels in Pinot Noir were higher than in Xarel lo. ANOVA studies showed that these differences were significant. For instance, for some representative amines such as ethanolamine and putrescine, *p*-values of 1.3×10^{-11} and 3.1×10^{-15} were respectively obtained. Except for ethanolamine, this difference is more noticeable among base and stabilized wines as can be seen in Figure S3 showing the boxplots of the most important amines (ethanolamine, putrescine, histamine, and tyramine).

The first comparison of sample classes revealed that musts contained the lowest contents of biogenic amines, among which ethanolamine and putrescine occurred at levels higher than 1 mg L^{-1}, while cadaverine, histamine, spermine, or agmatine if detected, were approximately 0.1 mg L^{-1}. Samples of monovarietal base wines and stabilized wines displayed quite similar chromatographic profiles, with 2- to 4-fold higher concentrations of ethanolamine and putrescine, compared with musts with remarkable levels of histamine and tyramine (above 2 mg L^{-1}) and smaller concentrations of the others. The rise in the concentrations of BAs in these two types of samples was attributed to the influence of alcoholic fermentation. In addition, samples elaborated with C and D qualities were subjected to malolactic fermentation (MLF) to reduce the tart character of wines while providing more creamy notes. During the transformation of malic acid to lactic acid, lactic bacteria also produce the decarboxylation of a percentage of free amino acids, thus MLF being one of the main processes contributing to the formation of BAs.

With regard to the further stages of cava elaboration, two different behaviors were observed depending on the amine. For ethanolamine, a slight increase in the amine levels after the second fermentation and during the aging process was found. In contrast, concentrations of putrescine, tyramine, putrescine, and histamine slightly decreased with aging after the second fermentation. Hence, a certain degradation of these amines with time was detected after the rise in the maximum levels at the stage of base wine.

As indicated above, apart from BAs, chromatograms of dansyl derivatives also showed several peaks corresponding to amino acids. The more polar nature of amino acid derivatives with respect to amines makes their elution faster, so most of them corresponded to those peaks in the time range from 2 to 10 min. Interestingly, as can be seen in Figure 1, one of the most outstanding peaks in the chromatograms, very abundant in all of the samples, corresponded to lysine (retention time 10.1 min) which is the precursor of cadaverine. Due to the importance of lysine, we decided to include this compound in our datasets when PCA was performed.

Beyond the influence of the winemaking stages on BA content, amine concentrations may also depend on the quality of the grapes in origin as well as the precautions during their manipulation, from harvesting and transport to must pressing. For instance, Figure 2 shows the comparison of two base wines of Pinot Noir of A and D qualities. All of the peaks in the chromatogram of D base wine are more intense than those in A base wine, except for lysine.

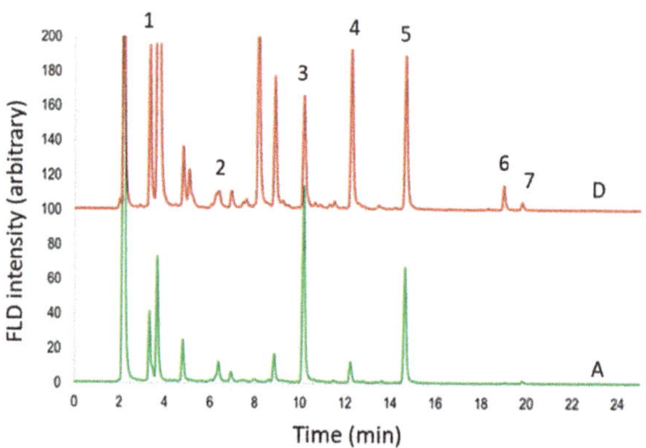

Figure 2. FLD chromatograms of two base wines of A and D qualities from Pinot Noir variety. Peak assignment: Peak assignment: 1 = Ethanolamine; 2 = Agmatine; 3 = Lysine; 4 = Cadaverine; 5 = Internal standard; 6 = Tyramine; 7 = Spermidine.

Figure 3 shows the boxplots of concentrations of two representative amines—ethanolamine and putrescine—as a function of sample product type, from must to sparkling wines and grape qualities. It can be seen that there is a significant rise in concentration from must to base wine. In the case of ethanolamine, the maximum concentration is achieved after the second fermentation while for putrescine it is found in the base wine. Regarding the wine quality, A samples show the lowest levels, and C and D samples show much higher concentrations. For the B type, the behavior of the two amines is quite different.

In order to try to isolate the contribution of the quality from that of the MLF on the formation of BAs an ANOVA study was applied. Samples from each variety (Pinot Noir and Xarel lo) were evaluated separately. For each amine, concentrations of equivalent samples of A and B qualities were compared with those from C and D counterparts. When MLF was not applied (i.e., pre-fermentation samples such as musts), no significant differences among A/B and C/D samples were found in most of the cases, with the only exception being ethanolamine in Pinot Noir and putrescine in Xarel lo. However, differences among A/B and C/D samples were always significant for post-fermentation products, thus proving that MLF was a much more crucial factor than quality on the formation of BAs. Results from this study can be found in Table S3 in the Supplementary Material.

However, some subtle details and relationships may have escaped our consideration so this approach may fail in complex situations, such as those presented here. At this stage, chemometric methods have proved their great performance in the analysis of large sets of multivariate data involving the simultaneous study of many samples of different characteristics.

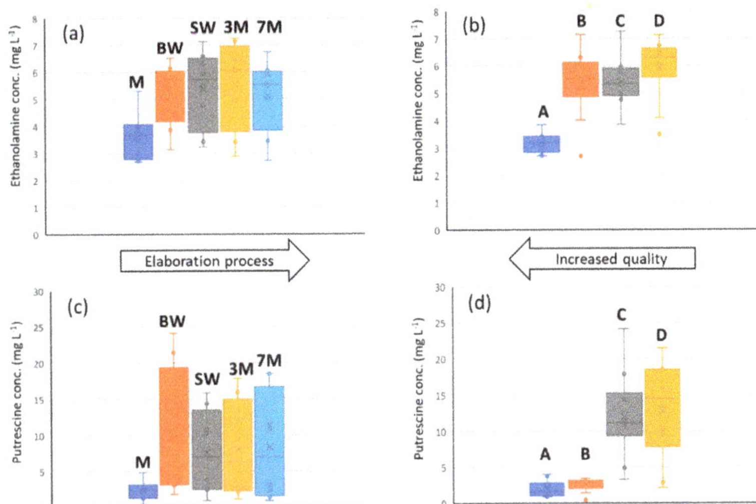

Figure 3. Boxplots with whiskers representing the concentration of ethanolamine and cadaverine in the samples. (**a**) Ethanolamine as a function of the sample types; (**b**) Ethanolamine as a function of the qualities; (**c**) Putrescine as a function of the sample types; (**d**) Putrescine as a function of the qualities. Sample assignment: M = must; BW = base wine; SW = stabilized wine; 3M = sparkling wine of 3 months aging; 7M = sparkling wine of 7 months aging. A = quality A; B = quality B; C = quality C; D = quality D. Error bars indicate the minimum and maximum values.

3.3. Principal Component Analysis

PCA was applied to explore the dependence of BA (and lysine) contents with fermentations and other oenological practices. The influence of the grape quality was assessed as well. For such a purpose, the X matrix of responses was built, which consisted of concentrations of BAs in the set of samples. Prior to PCA, data were autoscaled to equalize the descriptive relevance of all (major and minor) amines.

First, a model was calculated considering all of the samples under study to follow the evolution of BAs throughout the vinification. Two PCs were able to retain more than 66% of the total variance so the scatter plot of the scores of PC1 vs PC2 displayed a great portion of relevant information contained in the data. As shown in Figure 4, samples were clearly structured according to the main types, with musts predominating at the bottom left sector, sparkling wines at the top left sector, and base and stabilized wines on the right. Regarding sparkling wines, 3- and 7-month aged classes were, in general, well-separated, thus indicating that aging was a relevant factor in the evolution of the BA content. Conversely, base and stabilized wines were mixed in the same area and their distribution depended on additional features such as grape quality and variety. Moreover, all QCs were grouped in a compact group in the center of the model which supported the reproducibility of the chromatographic data and the significance of the chemometric results.

Concerning the map of variables, the plot of loadings showed that PC1 mainly explained the overall contribution of BAs, with some of the most abundant amines located to the right side (including putrescine, cadaverine, tyramine, spermidine, and histamine). Concentrations of these amines were correlated, thus indicating that, in general, they followed similar patterns. Determination coefficients R2 were better than 0.7 and correlations were significant statistically ($p < 0.05$), with the correlation among putrescine and histamine (R2 = 0.96 and $p = 2 \times 10^{-28}$) being especially remarkable. In contrast, they were not correlated with the polyamines, ethanolamine, and agmatine, thus suggesting that they followed different formation pathways. PC2 mainly explained the behavior of ethanolamine.

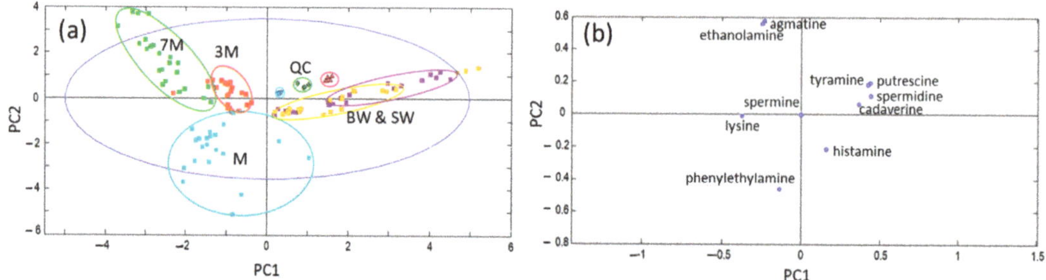

Figure 4. Characterization of product samples by PCA using the compositional profiles of amines as the data. (**a**) Scatter plot of scores of PC1 vs PC2; (**b**) Scatter plot of loadings of PC1 vs PC2. Sample assignment: M = must (light blue); BW = base wine (light ochre); SW = stabilized wine (purple); 3M = sparkling wine of 3 months aging (red); 7M = sparkling wine of 7 months aging (green); QC = quality control.

The simultaneous study of scores and loadings provided interesting conclusions on the descriptors or biomarkers which were up- or down-regulated depending on the different classes: musts, wines, or cava samples. It was found that base and stabilized wines, especially those subjected to MLF, reached the highest overall levels of BAs, such as putrescine, tyramine, and histidine. Sparkling wines resulting from the second fermentations showed a small decay in diamine and monoamine levels, except for ethanolamine, which slightly increased in parallel to the rise in the alcoholic content undergone after the second fermentation. In contrast, musts were characterized by low levels of all of the types of BAs.

When other sample features such as grape quality or grape type were explored in a similar way considering the full set of samples, no clear conclusions on the sample behavior were obtained, thus suggesting that the vinification stage was the most dramatic factor influencing the evolution of the content of BAs in the samples. However, it was evidenced that the initial quality of the grapes, assigned in origin as A, B, C, and D types (from best to worse quality), and the care in their further manipulations (maximum neatness for A type) were other key aspects in the BA content. In the preliminary statistical analysis, a noticeable overall increase in BA levels was evidenced from A to D qualities. Thus, to study the influence of each quality more thoroughly, various sample subsets focused on musts, base and stabilized wines, and sparkling wines separately were created to be submitted to PCA. Hence, specific PCA models were evaluated. The example depicted in Figure 5 shows the graph of scores of PC1 vs PC2 for both Pinot Noir and Xarel lo sparkling wines aged for a period of seven months. Samples were distributed along PC1, from left to right, as a function of the grape quality with a compact group on the left side containing both A and B types while C and D sparkling wines were spread on the right section. As deduced from ANOVA, this discrimination was mainly due to amines such as putrescine ($p = 6.4 \times 10^{-14}$), while others such as ethanolamine did not influence this behavior. PC2 especially described differences among Xarel lo and Pinot Noir, in which samples from each variety were mainly located at the upper and lower areas, respectively. This behavior was attributed to the significant differences in the concentrations of some important amines, also confirmed by ANOVA. The most significant differences were found for cadaverine, ethanolamine, and agmatine with p-values of 2.1×10^{-5}, 0.0043 and 0.0088, respectively. Regarding the variables, it was again confirmed that the lower the quality, the higher the BA content. Similar conclusions were obtained from models for the exploratory analysis of the other sample classes.

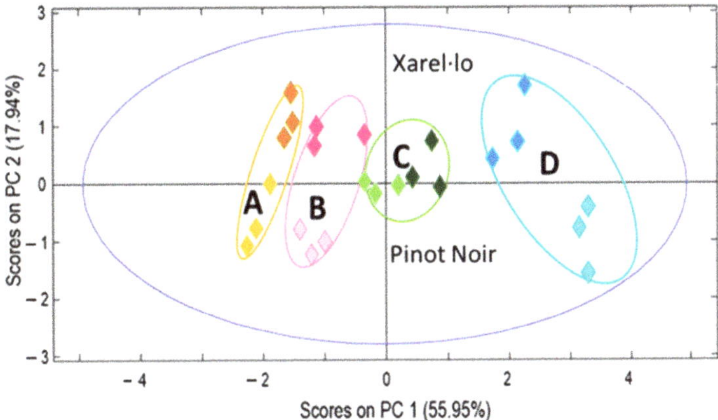

Figure 5. Characterization of sparkling wines aged for a period of seven months by PCA using the compositional profiles of amines as the data. Scatter plot of scores of PC1 vs PC2. Dark colors correspond to Xarel lo.

4. Conclusions

This paper explores the influence of grape/wine quality and vinification processes of sparkling wines (wines of Cava PDO—Protected Designation of Origin) on the occurrence of BAs. Samples of different characteristics (musts, base wines, stabilized base wines, and sparkling wines with 3 and 7 months of aging), different product qualities, and different grape varieties have been analyzed by HPLC–FLD. Conclusions from ANOVA and Principal Component Analysis (PCA) have shown that the concentrations of BAs increase significantly from musts to base wines. This rise is even more noticeable for wines subjected to malolactic fermentation. Subsequently, in general, BA levels remain constant or slightly decrease with aging after the second fermentation. From PCA results, the plot of scores has shown well-separated clusters corresponding must, base and stabilized wines, and sparkling wines while the plot of loadings has revealed that compounds such as putrescine, ethanolamine, and lysine are important descriptors. Regarding the quality, products of higher quality contain much lower amounts, meaning BA levels are notably lower if care and neatness are maintained throughout the process. This finding is even more noticeable for products subjected to malolactic fermentation, which is identified as one of the main sources responsible for the formation of BAs. When comparing the two varieties, Pinot Noir products, in general, display higher BA levels than Xarel lo ones.

BA profiles (as well as their precursors and related species) are potential biomarkers to determine wine quality and control the manufacturing processes. In this sense, the descriptive ability of lysine should be noted, thus suggesting that other amino acids, as BA precursors, could also be a complementary source of information. As indicated, at the must stage, high levels of amines can be associated with lower grape quality, and less careful harvesting and processing conditions. This trend regarding quality can be generalized to the other winemaking steps, therefore, for similar types of cava wines, the lower the amine levels, the better the products. Therefore, BA data can be essential for both producers and consumers to recognize the qualities of wine and cava samples, especially since this type of information is difficult to achieve by other means such as from sensory parameters.

Supplementary Materials: The following are available online at https://www.mdpi.com/article/10.3390/fermentation7030144/s1, Figure S1: Structures of the most common biogenic amines in fermented foodstuffs, Figure S2: FLD chromatograms of a derivatized standard solution of BAs at 2 mg L^{-1} each. Peak assignment: (a) ethanolamine; (b) agmatine; (c) tryptamine; (d) phenylethylamine; (e) putrescine; (f) cadaverine; (g) histamine; (h) hexylamine (IS); (i) octopamine; (j) tyramine; (k) spermidine; (l) spermine, Figure S3. Boxplots with whiskers representing the concentration of ethanolamine, putrescine, histamine, and tyramine in base and stabilized wines for Pinot Noir (light color) and Xarel lo (dark color) varieties, Table S1: MRM transitions for the detection of dansyl derivatives of biogenic amines by LC-MS/MS, Table S2: Concentration (mg L^{-1}) of biogenic amines in the different type of samples, Table S3: ANOVA applied to the study of quality and MLF factors.

Author Contributions: Conceptualization, A.I.-L. and J.S.; methodology, A.M.-C., A.I.-L. and S.S.; software, A.M.-C.; formal analysis, J.S. and S.S.; investigation, A.M.-C.; writing—original draft preparation, S.S. and J.S.; writing—review and editing, S.S. and J.S.; supervision, A.I.-L. and J.S. All authors have read and agreed to the published version of the manuscript.

Funding: This research was funded by the "Agencia Estatal de Investigación", grant number PID2020-114401RB-C22.

Institutional Review Board Statement: Not applicable.

Informed Consent Statement: Not applicable.

Acknowledgments: Authors thank the Raventós Codorníu Group for providing the samples.

Conflicts of Interest: The authors declare no conflict of interest.

References

1. Jaguey-Hernández, Y.; Aguilar-Arteaga, K.; Ojeda-Ramirez, D.; Añorve-Morga, J.; González-Olivares, L.G.; Castañeda-Ovando, A. Biogenic amines levels in food processing: Efforts for their control in foodstuffs. *Food Res. Int.* **2021**, *144*, 110341. [CrossRef]
2. Sivamaruthi, B.S.; Kesika, P.; Chaiyasut, C. A narrative review on biogenic amines in fermented fish and meat products. *J. Food Sci. Technol.* **2021**, *58*, 1623–1639. [CrossRef] [PubMed]
3. Costantini, A.; Vaudano, E.; Pulcini, L.; Carafa, T.; Garcia-Moruno, E. An Overview on Biogenic Amines in Wine. *Beverages* **2019**, *5*, 19. [CrossRef]
4. Ruiz-Capillas, C.; Herrero, A.M. Impact of biogenic amines on food quality and safety. *Foods* **2019**, *8*, 62. [CrossRef] [PubMed]
5. Simon Sarkadi, L. Amino acids and biogenic amines as food quality factors. *Pure Appl. Chem.* **2019**, *91*, 289–300. [CrossRef]
6. Wójcik, W.; Łukasiewicz, M.; Puppel, K. Biogenic amines: Formation, action and toxicity—A review. *J. Sci. Food Agric.* **2021**, *101*, 2634–2640. [CrossRef]
7. Önal, A. A Review: Current Analytical Methods for the Determination of Biogenic Amines in Foods. *Food Chem.* **2007**, *103*, 1475–1486. [CrossRef]
8. Marcobal, A.; Polo, M.C.; Martín-Álvarez, P.J.; Moreno-Arribas, M.V. Biogenic Amine Content of Red Spanish Wines: Comparison of a Direct ELISA and an HPLC Method for the Determination of Histamine in Wines. *Food Res. Int.* **2005**, *38*, 387–394. [CrossRef]
9. Marcobal, Á.; Martín-Álvarez, P.J.; Polo, M.C.; Muñoz, R.; Moreno-Arribas, M.V. Formation of Biogenic Amines throughout the Industrial Manufacture of Red Wine. *J. Food Prot.* **2006**, *69*, 397–404. [CrossRef] [PubMed]
10. Guo, Y.Y.; Yang, Y.P.; Peng, Q.; Han, Y. Biogenic Amines in Wine: A Review. *Int. J. Food Sci. Technol.* **2015**, *50*, 1523–1532. [CrossRef]
11. Zhang, Y.J.; Zhang, Y.; Zhou, Y.; Li, G.H.; Yang, W.Z.; Feng, X.S. A review of pretreatment and analytical methods of biogenic amines in food and biological samples since 2010. *J. Chromatogr. A* **2019**, *1605*, 360361. [CrossRef] [PubMed]
12. Ordóñez, J.L.; Troncoso, A.M.; García-Parrilla, M.D.C.; Callejón, R.M. Recent trends in the determination of biogenic amines in fermented beverages—A review. *Anal. Chim. Acta* **2016**, *939*, 10–25. [CrossRef]
13. Sentellas, S.; Núñez, O.; Saurina, J. Recent Advances in the Determination of Biogenic Amines in Food Samples by (U)HPLC. *J. Agric. Food Chem.* **2016**, *64*, 7667–7678. [CrossRef] [PubMed]
14. Hernández-Cassou, S.; Saurina, J. Derivatization strategies for the determination of biogenic amines in wines by chromatographic and electrophoretic techniques. *J. Chromatogr. B Anal. Technol. Biomed. Life Sci.* **2011**, *879*, 1270–1281. [CrossRef]
15. AECAVA. Available online: https://aecava.com/en/ (accessed on 4 August 2021).
16. Izquierdo-Llopart, A.; Saurina, J. Liquid chromatographic approach for the discrimination and classification of cava samples based on the phenolic composition using chemometric methods. *Beverages* **2020**, *6*, 54. [CrossRef]
17. Izquierdo-Llopart, A.; Carretero, A.; Saurina, J. Organic Acid Profiling by Liquid Chromatography for the Characterization of Base Vines and Sparkling Wines. *Food Anal. Methods* **2020**, *13*, 1852–1866. [CrossRef]
18. Izquierdo-Llopart, A.; Saurina, J. Characterization of sparkling wines according to polyphenolic profiles obtained by HPLC-UV/Vis and principal component analysis. *Foods* **2019**, *8*, 22. [CrossRef] [PubMed]
19. Pérez-Magariño, S.; Ortega-Heras, M.; Martínez-Lapuente, L.; Guadalupe, Z.; Ayestarán, B. Multivariate analysis for the differentiation of sparkling wines elaborated from autochthonous Spanish grape varieties: Volatile compounds, amino acids and biogenic amines. *Eur. Food Res. Technol.* **2013**, *236*, 827–841. [CrossRef]
20. Ivit, N.N.; Loira, I.; Morata, A.; Benito, S.; Palomero, F.; Suárez-Lepe, J.A. Making natural sparkling wines with non-Saccharomyces yeasts. *Eur. Food Res. Technol.* **2018**, *244*, 925–935. [CrossRef]
21. Martínez-Lapuente, L.; Apolinar-Valiente, R.; Guadalupe, Z.; Ayestarán, B.; Pérez-Magariño, S.; Williams, P.; Doco, T. Polysaccharides, oligosaccharides and nitrogenous compounds change during the ageing of Tempranillo and Verdejo sparkling wines. *J. Sci. Food Agric.* **2018**, *98*, 291–303. [CrossRef] [PubMed]

22. Martínez-Lapuente, L.; Guadalupe, Z.; Ayestarán, B.; Pérez-Magariño, S. Role of major wine constituents in the foam properties of white and rosé sparkling wines. *Food Chem.* **2015**, *174*, 330–338. [CrossRef] [PubMed]
23. Pérez-Magariño, S.; Martínez-Lapuente, L.; Bueno-Herrera, M.; Ortega-Heras, M.; Guadalupe, Z.; Ayestarán, B. Use of Commercial Dry Yeast Products Rich in Mannoproteins for White and Rosé Sparkling Wine Elaboration. *J. Agric. Food Chem.* **2015**, *63*, 5670–5681. [CrossRef] [PubMed]
24. Saurina, J. Characterization of wines using compositional profiles and chemometrics. *TrAC Trends Anal. Chem.* **2010**, *29*, 234–245. [CrossRef]
25. Ivit, N.N.; Kemp, B. The impact of non-Saccharomyces yeast on traditional method sparkling wine. *Fermentation* **2018**, *4*, 73. [CrossRef]
26. Restuccia, D.; Loizzo, M.R.; Spizzirri, U.G. Accumulation of biogenic amines in wine: Role of alcoholic and malolactic fermentation. *Fermentation* **2018**, *4*, 6. [CrossRef]
27. García-Villar, N.; Hernández-Cassou, S.; Saurina, J. Characterization of wines through the biogenic amine contents using chromatographic techniques and chemometric data analysis. *J. Agric. Food Chem.* **2007**, *55*, 7453–7461. [CrossRef] [PubMed]
28. Dugo, G.; Vilasi, F.; La Torre, G.L.; Pellicanò, T.M. Reverse phase HPLC/DAD determination of biogenic amines as dansyl derivatives in experimental red wines. *Food Chem.* **2006**, *95*, 672–676. [CrossRef]
29. Proestos, C.; Loukatos, P.; Komaitis, M. Determination of biogenic amines in wines by HPLC with precolumn dansylation and fluorimetric detection. *Food Chem.* **2008**, *106*, 1218–1224. [CrossRef]
30. Pineda, A.; Carrasco, J.; Peña-Farfal, C.; Henríquez-Aedo, K.; Aranda, M. Preliminary evaluation of biogenic amines content in Chilean young varietal wines by HPLC. *Food Control* **2012**, *23*, 251–257. [CrossRef]
31. Tuberoso, C.I.G.; Congiu, F.; Serreli, G.; Mameli, S. Determination of dansylated amino acids and biogenic amines in Cannonau and Vermentino wines by HPLC-FLD. *Food Chem.* **2015**, *175*, 29–35. [CrossRef]
32. Massart, D.L.; Vandeginste, B.G.M.; Buydens, L.M.C.; de Jong, S.; Lewi, P.J.; Smeyers-Verbeke, J. *Handbook of Chemometrics and Qualimetrics*; Elsevier: Amsterdam, The Netherlands, 1997.

 *fermentation*

Article

Sensory Profile of Kombucha Brewed with New Zealand Ingredients by Focus Group and Word Clouds

Hazel Alderson, Chang Liu, Annu Mehta, Hinal Suresh Gala, Natalia Rutendo Mazive, Yuzheng Chen, Yuwei Zhang, Shichang Wang and Luca Serventi *

Department of Wine, Food and Molecular Biosciences, Lincoln University, Lincoln 7647, New Zealand; Hazel.Alderson@lincolnuni.ac.nz (H.A.); Yvonne.Liu@lincolnuni.ac.nz (C.L.); Annu.Mehta@lincolnuni.ac.nz (A.M.); HinalSuresh.Gala@lincolnuni.ac.nz (H.S.G.); Natalia.Mazive@lincolnuni.ac.nz (N.R.M.); Yuzheng.Chen@lincolnuni.ac.nz (Y.C.); Yuwei.Zhang@lincolnuni.ac.nz (Y.Z.); kay.wang@lincolnuni.ac.nz (S.W.)

* Correspondence: Luca.Serventi@lincoln.ac.nz; Tel.: +64-3423-0860

Abstract: Kombucha is a yeast and bacterially fermented tea that is often described as having an acetic, fruity and sour flavour. There is a particular lack of sensory research around the use of Kombucha with additional ingredients such as those from the pepper family, or with hops. The goal of this project was to obtain a sensory profile of Kombucha beverages with a range of different ingredients, particularly of a novel Kombucha made with only Kawakawa (*Piper excelsum*) leaves. Other samples included hops and black pepper. Instrumental data were collected for all the Kombucha samples, and a sensory focus group of eight semi-trained panellists were set up to create a sensory profile of four products. Commercially available Kombucha, along with reference training samples were used to train the panel. Kawakawa Kombucha was found to be the sourest of the four samples and was described as having the bitterest aftertaste. The instrumental results showed that the Kawakawa Kombucha had the highest titratable acidity (1.55 vs. 1.21–1.42 mL) as well as the highest alcohol percentage (0.40 vs. 0.15–0.30%). The hops sample had the highest pH (3.72 vs. 3.49–3.54), with the lowest titratable acidity (1.21), and, from a basic poll, was the most liked of the samples. Each Kombucha had its own unique set of sensory descriptors with particular emphasis on the Kawakawa product, having unique mouthfeel descriptors as a result of some of the compounds found in Kawakawa. This research has led to a few areas that could be further studied, such as the characteristics of the Piperaceae family under fermentation and the different effects or the foaminess of the Kawakawa Kombucha, which is not fully explained.

Keywords: black pepper; focus group; hops; Kawakawa

Citation: Alderson, H.; Liu, C.; Mehta, A.; Gala, H.S.; Mazive, N.R.; Chen, Y.; Zhang, Y.; Wang, S.; Serventi, L. Sensory Profile of Kombucha Brewed with New Zealand Ingredients by Focus Group and Word Clouds. *Fermentation* **2021**, *7*, 100. https://doi.org/10.3390/fermentation7030100

Academic Editors: Claudia Gonzalez Viejo and Sigfredo Fuentes

Received: 10 June 2021
Accepted: 22 June 2021
Published: 23 June 2021

1. Introduction

Consumers are pushing a phenomenal surge for the potentially probiotic beverage known as Kombucha. The global sales volume reached USD 1779 million in 2020, with a compound annual growth rate (CAGR) projected at an astonishing 28.9% from 2020 to 2026, according to 360ResearchReports [1]. Kombucha is a sweetened tea that is fermented using a Symbiotic Culture of Bacteria and Yeast (or SCOBY). Examples of SCOBY are a combination of numerous lactic acid bacteria (*Acetobacter*, Brettanomyces, and *Gluconacetobacter*) and yeast (*Saccharomyces*, *Zygosaccharomyces*). These microorganisms form a biofilm-like structure at the top of the fermenting vessel. Kombucha is said to be well accepted in a sensory way by providing a unique drink that has an element of sweetness, while giving fruitiness and sour and acidic flavours. Ivanišová et al. [2] found that their Kombucha had a slightly better sensory preference overall compared to the sweetened black tea control, as well as improvements in taste and flavour intensity. Common flavours that are described by sensory panels seem to include lemon, vinegar, sour, and yeast flavours and aromas [3,4]. Mouthfeel is another sensory element to be considered. Gramza-Michałowska and collaborators [3] made Kombucha with different tea types (e.g., green, black, or yellow) and found that black tea had the poorest sensory acceptability, perhaps due to its lack of clarity and smoothness/mouthfeel in comparison to the other three tea types used, which were white, yellow, and green teas.

Currently, there is a lack of research around the sensory profile of Kombucha using different ingredients, such as the New Zealand native Kawakawa, or even simple Kombucha [5]. This has driven the need for research into the instrumental quality and flavour profiles of Kombucha made with New Zealand ingredients or other additives.

Black pepper (*Piper nigrum* L.) and the New Zealand plant Kawakawa (*Piper excelsum*) are both of the Piperaceae or pepper family and, therefore, have a unique spiciness. In New Zealand, Kawakawa has a history of being used as a medicinal plant for a range of small ailments [6]. In comparison, black pepper is a commonly found seasoning that also has a range of health benefits attributed to it, such as improvement to digestion and the ability to cure colds [7].

Hops are the flowers of the *Humulus lupulus* L. vine that are most commonly known for their role in beer production and flavour. Hops have concentrations of alpha acids that give bitterness to beer when they are boiled and isomerised. Other components of note are essential oils, resins, and phenolics that can be imparted during a process called dry hopping or during boiling [8]. Beta acids, found in hops, may be an issue for SCOBY growth since they have antibacterial properties. Hops can impart citrus flavours when dry-hopped (added after boiling), which may or may not have similar effects in Kombucha [9]. The hops variety used for this experiment was Riwaka™, which has a moderately low level of alpha acids and is described as having grapefruit and passionfruit aromas.

Word clouds, also known as tag clouds, are a valid visual tool to highlight the results of consumer studies such as focus groups. Word clouds are applied to qualitative analyses. The principle is to display those words that occur more often in a given description. The occurrence and size of each word reveals the relevance of each specific result, allowing observers to quickly notice differences among samples and the reasons behind them.

The hypothesis of this research is that each ingredient will produce a different response in the panellists, resulting in unique descriptors for each type of Kombucha. These may include acetic tastes, cloudiness or clarity, etc. The different ingredient additions may impact the growth of the microorganisms within, thereby impacting the instrumental quality of the Kombucha.

The aims of this research include testing the sensory characteristics of the Kombucha through the use of a semi-trained focus group consisting of at least eight people. This panel will test four samples of Kombucha, including a control Kombucha. The panellists will look at the various sensory qualities of the Kombucha, including areas such as clarity, bubble and foam formation, and colour. The second aim includes collecting instrumental data, including °Brix, pH, titratable acidity, and alcohol percentage. Instrumental data can be useful to support the discussion of sensory profiles, providing objective information on sensory-affecting parameters such as acidity and sweetness. Word clouds were used to visually highlight the results of the focus group, providing a novel, clear tool to describe food quality.

2. Materials and Methods

2.1. Materials

The Kombucha beverages were prepared using 1 L of boiling water, 70 g of white sugar (Pams, Auckland, New Zealand), 5 g of tea, 75 g of mother liquid, and 30 g of SCOBY (Get Cultured, Tauranga, New Zealand). The composition of this SCOBY can vary based on batches, but it usually comprises the following microorganisms: *Acetobacter aceti, A. intermedius, A. nitrogenifigens, A. pasteurianus, A. xylinum, Bacterium gluconicum, B. xylinum, Gluconacetobacter kombuchae, G. oxydans, Brettanomyces bruxellensis, B. claussenii, B. intermedius, Candida collecolusa, C. famata, C. guilliermondii, C. kefyr, C. obtusa, Koleckera apiculata, Mycoderma* sp., *Mycotorula* sp., *Pichia membranefaciens, Saccharomyces bisporus*, *S. cerevisiae*, *S. ludwegii*, *Saccharomycodes* sp., *Schizosaccharomyces pombe*, *Torulaspora delbrueckii*, *Zygosaccharomyces bailii*, *Z. kombuchaensis*, *Z. lentus*, and *Z. rouxii*, The adjuncts were 'Empire' cracked black pepper and Riwaka hops (brewshop.co.nz). The Kawakawa leaves used were from PureNature NZ (Auckland, New Zealand) and the black tea was from Bell Tea (Dunedin, New Zealand). The composition of Kawakawa dry leaves was determined in our laboratories, expressed in g/100 g of dry weight, and is as follows: protein, 17.1; minerals, 13.8; lipids, 8.41; soluble carbohydrates, 1.84; and insoluble carbohydrates, 48.2. The mineral profile was the following, expressed in mg/100 g dry weight: potassium, 3433; calcium, 1664; phosphorous, 603; magnesium, 557; sodium, 90.1; and iron, 11.9.

2.2. Kombucha Production

The SCOBY (30 g) and mother liquid (75 g) were from the same batch date, and the mother liquids were mixed, trying to maintain consistency through the four types. The boiling water, tea, and sugar were added to a pot, weighed, and kept boiling on a stove for 15 min, being stirred throughout. After boiling, boiling water was added to make up the initial weight. The tea was then filtered using a cheesecloth in a sieve before being placed in a pre-prepared ice bath to cool to approximately 22 °C. The tea was then transferred to a clean jar that had been sprayed with 70% ethanol, and the mother liquid and SCOBY were then added. The lid of the jar was placed on top but not screwed on in order to let any gas escape. The jars were transferred to a temperature-controlled room at 22 °C for seven days. The newly formed SCOBY as well as the original SCOBY were then removed, and any adjuncts added. The Kombucha was then kept in the fridge (+4 °C) overnight and in the morning was filtered using a sieve and cheesecloth.

2.3. pH Testing

The pH testing was performed using a Metrohm 730 Sample Changer Auto Titrator from MEP instruments, Newmarket (Auckland, New Zealand). The titrant used was 0.1 N NaOH from LabServ (Auckland, New Zealand), and this was administered using a Metrohm 702 SM Tritrino (Auckland, New Zealand). A sample size of 5 mL was used to perform the analysis. A standard pH meter was originally used, but this gave different results to the Auto Titrator; therefore, the Auto Titrator was chosen as the most accurate reading.

2.4. Titratable Acidity

Titratable acidity was performed using the Metrohm 730 Sample Changer Auto Titrator (Auckland, New Zealand). The titrant was 0.1 N NaOH from LabServ, and this was dispensed using a 702 SM Tritrino. A 5-millilitre sample was used. The corresponding software used was Tiamo 1.2.

2.5. Degrees Brix

The °Brix was obtained using a refractometer (Keg King, Springvale, Australia) calibrated with tap water. A drop of room temperature sample was placed on the testing panel, read, and was then cleaned from the panel using soft tissues. During preliminary testing, the °Brix test was performed daily, while in the final brew, °Brix was only obtained from day 0 and day 8, after the adjuncts were added.

2.6. Alcohol Testing

Alcohol testing was performed using a Dujardin-Salleron ebulliometer (Paris, France) that was calibrated using tap water. The boiling point of the tap water set the ebulliometer scale. A measure of the sample was added based on the ebulliometer equipment, and the temperature was read when the sample reached the boiling point. The boiling point was correlated to an equivalent alcohol percentage using the scale. The ebulliometer was cleaned between samples and was rinsed with the new sample.

2.7. Microbial Enumeration

Microbial enumeration was performed to establish the quantity of lactobacilli (probiotic bacteria) present in the beverages. Two serial dilutions were performed to 10^{-2}. The MRS agar plates (Fort Richard, Auckland, New Zealand) were incubated for 48 h in an anaerobic environment with Thermo Scientific™ Oxoid AnaeroGen 2.5 L sachets (Auckland, New Zealand). Then, colony forming units (log CFU/mL) were counted [10].

2.8. Sensory Panel

Eight master's students volunteered to be part of a sensory focus group. They were selected based on the following three criteria: experience in food science and sensory analysis, habit of drinking fermented and non-fermented beverages, and a multicultural background. The students were of different nationalities, including Chinese, Indian, and Vietnamese. They were of young age (twenties), 6 were female and 2 were male. This selection represented the consumer target of Kombucha, being mostly young females according to a recent survey by Mintel in 2019 [11]. The participants were asked to sign a consent form that included a list of potential allergens that may be in the samples. Sensory descriptors for four different categories, namely appearance, aroma, flavour, and mouthfeel, were collected for three commercial Kombucha products during a 1-h training session. The commercial samples were a control, original Kombucha by Daily Organics (Matakana Village, New Zealand), a spiced Kombucha by Daily Organics called 'Winter' that included a chai tea and ginger blend, and a Ginger and Kawakawa Kombucha by Mauriora (Christchurch, New Zealand) that provided a Kawakawa flavour. The most significant descriptors were chosen to use as a reference for the Kombucha samples. A set of sensory reference samples were created to use for training; these included carbonation, bitterness, sourness, sweetness, and colour. The carbonation training was performed using sparkling water, the bitterness using a hop pellet dissolved in roughly 25 mL of water, the sourness using a lemon slice, the sweetness using a 7% sugar solution, and the colour using black tea. With the reference samples as a guide, the semi-trained panel chose and agreed upon different sensory descriptors for each Kombucha sample. One sample of each product was provided in the sessions. In the second session, samples were labelled with 3-random-digit codes. Samples were presented in the following order: control, hops, Kawakawa, black pepper. This session lasted for about one hour. A 10-minute break was allowed between the two sessions (training, samples) to avoid sensory fatigue. Water was provided as a palate cleanser. All data were recorded on a whiteboard to ensure that the panellists could change their answers and could compare to previous samples. The resulting data were tabulated for discussion. The study was approved by the Ethics Committee of Lincoln University (protocol code 2020-60, approved on 11 December 2020).

2.9. Statistical Analyses

Instrumental measurements were taken in duplicate and an average was calculated with Excel 2016. Minitab 18 software was used to calculate statistical significance using a one-way analysis of variance (ANOVA) and a post hoc Tukey test, with an α value of 0.05. An online word cloud generator (www.wordclouds.com by Zigomatic, Vianen, The Netherlands, accessed on 3 May 2021 was used to summarise the results of the focus group and display the most agreed attributes for each sample.

3. Results and Discussion

3.1. Instrumental Results

The °Brix results for the four different Kombucha samples, both before and after fermentation, showed fairly similar results, with only one odd result for day zero for the hops sample, which came out at 6.8 °Brix. This result is odd since the control, black pepper, and hops all should have started the same since, in essence, they were all controls until day seven when their extra ingredients were added (Table 1).

Table 1. Parameters used to monitor the fermentation performance of different Kombucha samples. TA stands for titratable acidity. Different superscript letters refer to statistically significant samples (α 0.05).

		Monitoring of Fermentation			
Day	**Tea Type**	**°Brix**	**pH**	**TA (mL)**	**Alcohol (%)**
1	Control	6.2 ± 0 [b]	3.90 ± 0.03 [b]	0.77 ± 0.07 [a]	
	Black Pepper	6.0 ± 0 [d]	3.82 ± 0.00 [c]	0.76 ± 0.07 [a]	
	Hops	6.8 ± 0 [a]	3.80 ± 0.01 [c]	0.79 ± 0.05 [a]	
	Kawakawa	6.1 ± 0 [c]	3.98 ± 0.01 [a]	0.67 ± 0.08 [a]	
8	Control	6.2 ± 0 [a]	3.54 ± 0.02 [b]	1.28 ± 0.05 [a]	0.15 ± 0.07 [c]
	Black Pepper	6.0 ± 0 [b]	3.49 ± 0.00 [b]	1.42 ± 0.03 [a]	0.30 ± 0.00 [ab]
	Hops	6.2 ± 0 [a]	3.72 ± 0.04 [a]	1.21 ± 0.18 [a]	0.20 ± 0.00 [bc]
	Kawakawa	6.0 ± 0 [b]	3.50 ± 0.04 [b]	1.55 ± 0.01 [a]	0.40 ± 0.00 [a]

The measurements of the day eight samples were taken after roughly a day of the adjuncts being added, i.e., the hops, and black pepper. The initial pH values for the four samples were all fairly similar and were within 0.2 pH of each other. This almost negligible difference was to be expected since the SCOBY/mother liquid batch for all the samples was the same. Three of the samples reached a fairly similar pH after fermentation, including Kawakawa, which, in preliminary results, tended to have a higher pH throughout the whole fermentation. The one interesting result from the pH measurements is the starting point of 3.8 for the hops only dropping to 3.7, which is an unexpected result. In beer, which has a somewhat similar fermentation process to Kombucha, the hops can take various components out of solution through the formation of complexes during hop filtration. One such complex that is very common is the protein–phenol complex, which may have also occurred in our hops Kombucha sample from proteins in/from the yeast and phenolic compounds in the hops. Phenolic compounds, such as tannic acid, are acidic, and therefore, when they fall out of a solution, they may increase the pH of the Kombucha and partially explain the results. Overall, our pH results are a little high compared to some other studies, where a pH of 3 was more expected; it may be that the Kombucha beverages required one to two more days to ferment [12,13].

The results for the titratable acidity (TA) for day zero and day eight are shown in Table 1. As can be seen from the table, all four samples had a very similar TA at the beginning, with no statistically significant differences (p value 0.448 for TA at Day 1), while deviations occurred after the completion of the fermentation. The first TA would likely only be from the mother liquid, which would have previously produced acids, mainly organic acids. However, as the fermentation progressed, the current microorganisms would have produced new acids, partially dependent on the growing conditions and whether they were favourable. The lactic acid bacteria are responsible for the production of acids, meaning that the samples with a higher TA had better growing conditions for bacteria to grow and produce acids. In this case, the highest TA was present in the Kawakawa Kombucha, which may have been caused by a favouring of all the bacteria or perhaps of one particular type; this requires some further investigation as there is no literature in this area. Regardless of the larger Kawakawa value, the TAs for all the samples were fairly similar (within 0.3 mL), which was not unexpected since three of the four recipes all started the same. The difference between the start and end of the fermentation was reasonably large, particularly for Kawakawa, which had a change of approximately 0.9 mL for only a 5-millilitre sample, suggesting that the bacteria produced a significant amount of acid during fermentation. The Kombucha with the smallest change in TA was the hops sample that had a 0.4 mL increase for a 5-millilitre sample. Nonetheless, these differences were not significant (p value 0.070 for TA at Day 8), thus having minimal implications on the quality of the beverages. Some of these results can be explained with the help of microbial data. The experiment measured colony forming units (CFUs) from day zero and day eight. The results from day zero ranged from 7.18 log CFU/mL for Kawakawa to 7.40 log CFU/mL for hops, these initial differences may be related to the different distribution of bacteria in the SCOBY. The results for day eight are much more useful, where there are distinct increases and decreases in the bacterial levels. The amount that the control increased, the hops Kombucha decreased as much (7.34 log CFU/mL); this shows that the bacteria, including the acetic acid bacteria, significantly drop after the hop addition, which may partially explain why the pH of the hops Kombucha was higher than the rest, as there are not any acetic acid bacteria to convert sugar or ethanol to acetic acid. The bacterial count for both Kawakawa and black pepper was higher than the increase in the control. The Kawakawa had a 7.47 log CFU/m increase, and black pepper had a 7.67 log CFU/mL increase, which was over double the control. This may explain why both the black pepper and Kawakawa were low in pH and high in TA. These differences in results must be a result of only one day, as the ingredients were added on day seven, right before the Kombucha beverages were refrigerated, which is a quick change. The research around the Piperaceae family is mixed in conclusions, as one article describes many species as having antibacterial actions, particularly against Gram-negative bacteria, which include acetic acid bacteria, while another article describes black pepper as having some prebiotic qualities [14]. This area could perhaps use some deeper research to understand the underlying factors.

The alcohol percentages for all four samples were well below 1%, with the highest being Kawakawa at an average of 0.4% and the lowest being the control at an average of 0.15%. These results suggest that the yeast in the Kawakawa Kombucha were particularly active compared to the other samples, or it may be that particular yeast strains that tend to produce significantly higher alcohol levels are favoured by the conditions that the Kawakawa Kombucha gives. In comparison to the current literature, our control alcohol percentage for eight days of fermentation is reasonably low, though this is difficult to discern as many studies measure ethanol only through chromatography methods and even then, the results cover a large range of ethanol contents, such as nearly 0.3 to 0.6 mg/mL [15,16]. However, in comparison to some commercial ethanol contents, our ethanol contents are very low [16]. This may be due to different fermentation parameters, such as temperature ranges, as higher temperatures promote yeast growth [13]. This is interesting to compare to the previously discussed bacterial levels on day eight, as they suggested that the bacteria were favoured in all the samples but the hops Kombucha. Therefore, one would almost expect that the fermented hop beverage might be the highest in alcohol. However, this is not the case; therefore, perhaps the conditions provided by black pepper and Kawakawa promote the growth of both bacteria and yeast.

No samples were taken on day seven, meaning that comparisons before the flavourings were added could not be made; this would have been a useful comparison, and in future should be added to the design. This would allow for the comparison of the samples without the flavourings, as well as the impact that the flavourings have on the instrumental results.

3.2. *Training of Focus Group with Commercial Samples and Standards*

The descriptors that were agreed on during the focus group training are shown in Table 2. A range of descriptors were of particular note as well as potential explanations to correspond. One of the main descriptors for flavour was sourness with each Kombucha having a sour taste either above or below the reference of lemon and having a number of agreeing panellists. Sour and acetic flavours are some of the main descriptors for plain Kombucha, along with fruitiness and freshness [3]. This can be seen in the control descriptors of sour, tomato and honey aromas, and sour and Hawthorn flavours. The sour descriptors align with the instrumental results that showed a pH of 3.5. The fruity aromas and flavours are likely due to esters that are formed by the yeast. For example, it is possible that the phenyl ethyl acetate ester is responsible for the honey aroma [17,18].

Table 2. The training table for the focus group using three commercially available Kombucha beverages. The words in bold refer to those used as base descriptors for the final Kombucha samples.

Commercial Sample Sensory Training			
Sample	**1. Mauriora 'Ginger & Kawakawa'**	**2. Daily Organics 'Original'**	**3. Daily Organics 'Winter'**
Appearance	**Clear**	Hazy	Darker than 2
	Orange/**yellow**	**Foamy**	Bright orange/**yellow**
	Piece of **SCOBY**	Light colour	Hazier than 2
	Less foamy	Pale ale colour	Foamy
Aroma	**Ginger**	**Fermented** (beer)	**Sour** (more than 2)
	Tea	**Apple** cider	Less 'fermented' than 2
	Ginger beer	Malty	Fermented juice
	Spice	Not tea	Tea
	Cinnamon	**Pear**	Not like tea
	Sweet/sour (in between)		Less fruity, more smoky
			Rotten apple (in a nice way)
Flavour and Taste	**Sour**!	**Acetic** acid	**Bitter**
	Bad aftertaste (bitter)	Bitter aftertaste	Sour
	Astringent	Beer	Pear
	Acetic acid	Slightly sweet	Slightly sweet
	Metallic	**Alcohol**	Fruity
	Bread aftertaste		Acrid
Mouthfeel	Light	**Carbonated**	Carbonated
	Not very rich/thick		Not as fizzy

No astringency is mentioned for the control mouthfeel as it is for all other samples, this aligns with Tran and collaborators [13], who say that astringency is never mentioned as a Kombucha descriptor. This means that for all the other samples, the astringency likely comes from the additional ingredients.

3.3. Descriptive Analysis of Samples by Focus Group

The control Kombucha was described as orange and sour (in aroma and taste) (Table 3).

A result that is of particular interest is that seven of the eight panellists agreed that the black pepper sample was less sour than the lemon reference (Table 4), yet the TA was the second highest and the pH was one of the lowest, meaning that, theoretically, it should sit in a similar place to Kawakawa, yet this was not the case. This is further expounded by the agreeance of all panellists that the black pepper was sweeter than the 7% sugar reference. The sweetness result may be due to a few different reasons, such as a difference in fermentation since two different SCOBYs were divided between the four types. The difference may have been due to slightly different conditions since each Kombucha was made individually rather than separated out from one big batch. Another possible reason is that components of black pepper can promote the breakdown of proteins into amino acids [19]. Amino acids can sometimes have sweet flavours and these amino acids may also reduce the perception of acidity, though this requires further research. There are also other components in the pepper that may improve the perception of sweetness, such as the essential oils and other flavour components, including esters.

Table 3. The results from the sensory focus group of eight panellists. The descriptors were based on the commercial Kombucha beverages, and the new descriptors were from the control Kombucha sample.

Sensory	Control			
	Descriptors		New Descriptors	
Appearance	Hazy	7	Thick	5
	Orange	8		
Aroma	Fermented	6	Tomato	8
	Sweet	3	Honey (flowery)	6
	Sour	8		
Flavour and Taste	Less sweet	8	Hawthorn (fresh, leafy)	3
	Sour	5		
	Alcohol	2		
Mouthfeel				

Table 4. The results from the sensory focus group of eight panellists. The descriptors were based on the commercial Kombucha beverages, and the new descriptors were from the black pepper Kombucha sample.

Sensory	Black Pepper			
	Descriptors		New Descriptors	
Appearance	Hazy	7		
	Orange	7		
Aroma	Spice	3	Tomato	1
	Tea	4		
	Sweet	1		
	Fermented	6		
	Sour	5		
Flavour and Taste	Sweet	8	Apple	6
	Bitter	1		
	Less sour	7		
	Metallic	1		
	Acetic	3		
Mouthfeel	Astringent	3		

Apart from the interesting comparisons made above, there are very few other descriptors of note, except that there were three panellists that described a 'spice' aroma, which would be expected as a result of the black pepper. However, it would have been expected that the black pepper would produce aromas such as 'flowery', 'green', 'intense', or even 'peppery' [20]. This was not the case, however; therefore, it is likely that the components that cause these descriptors were broken down by microorganisms or other chemical processes during the cooling and refrigeration step.

The hops Kombucha (Table 5) had fewer descriptors than the other three samples. The whole panel found there to be hop aromas and flavours. Only three people described a sour flavour, which is to be expected given the instrumental results of a high pH and a low titratable acidity compared to the other Kombucha products. In beer, hops are added before or during boiling to isomerise the alpha acids to give a bitter flavour. Hops may also be added during or after fermentation to provide aromas—called dry hopping. The bitter flavour that two of the panellist's mentioned may be related to the original Kombucha, the actual hop bitterness, or even the effect of the hop aromas on perceived bitterness [21]. The hops are also likely to be responsible for the 'leafy' aroma. The astringency could also be a result of the hops, as polyphenols from the hops form complexes with the proteins in saliva, giving a feeling of astringency [22].

Table 5. The results from the sensory focus group of eight panellists. The descriptors were based on the commercial Kombucha beverages, and the new descriptors were from the hops Kombucha sample.

Sensory	Hops			
	Descriptors		New Descriptors	
Appearance	Orange	8		
	Hazy	8		
Aroma	Tea	4	Leafy	3
	Sweet	3	Hops	8
			Grape	1
Flavour and Taste	Less sweet	2	Hops	8
	Bitter aftertaste	2		
	Sour	3		
Mouthfeel	Astringent	4		

The Kombucha that was agreed upon as being the sourest was Kawakawa (Table 6), and this corresponds well to both the pH and the TA of the Kombucha, which were comparatively low and high, respectively. An interesting descriptor to note is the alcohol descriptor for flavour, which was mentioned by one panellist. The only other mention of an alcohol flavour was for the control, where two panellists agreed. This is interesting, as the control had the lowest alcohol percentage, and the Kawakawa had the highest. Tran and collaborators [13] mentioned that the flavour threshold for ethanol is around 1–2; therefore, it is likely that the panellists were describing another flavour that is similar to ethanol, as the Kawakawa contained only 0.4% alcohol, though it may have played a slight role in the bitter flavour described. Many of the flavour and mouthfeel characteristics could be explained by the presence of compounds such as piperidides, isobutyl amides, and esters [23]. Furthermore, compounds of essential oils such as myristicin might contribute to the strong aroma and flavour found in the Kawakawa Kombucha. It was suggested that essential oils from plant secondary metabolites have a wide application in food flavouring and in fragrance industries [24].

These different compounds are found in Kawakawa and are described as being hot, astringent, bitter, tingly, dry, irritating, numbing, and cooling. The descriptors mentioned correspond well to the spice aroma, bitter taste and aftertaste, as well as the astringency and minty/cooling feeling described for mouthfeel. One other descriptor of note that is different from the control is the foamy appearance descriptor. There is little Kombucha research in this area, but the theory should be somewhat similar to beer. In beer, foam stability is often the result of protein presence as well as carbonation [25]. Therefore, in this case, it may be that there were more proteins present, perhaps from the yeast. This aligns with the higher alcohol percentage found in the Kawakawa Kombucha. The foaminess may also be a result of some other compound from the Kawakawa; this could use more research.

Table 6. The results from the sensory focus group of eight panellists. The descriptors were based on the commercial Kombucha beverages, and the new descriptors were from the Kawakawa Kombucha sample.

Sensory	Kawakawa			
	Descriptors		New Descriptors	
Appearance	Foamy	8		
	Hazy	8		
Aroma	Spice	8	Leafy	2
	Tea	4	Dried orange peel	2
	Less sweet (ref)	1		
	Sour	8		
	Fermented	3		
Flavour and Taste	Bitter	2	Hops	7
	Sour	7	Preserved plum	2
	Bitter aftertaste	5	Sweet and bitter	6
	Alcohol	1	Spice	7
Mouthfeel	Astringent	7	Mint (cooling feeling)	6

3.4. Word Clouds of Samples

The focus group has shown differences in the sensory profiles of the Kombucha samples tested. Each recipe was characterized by distinctive attributes for appearance, aroma, flavour, and mouthfeel. While Tables 3–6 might provide an overwhelming amount of information, only some of these differences resulted in a large change in perception of the overall quality. Considering that eight panellists participated in the session, only the attributes agreed to by at least seven panellists were considered representatives of high consensus. It must be noted that word clouds do not represent a sensory profile, but rather a depiction of sensory results. Therefore, scientific conclusions were drawn based on the results listed in Tables 3–6. Word clouds were a useful tool to display such findings.

With this logic in mind, the control Kombucha was hazy and orange in look, had a sour and tomato aroma, and was less sweet than the others in flavour. Nonetheless, the word cloud (Figure 1) highlighted hazy as the main attribute. This can be explained by the occurrence of this attribute in all the samples, which is likely due to the high biomass produced upon fermentation. The black pepper Kombucha's sensory profile was mostly similar to the control (Figure 1). Tables 3 and 4 indicated that the black pepper Kombucha was less sour and sweeter than the control, with a spice aroma, but no unanimous agreement was reached. This was likely due to small differences; consequently, the word cloud displayed this lack of significant difference by producing a similar representation as that for the control. Therefore, it can be deduced that these two products did not differ in sensory quality.

Alternatively, the hops and Kawakawa recipes were characterized by one key difference: an orange and less hazy appearance (hops), a foamy appearance (Kawakawa) (Figure 1). Once again, the word clouds highlighted significant differences among the samples. Interestingly, Table 5 showed a unanimous agreement on the hops Kombucha for hazy, but this attribute was only a minor component of the word cloud. This difference can be ascribed by the phenolic content of hops (orange colour) as well as a comparable amount of biomass produced (hazy appearance), resulting in a perception of a less hazy product. Finally, the Kawakawa Kombucha was described as both hazy and foamy (Table 6) due to biomass production and protein-induced foam development. Spicy and sour aromas were agreed on, as well as astringent mouthfeel, but all of these attributes were overshadowed by a visible difference in the foam volume (Figure 1).

Figure 1. Word clouds originated from the focus group, showing the most agreed attributes for the Kombucha samples.

4. Conclusions

The hypothesis was that each Kombucha would have a unique sensory profile depending on the different ingredients added. This turned out to be true, with distinct differences for the Kawakawa and hops Kombucha beverages, and subtler differences for the black pepper Kombucha. The differences were also shown in the instrumental analyses, with minor differences in the pH and alcohol levels among the fermented products, some of which were picked up in the sensory profiling. For example, the hops Kombucha was described as sour in taste by fewer panellists (3 vs. 5–7 people). This agreed with the significantly higher pH than the other beverages (3.72 vs. 3.49–3.54). In addition, an alcohol flavour was noted only for the Kawakawa Kombucha, which, in fact, was significantly more alcoholic than the other beverages (0.40 vs. 0.15–0.30%).

This experiment provided a range of information and potential explanations as to why the results occurred, though there are many areas that require more research for a surer knowledge. Some of these areas include the discrepancy between the professed antimicrobial, yet prebiotic nature of the *Piperaceae* family, particularly for Kawakawa, that have not been investigated in depth. The sweetness that black pepper gives could also benefit from a confirmation as to its origin. Finally, the foaminess of the Kawakawa Kombucha could use with further investigation, as there would likely be fairly low protein levels; therefore, a different compound may have a part to play.

What the focus group could not display was the extent of the differences. Word clouds provided this information and put it into context, allowing a visual description of the sensory profiles. Focus groups provide detailed information of scientific significance, while word clouds offer an easily readable overall view. An integration of both methods is a comprehensive approach to descriptive sensory analysis.

Author Contributions: Conceptualization, L.S.; the training session and the focus group—planning, organization, and management, C.L.; methodology planning, L.S. (overall), Y.Z. (fermentation), and C.L. (sensory analysis); software use, L.S. and H.S.G.; validation, L.S. and A.M.; formal analysis, L.S.; investigation, H.A., N.R.M., and Y.C.; resources, L.S.; data curation, L.S. and H.S.G.; writing—original draft preparation, H.A., S.W., and L.S.; writing—review and editing, L.S. and A.M.; visualization, L.S.; supervision, L.S.; project administration, L.S.; funding acquisition, L.S. All authors have read and agreed to the published version of the manuscript.

Funding: This research was funded by Department of Wine, Food, and Molecular Biosciences, Faculty of Agriculture and Life Sciences, Lincoln University, New Zealand. The budget was allocated to teaching the course named "FOOD399—Research Placement".

Institutional Review Board Statement: The study was conducted according to the guidelines of the Ethics Committee of Lincoln University for sensory analysis (Application No: 2020-35, Approved on 14 August 2020).

Informed Consent Statement: Informed consent was obtained from all subjects involved in the study.

Acknowledgments: Authors would like to thank Shiyun Li for her preliminary work on Kawakawa Kombucha and Jamel Elizabeth Barber for her study of black pepper as an aromatic compound in fermented beverages. In addition, the authors would like to acknowledge the technical work of Roger Cresswell (protein analysis of Kawakawa dry leaves) and Rosy Tung (dry matter, ash, lipid, and soluble carbohydrates of dry Kawakawa leaves). Finally, the authors would like to thank Leo Vanhanen for his help with the instrumental analysis of pH and titratable acidity.

Conflicts of Interest: The authors declare no conflict of interest.

References

1. 360 Research Reports. Available online: https://www.360researchreports.com/-global-kombucha-market-17759175 (accessed on 2 June 2021).
2. Ivanišová, E.; Meňhartová, K.; Terentjeva, M.; Harangozo, L'.; Kántor, A.; Kačániová, M. The evaluation of chemical, antioxidant, antimicrobial and sensory properties of kombucha tea beverage. *J. Food Sci. Technol.* **2019**, *57*, 1840–1846. [CrossRef]
3. Gramza-Michałowska, A.; Kulczyński, B.; Xindi, Y.; Gumienna, M. Research on the effect of culture time on the kombucha tea beverage's antiradical capacity and sensory value. *Acta Sci. Pol. Technol. Aliment.* **2016**, *15*, 447–457. [CrossRef]
4. Neffe-Skocińska, K.; Sionek, B.; Ścibisz, I.; Kołożyn-Krajewska, D. Acid contents and the effect of fermentation condition of Kombucha tea beverages on physicochemical, microbiological and sensory properties. *CYTA J. Food* **2017**, *15*, 601–607. [CrossRef]
5. Kim, J.; Adhikari, K. Current Trends in Kombucha: Marketing Perspectives and the Need for Improved Sensory Research. *Beverages* **2020**, *6*, 15. [CrossRef]
6. Butts, C.A.; van Klink, J.W.; Joyce, N.I.; Paturi, G.; Hedderley, D.I.; Martell, S.; Harvey, D. Composition and safety evaluation of tea from New Zealand kawakawa (*Piper excelsum*). *J. Ethnopharmacol.* **2019**, *232*, 110–118. [CrossRef]
7. Gülçin, İ. The antioxidant and radical scavenging activities of black pepper (*Piper nigrum*) seeds. *Int. J. Food Sci. Nutr.* **2009**, *56*, 491–499. [CrossRef]
8. Teghtmeyer, S. Hops. *J. Agric. Food Inf.* **2018**, *19*, 9–20. [CrossRef]
9. Lafontaine, S.R.; Shellhammer, T.H. Impact of static dry-hopping rate on the sensory and analytical profiles of beer. *J. Inst. Brew.* **2018**, *124*, 434–442. [CrossRef]
10. Dave, R.I.; Shah, N.P. Evaluation of media for selective enumeration of *Streptococcus thermophilus*, *Lactobacillus delbrueckii* ssp. *bulgaricus*, *Lactobacillus acidophilus*, and *bifidobacteria*. *J. Dairy Sci.* **1996**, *79*, 1529–1536. [CrossRef]
11. Mintel. Tea and RTD Teas—US—August 2019. Reasons for Drinking Kombucha. 2019. Available online: www.reports.mintel.com (accessed on 8 June 2021).
12. Goh, W.N.; Rosma, A.; Kaur, B.; Fazilah, A.; Karim, A.A.; Rajeev, B. Fermentation of black tea broth (Kombucha): I. Effects of sucrose concentration and fermentation time on the yield of microbial cellulose. *Int. Food Res. J.* **2012**, *19*, 109.
13. Tran, T.; Grandvalet, C.; Verdier, F.; Martin, A.; Alexandre, H.; Tourdot-Maréchal, R. Microbiological and technological parameters impacting the chemical composition and sensory quality of kombucha. *Compr. Rev. Food Sci. Food Saf.* **2020**, *19*, 2050–2070. [CrossRef] [PubMed]

14. Cvetković, D.; Markov, S.; Djurić, M.; Savić, D.; Velićanski, A. Specific interfacial area as a key variable in scaling-up Kombucha fermentation. *J. Food Eng.* **2008**, *85*, 387–392. [CrossRef]
15. Perigo, C.V.; Torres, R.B.; Bernacci, L.C.; Guimarães, E.F.; Haber, L.L.; Facanali, R.; Vieira, M.A.R.; Quecini, V.; Marques, M.O.M. The chemical composition and antibacterial activity of eleven *Piper* species from distinct rainforest areas in Southeastern Brazil. *Ind. Crop. Prod.* **2016**, *94*, 528–539. [CrossRef]
16. Narendra Babu, K.; Hemalatha, R.; Satyanarayana, U.; Shujauddin, M.; Himaja, N.; Bhaskarachary, K.; Dinesh Kumar, B. Phytochemicals, polyphenols, prebiotic effect of *Ocimum sanctum*, *Zingiber officinale*, *Piper nigrum* extracts. *J. Herb. Med.* **2018**, *13*, 42–51. [CrossRef]
17. Wang, S.; Zhang, L.; Qi, L.; Liang, H.; Lin, X.; Li, S.; Yu, C.; Ji, C. Effect of synthetic microbial community on nutraceutical and sensory qualities of kombucha. *Int. J. Food Sci. Technol.* **2020**, *55*, 3327–3333. [CrossRef]
18. Pires, E.J.; Teixeira, J.A.; Brányik, T.; Vicente, A.A. Yeast: The soul of beer's aroma—A review of flavour-active esters and higher alcohols produced by the brewing yeast. *Appl. Microbiol. Biotechnol.* **2014**, *98*, 1937–1949. [CrossRef]
19. Meghwal, M.; Goswami, T.K. *Piper nigrum* and Piperine: An Update. *Phytother. Res.* **2013**, *27*, 1121–1130. [CrossRef]
20. Narasimhan, S.; Rajalakshmi, D.; Chand, N. Quality of powdered black pepper (*Piper nigrum* L.) during storage: II. Principal Components Analyses of GC and sensory profiles. *J. Food Qual.* **1992**, *15*, 67–83. [CrossRef]
21. Oladokun, O.; James, S.; Cowley, T.; Dehrmann, F.; Smart, K.; Hort, J.; Cook, D. Perceived bitterness character of beer in relation to hop variety and the impact of hop aroma. *Food Chem.* **2017**, *230*, 215–224. [CrossRef] [PubMed]
22. François, N.; Guyot-Declerck, C.; Hug, B.; Callemien, D.; Govaerts, B.; Collin, S. Beer astringency assessed by time—Intensity and quantitative descriptive analysis: Influence of pH and accelerated aging. *Food Qual. Prefer.* **2006**, *17*, 445–452. [CrossRef]
23. Obst, K.; Lieder, B.; Reichelt, K.V.; Backes, M.; Paetz, S.; Geißler, K.; Krammer, G.; Somoza, V.; Ley, J.P.; Engel, K.-H. Sensory active piperine analogues from *Macropiper excelsum* and their effects on intestinal nutrient uptake in Caco-2 cells. *Phytochemistry* **2017**, *135*, 181–190. [CrossRef] [PubMed]
24. Richardson, A.T. Hot new chemistry from a kiwi pepper tree. *Chem. N. Z.* **2015**, *79*, 91–93.
25. Steiner, E.; Gastl, M.; Becker, T. Protein changes during malting and brewing with focus on haze and foam formation: A review. *Eur. Food Res. Technol.* **2011**, *232*, 191–204. [CrossRef]

fermentation

MDPI

Article

Application of Augmented Reality in the Sensory Evaluation of Yogurts

Yanyu Dong, Chetan Sharma, Annu Mehta and Damir D. Torrico *

Centre of Excellence—Food for Future Consumers, Department of Wine, Food and Molecular Biosciences, Faculty of Agriculture and Life Sciences, Lincoln University, Lincoln 7647, New Zealand; Yanyu.Dong@lincolnuni.ac.nz (Y.D.); Chetan.Sharma@lincoln.ac.nz (C.S.); Annu.Mehta@lincolnuni.ac.nz (A.M.)
* Correspondence: Damir.Torrico@lincoln.ac.nz; Tel.: +64-3-423-0641

Abstract: Augmented reality (AR) applications in the food industry are considered innovative to enrich the interactions among consumers, food products, and context. The study aimed to investigate the effects of AR environments on the sensory responses of consumers towards different yogurts. AR HoloLens headsets were used to set up two AR environments: (1) AR coconut view (ARC) and (2) AR dairy view (ARD). Hedonic ratings, just-about-right (JAR), check-all-that-apply (CATA) attribute terms, emotional responses, purchase intent, and consumer purchasing behaviors of three types of yogurts (dairy-free coconut, dairy, and mixed) were measured under ARC, ARD, and sensory booths (SB). The results showed that the liking scores of dairy and mixed yogurts were generally higher than the coconut yogurt regardless of the environment. The interaction effect of yogurts and environments was statistically significant in terms of appearance, taste/flavor, sweetness, mouthfeel, aftertaste, and overall liking. JAR and penalty analysis revealed that consumers penalized the coconut yogurt for being "too much" in sourness, "too little" in sweetness, and "too thin" in mouthfeel. For the CATA analysis, attribute terms positively associated with overall liking (such as "sweet", "smooth", and "creamy") were selected for dairy and mixed yogurts, whereas the attribute terms negatively associated with overall liking (such as "firm", "heavy", and "astringent") were only selected for coconut yogurts. Regarding yogurt-consumption behaviors, the purchase intent of dairy and mixed yogurts was higher than that of the coconut yogurt, and taste and health were considered to be the most critical factors for yogurt consumption.

Keywords: augmented reality; non-dairy yogurt; contexts; consumer acceptability; emotional responses

Citation: Dong, Y.; Sharma, C.; Mehta, A.; Torrico, D.D. Application of Augmented Reality in the Sensory Evaluation of Yogurts. *Fermentation* **2021**, 7, 147. https://doi.org/10.3390/fermentation7030147

Academic Editor: Chrysoula Tassou

Received: 17 June 2021
Accepted: 30 July 2021
Published: 9 August 2021

1. Introduction

Humans are social beings, and their actions depend on the social contexts. Building on this age-long traditional knowledge, sensory science, as a discipline, is actively integrating this overlooked dimension. In this regard, the changing scientific worldview [1] concerning the nature of individuals [2], the social agency of materials [3], and the theory that everything relates to everything else [4] stressed the scientific disciplines to realign themselves with these philosophies. The ongoing sensory science is actively seeking to connect and accommodate the social context in consumption settings. In the midst of this, a variety of novel techniques, such as virtual or augmented reality, are being tested for their efficacy in bringing context to the sensory booths. The impact of contextual information on sensory perception and acceptability should not be neglected [5]. Context and environments can have a significant effect on the perception, liking, and emotions of consumers when tasting different products [6]. In this succession, we tested the application of augmented reality (AR) for the inclusion of context.

The delivery of AR experience could be achieved by using AR head-mounted-display (HMD) devices and mobile phones capable of controlling users' visual and auditory experiences [7]. AR menus have been used in some restaurants to make the consumers see the real-size 3D model of the dishes which they would like to order [8]. Recent interests of sensory AR systems mostly focus on engaging human–computer interactions or enrich intuitions by facilitating engagement with multisensory dimensions from both the physical and virtual information spaces [9]. Therefore, more vivid and realistic contexts could be created by immersive technologies in sensory studies. The effect of immersive technologies on improving ecology validity and engagement has been proven by previous studies [10–12]. Sinesio et al. [13] reported that similar results of emotions and hedonic scores were acquired in real-life environments, immersive rooms, and virtual-reality conditions [11]. Moreover, AR could increase the interactivity of food products, such as modifying texture or color [14] and adding digital nutrition information [15]. Extended global food competition demands high-quality products. If AR technology could be appropriately used to enhance interactions and engagements during sensory-evaluation tests, the more reliable consumption contextual data would be obtained from the sensory-evaluation assessments.

In this study, non-dairy and dairy yogurts were selected as the food models to test the effect of AR environments on the sensory experiences. In this regard, yogurt has a trustworthy reputation for its desirable taste and health benefits (probiotics). In comparison with other plant-based foods, non-dairy yogurts have the higher potential to be accepted by the general populations rather than special groups (e.g., vegan/vegetarian.). The sensory characteristics of plant-based foods may be the main obstacle to achieve higher consumer likings and preferences.

So far, very few studies have been conducted about the use of immersive technologies in creating contexts regarding the sensory evaluation of food products. Although some previous virtual reality (VR)-immersive studies revealed that contexts had a marginal effect on sensory acceptability [6,10], both AR and VR sensory studies are still at the early stages of research. In this experiment, a vanilla-flavored dairy milk-based yogurt (dairy yogurt), a vanilla-flavored coconut-milk-based yogurt (coconut yogurt), and a mixed combination yogurt (50% dairy and 50% coconut) were selected for the sensory test. The HoloLens 2 headsets (Microsoft, Ltd., Redmond, WA, USA) were used to create immersive AR environments. Three contextual environments referring to the traditional sensory booth (SB), AR dairy view (ARD), and AR coconut view (ARC) were used for consumers. This study aimed to investigate three environments' effects on the sensory acceptance, emotional responses, purchase intent, and yogurt-consumption behaviors among three yogurt samples using augmented reality as a contextual factor of the evaluation.

2. Materials and Methods

2.1. Participants

A total of N = 63 untrained participants (23 males and 40 females), with 92% of them aging from 21 to 40 years, were recruited voluntarily for this study at Lincoln University, Lincoln, New Zealand. All participants were required to be not allergic to yogurt ingredients involved in the study and were asked to sign the consent form before the sensory evaluation. All work involving human participants was approved by the Lincoln University Human Ethics Committee (Approval No: 2019-68; date of approval 18-11-2019). A brief introduction was given to each participant about the sensory test procedure. Participants were also briefed about the use of the AR headsets. This study included three sensory sessions (sensory booth, AR dairy context, and AR coconut context), and all sessions were conducted in the sensory laboratory facility of Lincoln University, New Zealand. Each participant was required to complete the three sessions at once with a resting period of 5 min in between the sessions. The duration of all sessions was approximately 45 min. The order of three sessions for each participant was randomized.

2.2. Food Stimuli

A focus group of N = 7 was used to select dairy and non-dairy yogurt samples for the subsequent consumer study. Five coconut-milk-based yogurts and five dairy yogurts were selected from the local supermarket. After assessing the sensory attributes in terms of appearance, flavor, mouthfeel, texture, and overall liking, a dairy-based vanilla bean yogurt (Yoplait, Ltd., Auckland, Auckland, New Zealand) and a coconut-milk-based vanilla bean yogurt (Raglan, Ltd., Reglan, Waikato, New Zealand) were selected as the final stimuli to be tested for the context-effect experiment. The ingredients of the dairy yogurt included skim milk, milk solids, sugar, water, cream, thickener (modified starch), gelatin, natural flavors, acidity regulators (citric acid, sodium citrate), preservative (potassium sorbate), vanilla bean seeds (0.01 %), and cultures. The ingredients of the coconut yogurt were organic coconut cream, apple juice concentrate, natural starch, vanilla paste, and live vegan cultures (*Acidophilus and Bifidobacterium*). To avoid appearance bias and logical errors, a mixed vanilla bean yogurt (50% dairy-based vanilla bean yogurt and 50% coconut-milk-based vanilla bean yogurt) was used as a third sample for consumers. Yogurt products were purchased and kept in the refrigerator at 4 °C. The preparation and sampling process was conducted within one hour before the sensory session. About 15 mL samples were poured into 30 mL cups with lids. Samples were coded with 3 random digital numbers. Participants were randomly served with one of the three yogurt samples in each sensory session (the order of the samples was balanced and randomized to avoid positional bias).

2.3. *Sensory Procedure*

A brief explanation of the experimental procedures was given to each participant at the beginning of three sensory sessions. These instructions included the explanation of the proper wearing and operation of the AR headsets, the duration of the experiment, and how to fill out the questionnaire using the RedJade Sensory Solutions (Martinez, CA, USA) platform. Each participant was instructed to evaluate three yogurt samples in one of the three sensory sessions. There were three contextual environmental settings (SB, ARC and ARD).

2.3.1. The 9-Point Hedonic Scale Questioning

For each environment, the participant was asked to rate the acceptability of sensory characteristics of each yogurt sample based on the 9-point hedonic scale, where 0 means "dislike extremely", and 9 means "like extremely", with a neutral response at 5 [16]. Sensory attributes included appearance, color, aroma, taste/flavor, sweetness, sourness, mouthfeel, viscosity, aftertaste, and overall liking.

2.3.2. Just-about-Right (JAR) Scale Questioning

Following the 9-point questions, the intensities of the sweetness, sourness, dairy flavor, coconut flavor, and mouthfeel were assessed using the just-about-right (JAR) scale, where 1 = not at all sweet/sour enough or much too weak/thin; 2 = just about right; and 3 = much too sweet/sour or much too strong/thick [17].

2.3.3. Check-All-That-Apply (CATA) Questioning

A check-all-that-apply (CATA) question was utilized for the sensory attributes of the yogurt samples. The terms in the CATA question were referenced in previous studies [18–20], including fruity flavor, dairy flavor, coconut flavor, vanilla flavor, sweet, sour, plain, smooth, creamy, astringent, homogeneous, cohesive, thick, thin, light, firm, compact, and heavy.

2.3.4. Emotions

Besides, the CATA-based method was used to assess the emotional responses evoked by the yogurt-tasting experience. The 33 emotional terms were pre-selected [21–23], including 24 positive terms (adventurous, satisfied, active, affectionate, energetic, enthusiastic, free, friendly, glad, good, happy, interested, joyful, loving, merry, nostalgic, pleased, pleasant, secure, warm, daring, eager, polite, understanding), 4 neutral terms (calm, peaceful, steady, and wild) and 5 negative terms (bored, disgusted, worried, aggressive, and guilty).

2.3.5. Purchase Intent, Demographics, and Consumption Questions

The purchase intent of each yogurt sample was also measured in the questionnaire (1 = No, 2 = Yes). For each participant, the demographic information (gender, age, and ethnicity) and consumer yogurt-consumption behavior were collected once at the beginning of the sensory booth session. Each participant was asked to multi-select the reasons for the yogurt consumption (nutritional, taste, probiotics, health, as a habit, emotional pleasantness, and other) and what factors they consider most when purchasing yogurt (flavor/taste, price, brand, plant-based yogurt/dairy-based yogurt, organic, locally produced, packaging, other/explanation). Participants were asked to use crackers (Arnott, Ltd., Silverwater, NSW, Australia) or water in between samples to avoid possible carryover effects.

2.4. Contextual Settings (Environments)

Figure 1 shows the three contextual settings, including the traditional sensory booths (the control setting), AR coconut view setting (ARC), and AR dairy view setting (ARD). The sensory testing booth (1.5 m width × 2.1 m height) was individual and isolated, located at RFH building, Lincoln University, Lincoln, New Zealand. The room temperature was controlled around 21 °C. Both incandescent light and the white-color fluorescent light were constant during the whole evaluation process. There was a window-door within each booth unit, which connected the testing booth with the food preparation room for easily placing samples and tablets.

Figure 1. Contextual settings (environments) for sensory evaluation of yogurt samples. (**a**) Sensory booth, (**b**) AR coconut view, (**c**) AR dairy view, (**d**) HoloLens 2 headset.

The HoloLens 2 headset (Microsoft, Redmond, WA, USA) was applied for two AR testing settings to generate the two AR environments. Participants were able to see the appearance of the samples through the transparent visor. While assessing samples in the AR sessions, participants were asked to wear the AR headset for viewing the pre-set contextual settings. When it was time to fill out the sensory questionnaire after tasting the sample, participants were able to easily flip the visor up (no need to take off the headset) to step out of the AR environment. Participants repeated the process until they were finished evaluating all samples. Moreover, participants were required not to touch the 3D models to avoid the setting bias for each observation.

Two AR contextual settings were generated in the isolated focus room and displayed by the 3D Viewer application, which was run by the HoloLens 2 headset. Two HoloLens

2 devices were put on the table in the sensory focus group room, where one was used for the ARC and another for ARD. Both AR contextual settings included 3D models, photos, and music. The criterion for creating the AR settings was based on the actual environment and perceptions of milk or coconut production. One instructor helped participants to properly wear the headsets and explain the sensory procedure, while another instructor set the table for the tasting experience. When participants understood the entire process and saw the right contextual settings, instructors left the focus room to give participants a private space for the sensory evaluation.

2.4.1. Coconut View Environment

For ARC (Figure 1b), a coconut photo was used as the background while 3D palm trees and one 3D coconut were displayed, accompanied by beach sounds. The 3D models used for ARC were extracted from the Sketchfab 3D model's platform (New York, NY, USA), using free-to-download non-commercial 3D models. Free 3D models used in ARC included palm trees [24] and one coconut [25]. The palm-tree photo comes from the Pixabay photo platform (Berlin, Berlin, Germany) [26]. The beach sound for the ARC originated from a beach-view video on YouTube [27], which gave a natural beach sound with sea waves and bird sounds.

2.4.2. Dairy View Environment

The ARD settings are presented in Figure 1c, including the background photo of a dairy farm with cows, two 3D cow models, one 3D milk model, and the farm's natural cow sounds. All 3D models were free downloaded from the Sketchfab platform, including the cow [28] and the milk models [29]. The background photo of the farm view was downloaded free from the Pixabay platform [30]. The cow sounds with slight bird sounds were used from a cow-farm video on YouTube [31], which displayed the natural peaceful farm sound with cows' mooing.

2.5. *Statistical Analysis*

The normality of hedonic data of sensory attributes was assessed by the Shapiro–Wilk test in XLSTAT Statistical Software 2019 (Addinsoft, New York, NY, USA). The result showed that all attribute data were not normally distributed. Therefore, the non-parametric Freidman's test followed by a *post hoc* Nemenyi test in XLSTAT Statistical Software 2019 (Addinsoft, New York, NY, USA) was performed to assess if there were significant differences in hedonic ratings among the combinations of 3 yogurt types $\times$ 3 environments. The H_0 assumption of Friedman's test is that at least one treatment significantly differs from the other treatments. If H_0 is approved, the Nemenyi test will be applied for the multiple means comparisons. The data of acceptability was auto sorted and analyzed for mean ranks using the XLSTAT Statistical Software. Penalty analysis performed on JAR data was used to determine the effects of the sensory attributes on the yogurt samples' liking scores. Cochran's Q test, correspondence analysis (CA), and principal coordinate analysis (PCoA) were applied to assess the CATA emotional responses and the CATA attributes of yogurts under different contextual settings. Principal component analysis (PCA; associations bi-plot) was applied to investigate the correlations of liking scores of attributes and yogurt samples under SB, ARC, and ARD. Agglomerative hierarchical cluster (AHC) analysis was performed to categorize the nine yogurt x environment combinations. The dissimilarity of these treatments was analyzed, relying on the Euclidean distance and Ward's method for investigating the difference between the yogurts' sensory attributes compared to the overall liking. Purchase-intent frequencies were analyzed by the Cochran's Q test and simultaneously confidence interval test with multiple comparisons. Frequency percentages were reported for the demographic data. Data from all participants were collected automatically by RedJade Sensory Software (Martinez, CA, USA) and analyzed by the XLSTAT Statistical Software 2019 (Addinsoft, New York, NY, USA).

3. Results

3.1. Hedonic Results

Table 1 shows the Friedman analysis results of sensory acceptance regarding the three yogurt types (coconut vanilla, dairy, and mixed) under three contextual settings (SB, ARC and ARD). Realizing the significant interaction effects for appearance, taste/flavor, sweetness, mouthfeel, aftertaste, and overall liking, only interactions and no main effects were further analysed for these attributes in the discussion section.

Table 1. The Q- and *p*-values summary (α = 0.05) of the Friedman analysis from the sensory attributes on yogurt samples under different environments.

Effects	Acceptability Attributes									
	Appearance		Color		Aroma		Taste/Flavor		Sweetness	
	Q-Value	*p*-Value	Q-Value	*p*-Value	Q-Value	*p*-Value	Q-Value	*p*-Value	Q-Value	*p*-Value
Yogurt	8.20	0.02	12.27	<0.01	7.73	0.02	18.05	<0.01	15.00	<0.01
Environment	4.70	0.10	0.57	0.75	1.31	0.52	0.05	0.98	2.51	0.29
Yogurt × Env.	16.75	0.03	14.95	0.06	12.29	0.14	27.00	<0.01	22.02	<0.01
Effects	**Acceptability Attributes**									
	Sourness		Mouthfeel		Viscosity		Aftertaste		Overall Liking	
	Q-Value	*p*-Value	Q-Value	*p*-Value	Q-Value	*p*-Value	Q-Value	*p*-Value	Q-Value	*p*-Value
Yogurt	7.19	0.03	11.80	<0.01	12.07	<0.01	18.06	<0.01	21.53	<0.01
Environment	2.73	0.26	1.54	0.46	2.14	0.34	4.31	0.12	4.70	0.10
Yogurt × Env.	10.84	0.21	19.88	0.01	14.93	0.06	22.35	<0.01	25.09	<0.01

Table 2 shows the Nemenyi test results comparing the mean ranks at alpha 5% for the sensory attributes, including appearance, taste/flavor, sweetness, and overall liking. Only attributes that had shown significance ($p < 0.05$) at the multiple pairwise comparisons using Nemenyi's procedure were summarized in Table 2. The liking score of the dairy yogurt under ARD significantly differed from the coconut yogurt under SB for the overall liking. Regarding appearance and sweetness, mixed yogurt under ARD had higher liking scores than the coconut yogurt under SB. Besides, the hedonic ratings of taste/ flavor of the mixed and dairy yogurts under SB were significantly higher than the coconut yogurt under SB. However, there were no differences in the liking scores between M-SB and D-ARD. Overall, coconut-based yogurt was the most disliked yogurt among the three yogurt types under the three contextual settings.

3.2. JAR Results and Penalty Analysis

The JAR frequencies and mean drops based on the penalty analysis for five attributes (sweetness, sourness, dairy flavor, coconut flavor, and mouthfeel) of the yogurt samples (coconut-milk-based, dairy, and mixed) under three contexts (SB, ARC and ARD) are presented in Figure 2. Coconut-milk-based yogurt under ARC had the highest selections of JAR for sweetness (65%), sourness (65%), dairy flavor (62%), and mouthfeel (75%). In contrast, the highest selection of JAR for coconut flavor (44) was shown under ARD. The highest proportions of participants selected that coconut-milk-based yogurt was "too little/weak" in sourness and coconut flavor under the ARC. Similarly, the selection frequencies of "too little/weak/thin" in sweetness, dairy flavor, and mouthfeel, respectively, were the highest under the sensory booths compared to the AR environments. The coconut yogurt under sensory booths had the highest selection of "too much/strong" in coconut flavor (46%) and sourness (32%), while the attributes of sweetness (32%) and dairy flavor (19%) were selected to be "too much" under the ARC. Regarding the dairy yogurt under all environments, the JAR frequencies of sweetness, sourness, dairy flavor, and mouthfeel were not much different, which varied from 62% to 78%. Approximately, 56–59% of participants selected "too weak" for coconut flavor, and 25–32% selected "too much" for sweetness. For the mixed yogurt, JAR was selected more frequently for sweetness (73–79%), sourness (67–70%), dairy flavor (63–75%), coconut flavor (51–54%), and mouthfeel (68–79%) under all contextual settings. The selections of "too little/weak/thin" for all evaluated attributes of the mixed yogurts except sweetness were similar under the different environments, ranging from 19% to 33%. Only 14–22% of participants selected "too much/strong" in the sweetness and coconut flavor of the mixed yogurt regardless of the environment.

Table 2. Results of the Nemenyi analysis (α = 0.05) on four attributes for nine yogurt treatments (critical difference: 1.53).

Attribute	Yogurt	Environment		
		SB	ARC	ARD
Appearance	C	4.08 ^{b}	4.89 ^{ab}	4.88 ^{ab}
	D	5.03 ^{ab}	5.22 ^{ab}	5.06 ^{ab}
	M	4.90 ^{ab}	5.15 ^{ab}	5.79 ^{a}
Taste/Flavor	C	4.06 ^{b}	4.60 ^{ab}	4.22 ^{ab}
	D	5.71 ^{a}	4.98 ^{ab}	5.29 ^{ab}
	M	5.65 ^{a}	5.37 ^{ab}	5.10 ^{ab}
Sweetness	C	4.06 ^{b}	4.90 ^{ab}	4.28 ^{ab}
	D	5.15 ^{ab}	5.08 ^{ab}	5.44 ^{ab}
	M	5.37 ^{ab}	5.64 ^{a}	5.10 ^{ab}
Overall liking	C	4.08 ^{b}	4.87 ^{ab}	4.13 ^{ab}
	D	5.12 ^{ab}	5.24 ^{ab}	5.62 ^{a}
	M	5.25 ^{ab}	5.58 ^{ab}	5.11 ^{ab}

Data displayed as mean ranks; the different letters within each attribute mean significant difference ($p < 0.05$). C-SB, coconut-milk-based yogurt in the sensory booth; D-SB, dairy yogurt in the sensory booth; M-SB, Mixed yogurt (50% dairy yogurt and 50% coconut-milk-based yogurt) in the sensory booth; C-ARC, coconut-milk-based yogurt in AR coconut view; D-ARC, dairy yogurt in AR coconut view; M-ARC, mixed yogurt in AR coconut view; C-ARD, coconut-milk-based yogurt in AR dairy view; D-ARD, dairy yogurt in AR dairy view; M-ARD, mixed yogurt in AR dairy view.

Penalty analysis was based on the JAR frequencies and the hedonic scores of yogurt samples under different contexts. The threshold population size of 20% is considered to be the significant level for an attribute to be penalized in the analysis [32,33]. According to the explanation of the penalty plots [32], more than 20% of participants penalized coconut yogurt for being "too much/strong" in sourness and coconut flavor and "too little/weak" in sweetness and dairy flavor under SB and ARD, whereas "too much" sweetness and "too little/weak" sourness and coconut flavor were penalized under ARC. Dairy yogurts under all environments were penalized due to lack of sourness and coconut flavor and for having "too much" sweetness. For the mixed yogurts, having "too little/weak" sourness and coconut flavor affected the liking scores regardless of the environment. For the dairy yogurts, having "too thin" mouthfeel under SB and ARC as well as having "too much" sweetness under ARC and having "too weak" dairy flavor under ARD were penalized by over 20% of consumers. In addition, having a "too thin" mouthfeel influenced the hedonic scores of dairy yogurts under all contexts.

In general, the mixed yogurts had higher proportions of JAR scores for all sensory attributes compared to those of the dairy and coconut yogurts. Besides, coconut yogurts were slightly penalized for having the "too much" coconut flavor in SB and ARD. Comparing the environments, marginal differences were observed in the JAR responses, as participants had slightly different penalizations for all yogurts in each environment.

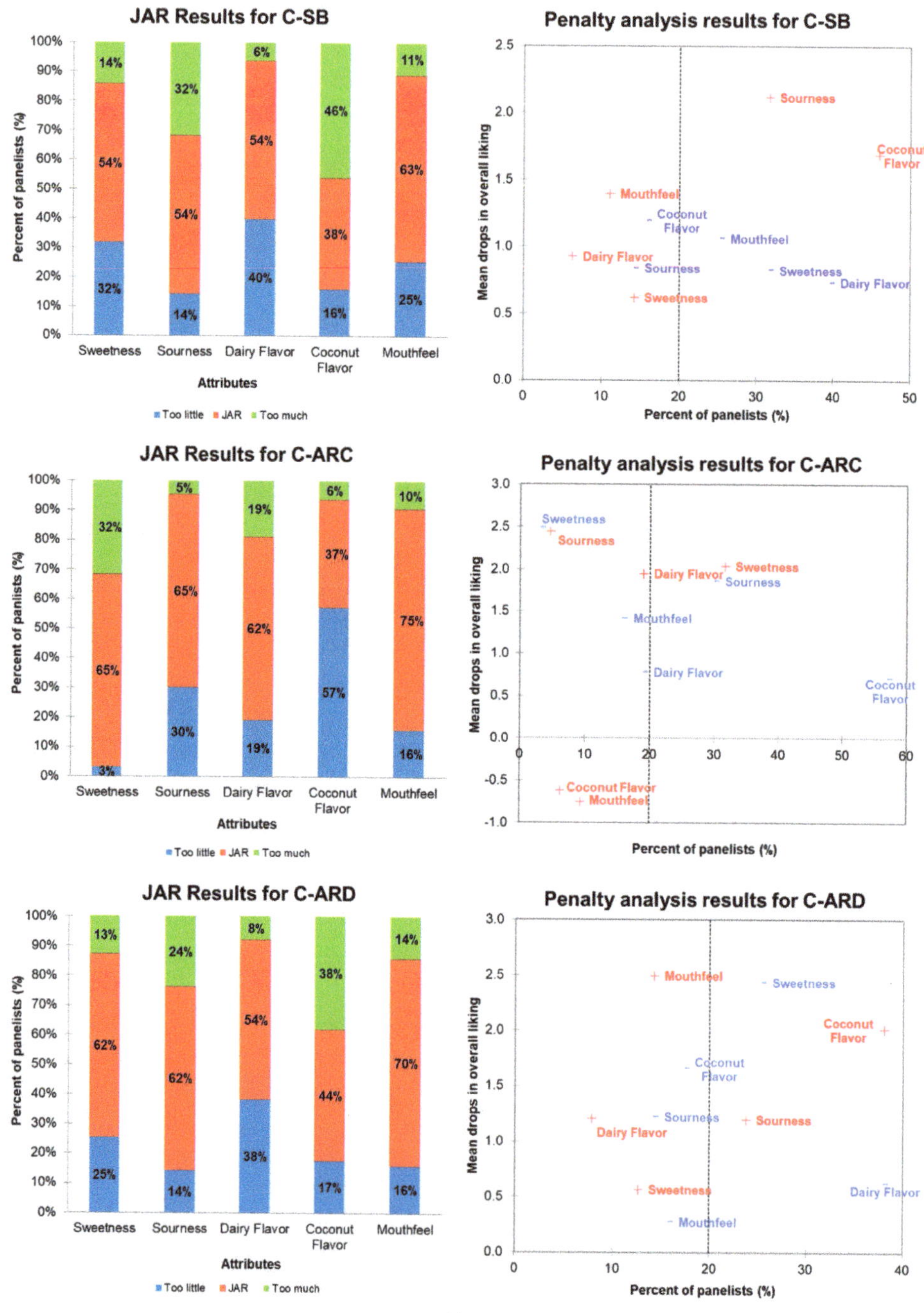

(a)

Figure 2. *Cont.*

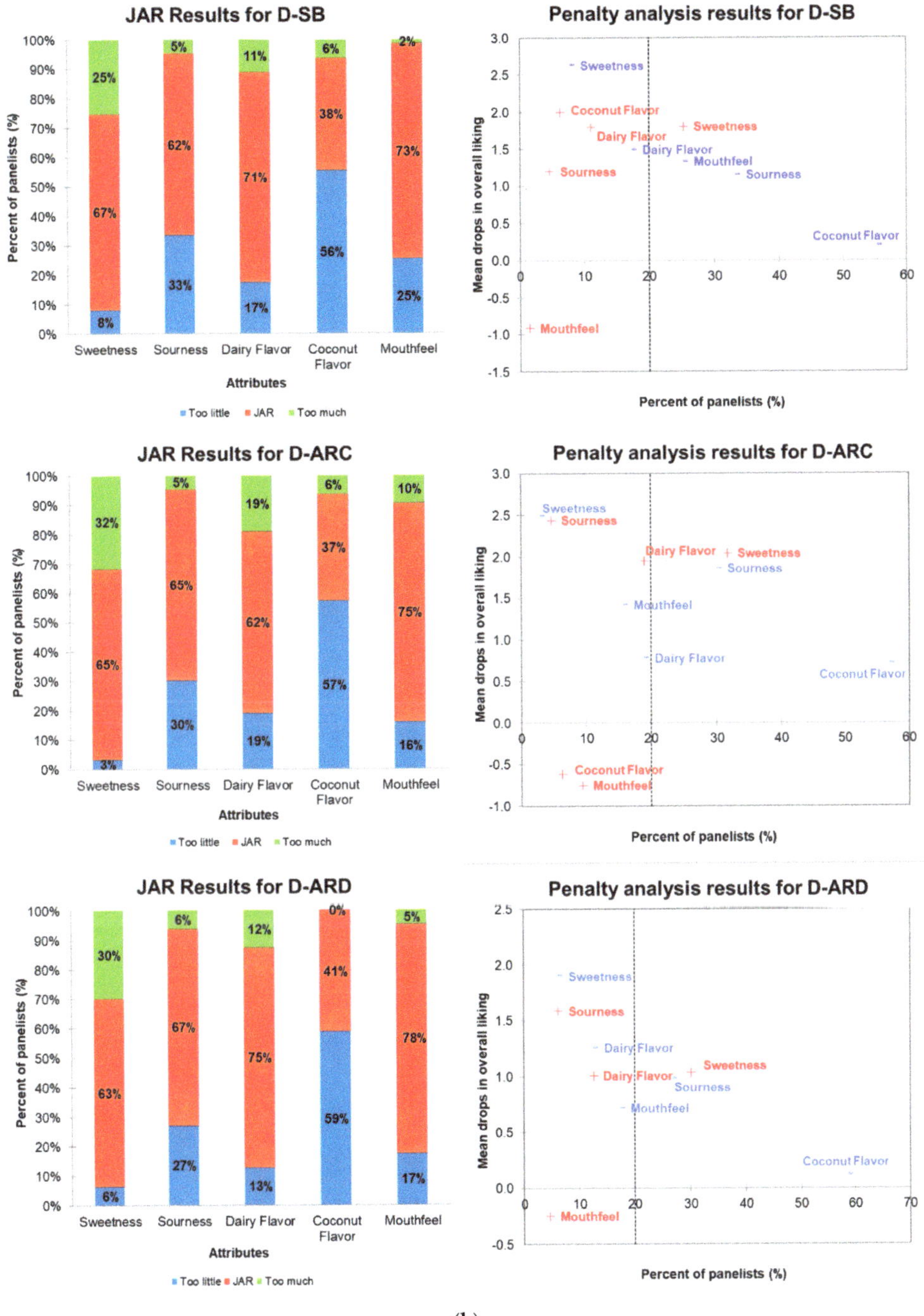

(b)

Figure 2. *Cont.*

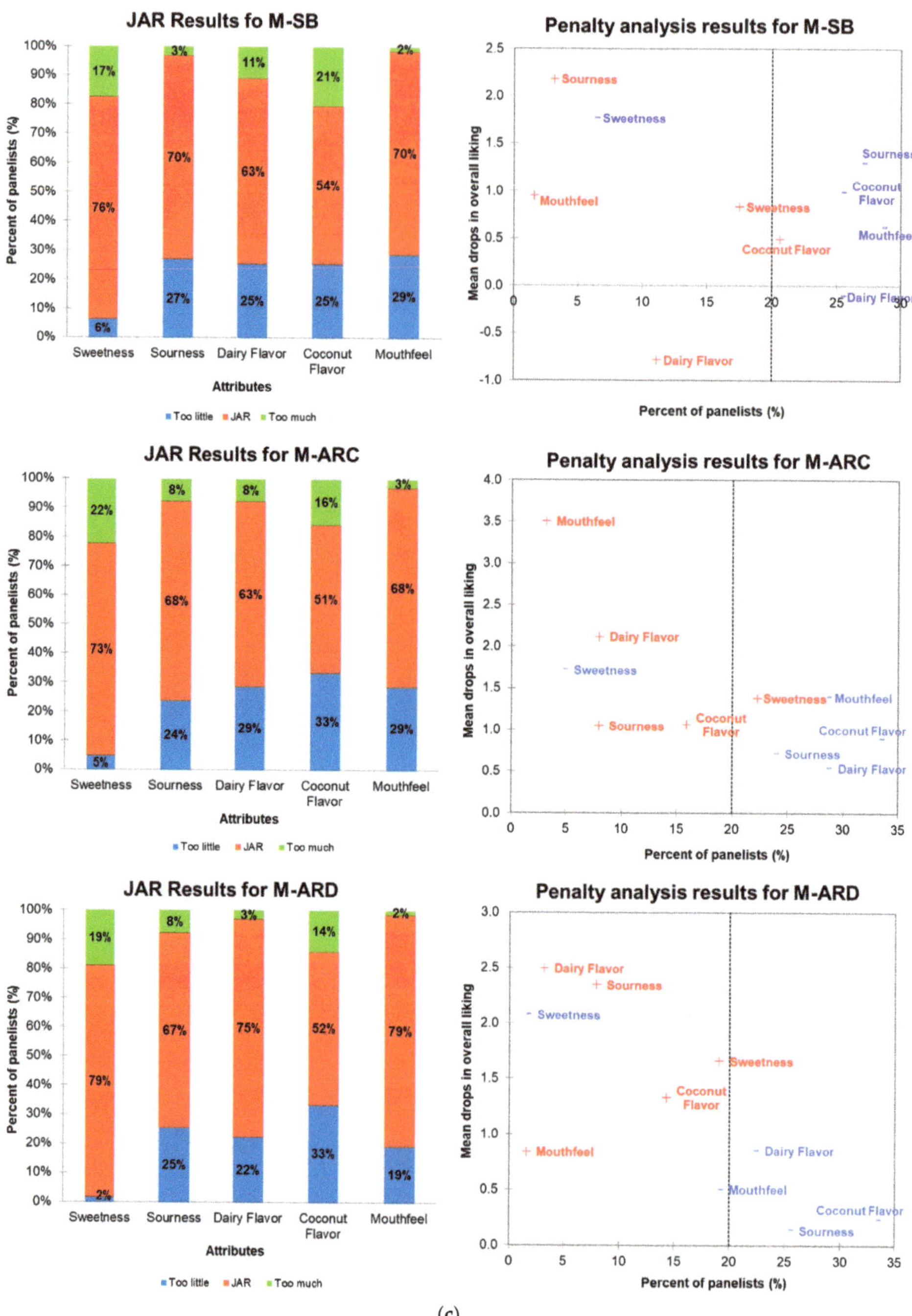

(**c**)

Figure 2. Just-about-right (JAR) frequencies and penalty analysis results of sensory attributes (-, too little;+, too much) of (**a**) coconut, (**b**) dairy, and (**c**) mixed yogurts under SB, AR1, and AR2. The description of the treatment's labels can be found in the footnote of Table 2.

3.3. CATA Analysis of Attribute Terms of Yogurt Samples in Different Environments

Figure 3 shows the correspondence analysis of the CATA attribute terms for the yogurts in the different environments and the principal coordinate analysis of the CATA attribute terms with the overall liking score. The principal component one (PC1) of the correspondence analysis accounted for 73.47%, and the principal component two (PC2) was 18.27%. Both components explained 91.77% of total data variability. According to the Cochran's *Q* test results, 15 terms were significantly different among the yogurt samples, including "fruit flavor", "dairy flavor", "coconut flavor", "vanilla flavor", "sweet", "sour", "smooth", "creamy", "astringent", "homogeneous", "thick", "thin", "light", "firm", and "heavy" (data not shown). The participants selected attribute terms toward the dairy yogurt samples under all environments, such as "homogeneous", "fruity flavor", "dairy flavor", "vanilla flavor", "sweet", "smooth", and "creamy". The mixed yogurt sample was related to the terms "thin", "light", "sweet", and "smooth". Besides, the mixed yogurt under ARD was associated with "creamy" and "vanilla flavor". On the contrary, the coconut yogurt was linked with "coconut flavor", "thick", "firm", "sour", and "astringent" (Figure 3a). According to the results of the principal coordinate analysis regarding the descriptive terms against the acceptability of all samples in the different contexts, the liking score was positively associated with "sweet", "smooth", "dairy flavor", "fruity flavor", "vanilla flavor", "creamy", "light", and "homogeneous", whereas it was negatively associated with "firm", "heavy", and "astringent" (Figure 3b).

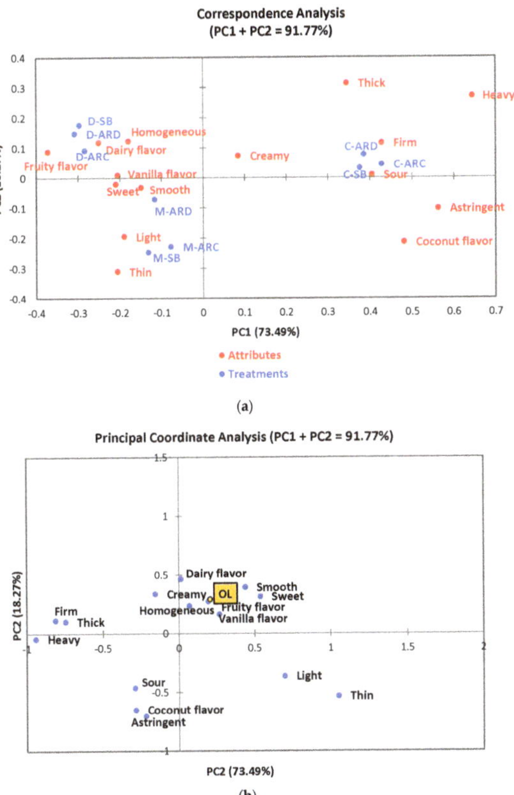

Figure 3. (**a**) Correspondence analysis of the CATA attribute terms for the yogurts under the different environments and (**b**) principal coordinate analysis of the CATA descriptive terms with the overall liking score (OL). The description of the treatment's labels can be found in the footnote of Table 2.

In general, samples were grouped according to the yogurt type (coconut, dairy, or mixed), with marginal effects of the environment in the discrimination of the attributes (Figure 3a). The dairy and mixed yogurts were more similar in their sensory attributes compared to those of the coconut yogurt. In the principal coordinate analysis, the overall liking was more related to the dairy yogurts compared to the other samples.

3.4. Emotional Responses

The results of the correspondence analysis of emotional terms for the yogurts in different environments were shown in Figure 4a. PC1 (50.33%) and PC2 (16.43%) explained 66.89% of the total variability. According to the Cochran's Q test results, 14 emotional terms showed significant differences among the treatments, including "satisfied", "calm", "free", "friendly", "good", "happy", "interested", "joyful", "loving", "peaceful", "pleased", "pleasant", "disgusted", and "aggressive" (data not shown). Specifically, the dairy yogurt under SB and ARD, the mixed yogurt under all contexts, and the coconut yogurt under SB and ARC were associated with nine positive emotional terms ("friendly", "satisfied", "happy", "joyful", "pleased", "good", "interested", "loving", and "peaceful") and one neutral emotional term ("calm"). In addition, the emotional term "free" was associated with the coconut-milk-based yogurt under the ARD treatment. On the other hand, negative terms, such as "aggressive" and "disgusted", were linked with the diary yogurt under ARC.

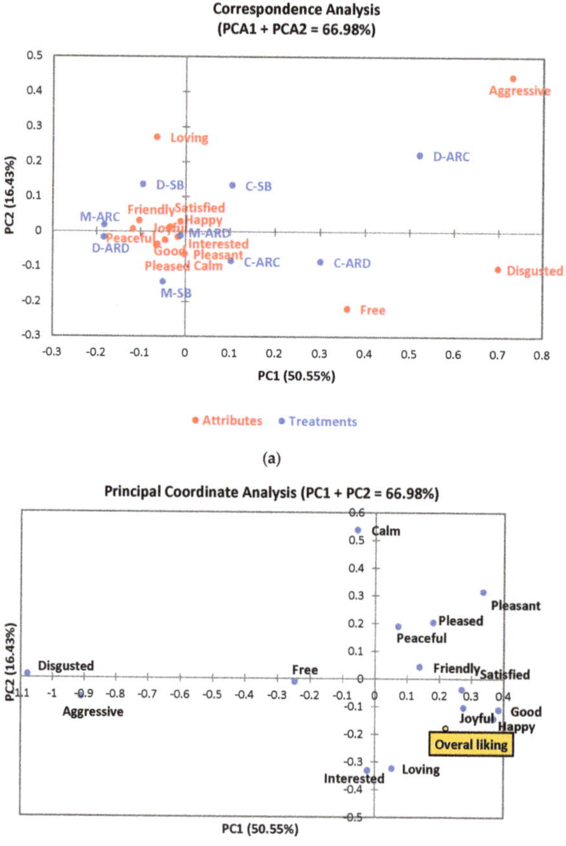

Figure 4. *Cont.*

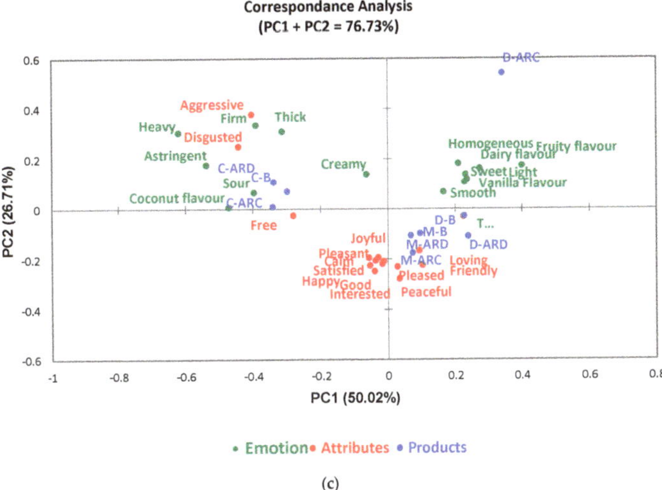

Figure 4. (**a**) Correspondence analysis of the emotional terms for the yogurts under the different environments, (**b**) principal coordinate analysis of the emotional terms with the overall liking score, and (**c**) correspondence analysis of the merged results from the attributes and emotions of the yogurt samples. The description of the treatment's labels can be found in the footnote of Table 2.

Figure 4b presents the results of the principal coordinate analysis of the emotional terms concerning the overall liking score. In general, the positive emotional terms, such as "satisfied", "happy", "good", "joyful", "pleasant", and "peaceful", were positively associated with the overall liking of yogurt products. On the other hand, negative emotional terms, such as "aggressive" and "disgusted", were negatively associated with overall liking.

In general, no clear separations among yogurt types and environments were shown in the correspondence analysis results. The dairy yogurt under the AR coconut environment was different in its emotional profile compared to the other samples in this experiment. Figure 4c shows the merged results from the attributes and emotions of the different yogurt samples under the different environments. Generally, the sensory attributes were separated from the emotional terms. However, the textural attributes "heavy", "firm", "thick", and "creamy" were related to the emotional terms "aggressive" and "disgusted". Besides, these negative emotions and textural attributes were closely related to the coconut yogurts.

3.5. Principal Component and Cluster Analyses of Yogurt Samples under the Three Contexts

PCA biplot visualized the associations between the nine yogurt × environment combinations and the hedonic scores of the yogurt samples' ten sensory attributes. According to Figure 5, the PC1 accounted for 91.70% of the data variability, while the PC2 accounted for 3.80%. Both PCs explained 95.49% of total data variability. Liking vectors of most attributes were aligned with the horizontal axis (PC1), with the squared cosines ranging from 0.80 to 0.96. However, the liking vectors of appearance and aftertaste were linked with the vertical axis (PC2), with squared cosines ranging from 0.01 to 0.18. Liking vectors of taste/flavor, sweetness, sourness, mouthfeel, and overall liking were close to each other. Liking vectors of aroma and color were closely associated with viscosity. Besides, the liking vector of appearance was not closely related to color, as both vectors were almost orthogonal. In the case of the different treatments, the dairy and mixed yogurts, regardless of the contextual effects, were highly associated with the liking of mouthfeel, sourness, and viscosity. Moreover, dairy and mixed yogurts were relatively linked with the liking of taste/flavor, sweetness, aroma, color, and overall liking. In contrast, coconut yogurt was negatively associated with the liking of all evaluated attributes under all environments.

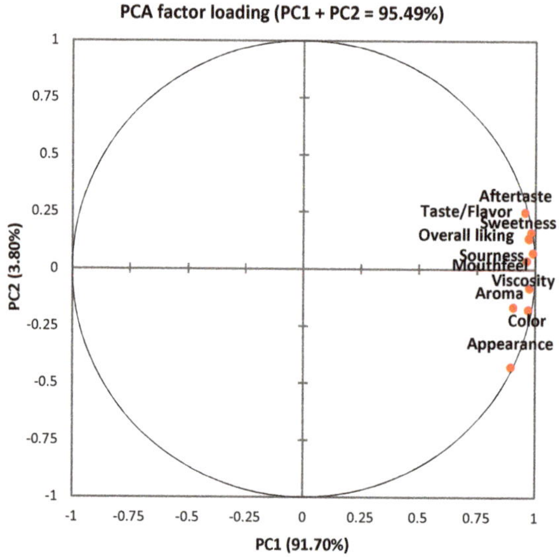

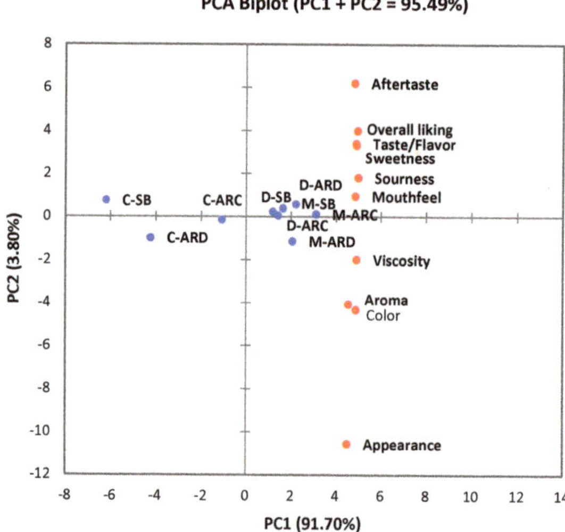

Figure 5. Principal component analysis (PCA) biplot regarding hedonic scores of yogurt attributes under the three contextual settings. The description of the treatment's labels can be found in the footnote of Table 2.

Figure 6 shows the dendrogram of the nine yogurt × environment combinations (3 × 3 factorial). Two main cluster groups are presented, which refer to (1) coconut-milk-based yogurt under the sensory booth and AR dairy view, (2) coconut yogurt under AR coconut view, and dairy and mixed yogurts under all environments.

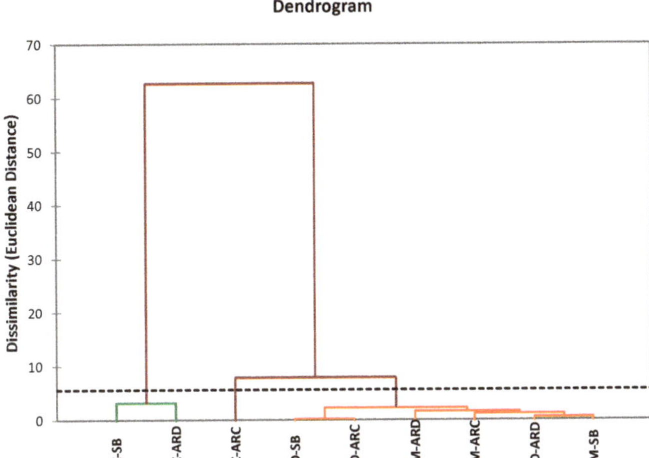

Figure 6. Dendrogram of agglomerative hierarchical clustering (AHC) grouping yogurt samples under the different environments. Different colored lines indicate different groups of yogurt samples given by AHC analysis. The description of the treatment's labels can be found in the footnote of Table 2.

3.6. The Purchase Intent of Yogurt Samples under Different Environments

Table 3 presents the yogurt samples' frequencies of purchase intent (dairy, coconut, and mixed) under three environments (SB, VRC and VRD). Both dairy yogurt (61.9–68.3%) and mixed yogurt (63.5–69.8%) showed higher purchase intent than coconut yogurt (41.3–55.6%) regardless of environment. The frequencies of purchase intent were not significantly different among the dairy and mixed yogurts under all contexts. In contrast, the purchase intent of mixed yogurt under ARD was significantly different compared to the coconut yogurt under the ARC and SB. Therefore, consumers would like to buy dairy and mixed yogurts in the market.

Table 3. The purchase-intent frequencies of yogurt samples under three contextual environments.

Variable	Yogurt [1]	Environment [1]		
		SB	ARC	ARD
Purchase Intent (%) [2]	C	41.3 [c]	42.9 [bc]	55.6 [abc]
	D	68.3 [ab]	68.3 [ab]	61.9 [abc]
	M	63.5 [abc]	66.7 [abc]	69.8 [a]

[1] Dairy yogurt (D), non-dairy yogurt (coconut-based yogurt; C), and mixed yogurt (50% dairy yogurt and 50% non-dairy yogurt; M) were tested under environments (SB, sensory booth; ARC, AR coconut view; ARD, AR dairy view). [2] Cochran's Q test with Marascuilo procedure was applied for multiple pairwise comparisons ($n = 63$); frequencies with different superscript (a, b, c) in the purchase-intent table indicate significant differences ($p < 0.05$).

3.7. The Results of Consumers Consumption Behaviour on Yogurt

Table 4 shows the results of yogurt-consumption frequencies, reasons for yogurt consumption, and factors that participants consider when purchasing yogurt. For the results of purchase frequencies, about 60.4% of participants selected "two or three times a week" or "sometimes in a week", while 14.3% of participants consume yogurt every day. The rest of the participants (25.4%) had relatively low yogurt-consumption frequencies. Regarding the reasons for yogurt consumption, 71% of consumers considered "health" and "taste" to be the main reason, followed by "nutrition" (60.3%), "probiotics" (49.2%), "as a habit" (28.6%), and "emotional pleasantness" (15.9%). Approximately 58.7% of consumers selected "price" as the most important factor that affected the purchase intent. "Brand", "type" (dairy or non-dairy yogurt), and "packaging" of the yogurt were the least selected factors affecting purchase intent, accounting for 28.6%, 20.5% and 15.9%, respectively. Only 9.5% of consumers were concerned with the factor "locally produced", and 6.3% of consumers considered "organic" as the most important information that influenced their purchase decision.

Table 4. Factors affecting consumers' consumption of yogurts and behaviors.

Frequencies of Yogurt Consumption	Percentage (%)
Everyday	14.3
Two or three times a week	30.2
Sometimes in a week	30.2
Two or three times a month	12.7
Sometimes in a month	11.1
Occasionally	1.6
Reasons for Yogurt Consumption	**Percentage (%)**
Health	71.4
Taste	71.4
Nutrition	60.3
Probiotics	49.2
As a habit	28.6
Emotional pleasantness	15.9
Factor Considered most When Purchasing Yogurt	**Percentage (%)**
Price	58.7
Brand	28.6
Type (dairy yogurt or non-dairy yogurt)	20.6
Packaging	15.9
Locally produced	9.5
Organic	6.3

4. Discussion

4.1. *The Effect of Contexts on Consumer Acceptability of Yogurt Products*

A significant interaction between environment and test samples signifies the importance of the stimulus context to the psychophysical judgments in food. Mostly, context affected all modalities or attributes except aroma and sourness. The overall mild, sweet impression of most yogurt aroma types could be responsible for this finding. The aroma as a stimulus may be lost in the high contextual effects of AR. No effect of context in the case of sourness may be attributed to the intensity of surprise bestowed by the acidity of sour stimuli. The sourness has been previously related to high-pitch sounds and angular shapes [34]. Physiologically, sourness has been found related to increased blood flow in the face [35], higher heart rate [36], and muscular concentration [37]. This strong context-less intensity of sourness leads towards stimulus-specific deliberations, which may also be related to survival instincts; for instance, conveying the detection of elements of microbe [38]. However, in general, consumers significantly liked dairy-based or mixed yogurts over coconut-based yogurts. Participants with a natural aversion to coconut smell or flavor perceived that coconut-flavored food products were unacceptable regardless of the different contexts [16].

Under sensory booths, the liking score of the taste, sweetness, mouthfeel, and viscosity differed significantly among yogurt types, whereas the liking scores of those attributes were not significantly different under ARD and ARC. This might be explained because sensory booths are isolated environments that allow participants to focus on the tasting task, and sensory differences are more evident in these contexts [16]. In this study, the liking scores of coconut yogurts for all sensory attributes under the two AR contexts (dairy and coconut) were higher compared to those scores under the sensory booths. This might reflect the enhancing effects of immersive technologies on the hedonic responses of consumers towards food products [6]. In contrast, Bangcuyo et al. [39] revealed that participants showed more discrimination in the immersive coffeehouse than the sensory booths. Overall, stimulus-specific contextual settings should not be overlooked; building on the significant interactions found between context and stimuli in this and other studies, critical insights into the intended consumer behaviors can be found.

4.2. JAR Results

Context primed consumers to desire more coconut and dairy flavors in their respective contexts in the case of coconut- and dairy-based yogurts (Figure 2a,b). This congruent priming of consumers' mental states has been previously reported by others [37,40], and it is frequently used by marketers through advertisement campaigns. The priming of consumers' mental states through context leads consumers to expect something relevant. Additionally, context-incongruent probing of coconut flavor in dairy-based yogurts was an unintentional act laid out by less coconut flavor (Figure 2b), which should be avoided in most if not all cases. Thanks to this, coconut flavor appeared as a polarized attribute, penalizing for both "too strong" and "too weak" in most cases (Figure 2). This may also be due to consumers having no agreement on their ideal level, i.e., one segment preferring a weaker level while another segment preferring a stronger level. For instance, about 6–46% of consumers perceived coconut yogurt as being "too strong" in coconut flavor, while 16–57% of consumers selected "too weak" in coconut flavor (Figure 2a). The mixed yogurts (half coconut and half dairy yogurts) obtained the highest JAR scores (varied from 51% to 79%) for all attributes (sweetness, sourness, dairy flavor, coconut flavor, and mouthfeel) under the three contexts, and this might be due to the balance of flavors in this yogurt.

4.3. Attribute Terms and Emotional Responses of Yogurts

The results of the correspondence analysis regarding the CATA attribute terms of the yogurt samples under different contexts showed that the same-type yogurts were close to each other when considering all sensory characteristics (such as C-SB, C-ARC, and C-ARD; D-SB, D-ARC, and D-ARD; M-SB, M-ARC, and ARD). The layout of the nine test samples in the plot was mainly dependent on the yogurt type rather than on the context. As shown in Figure 3, the dairy and mixed yogurts were linked to textural terms, including "homogeneous", "smooth", "sweet", and "creamy". Previously, those terms were also used to describe the desired texture in yogurts using premium and private labels [22]. In addition, the principal coordinate analysis proved the positive association between those attribute terms and overall liking. On the contrary, coconut yogurts were associated with "firm", "sour", and "astringent", which had a negative association with the overall liking. These findings confirm that textural and mouthfeel properties can have a significant effect on the acceptability of yogurt products [41].

In the context of emotional responses, the environments showed a more dissociative layout of yogurt samples (Figure 4). Before interpreting the superimposed representation of rows and columns in Figure 4, it would be worthy to mention that points closer to the origin contribute little to the derived space [42]. Ten positive, two negative, and two neutral emotional terms were the most frequent among consumers. All coconut-based samples are located in the right quadrants of the plot, whereas mostly all other samples are on the left side of the correspondence plot. The closeness of treatments indicates that those have either similar row or column profiles, meaning that they are related. Similarly, emotional terms clustered in the middle of the correspondence map have either similar row or column profiles. The position of D-ARC associates it with negative terms, such as "aggressive" and "disgusted". The incongruent context types, such as in the case of C-ARD and D-ARC, may be responsible for this layout. Previously, Bangcuyo et al. [39] used more comparable environments, such as "VR-dark coffee shop" and "VR-bright coffee shop", concerning one sample type, whereas here, we used two context types in two different sample types. Realizing the importance of context in food-consumption settings, it should be also kept in mind that yogurts are never consumed in dairy parlors as compared to other products, such as the case of coffee or beer, so the use of the context here was to demonstrate the effects of AR elements on the tasting experience.

4.4. Purchase Intent and Consumption Behaviors

Health, taste, and nutrition were the most frequent reasons for purchasing yogurt. Not surprisingly, the above findings reflect consumers' perceptions of different yogurts. Dairy and mixed yogurts are sweet, homogenous, creamy, smooth, and fruity, while the coconut yogurt is astringent, sour, and thick. For the most important factor when purchasing yogurts, interestingly, the majority of participants selected "price" in the first place. Besides, "brand" and "type of yogurt" were also important factors for consumers when selecting a particular yogurt. Familiarity with the products can have a significant effect on the purchase decision of consumers [43]. If the consumers do not belong to a special population cohort (for instance, vegan/vegetarian), they will probably purchase dairy yogurts with desirable sensory attributes. Therefore, the above reasons could potentially explain why coconut yogurt's purchase intent was lower than the dairy and mixed yogurts regardless of the effects of context.

5. Limitations

The limited number of partakers was a limitation of this study, which may pose constraints in generalizing these results. Still, this work gives an insight into the first-of-its-kind approach for the investigation of augmented reality in a fast-changing scientific worldview to include context. A more segregated sample size of higher number and congruent context type would be a further next step to follow.

6. Conclusions

The study investigated the consumer acceptability, JAR attributes, emotional response, purchase intent, and consumer purchase behaviors on coconut-milk-based, dairy, and mixed yogurts (half coconut and dairy) under three environments (sensory booth, AR coconut view, and AR dairy view). The augmented reality technique was explored in this study to create test environments for yogurts. A significant interaction effect between yogurt type and environment was observed in most sensory attributes. Dairy and mixed yogurts were significantly linked to attributes associated with liking scores, such as "sweet", "smooth", and "creamy", while coconut yogurt was described with attributes against overall liking in terms of "astringent", "thin", and "sour". Taste/flavor and health were the most important reasons to determine the purchase intent of yogurts. Moreover, dairy and mixed yogurts' purchase intents were higher than that of the non-dairy yogurt regardless of environment. Compared to virtual reality, one obvious advantage of AR technology was the seamless combination of the real environment with different virtual elements. Besides, AR has a significant potential to develop practical, advanced mixed-immersive applications in the food industry.

Author Contributions: Conceptualization, D.D.T. and Y.D.; methodology, D.D.T. and Y.D.; formal analysis, Y.D.; investigation, Y.D.; data curation, Y.D.; writing—original draft preparation, Y.D.; writing—review and editing, Y.D., D.D.T., A.M. and C.S.; supervision, D.D.T. and C.S.; project administration, D.D.T.; funding acquisition, D.D.T. All authors have read and agreed to the published version of the manuscript.

Funding: This research was funded by Lincoln University, New Zealand, through the Centre of Excellence-Food for Future Consumers.

Institutional Review Board Statement: The study was conducted according to the guidelines of the Declaration of Helsinki, and approved by the Human Ethics Committee of Lincoln University (Approval: 2019-68, date of approval 18-11-2019).

Informed Consent Statement: Informed consent was obtained from all subjects involved in the study.

Data Availability Statement: The data presented in this study are available on request from the corresponding author.

Acknowledgments: I would like to acknowledge all the subjects who participated in the sensory study.

Conflicts of Interest: The authors declare no conflict of interest. The funders had no role in the design of the study; in the collection, analyses, or interpretation of data; in the writing of the manuscript, or in the decision to publish the results.

References

1. Irzik, G.; Nola, R. Worldviews and their relation to science. In *Science, Worldviews and Education*; Matthews, M.R., Ed.; Springer: Dordrecht, The Netherlands, 2009; pp. 81–97.
2. Chapman, C.A.; Huffman, M.A. Why do we want to think humans are different? *Anim. Sentience* **2018**, *3*. [CrossRef]
3. Sillar, B. The social agency of things? Animism and materiality in the Andes. *Camb. Archaeol. J.* **2009**, *19*, 367–377. [CrossRef]
4. Rout, M.; Reid, J. Embracing indigenous metaphors: A new/old way of thinking about sustainability. *Sustain. Sci.* **2020**, *15*, 945–954. [CrossRef]
5. Jaeger, S.; Hort, J.; Porcherot, C.; Ares, G.; Pecore, S.; MacFie, H. Future directions in sensory and consumer science: Four perspectives and audience voting. *Food Qual. Prefer.* **2017**, *56*, 301–309. [CrossRef]
6. Torrico, D.D.; Sharma, C.; Dong, W.; Fuentes, S.; Viejo, C.G.; Dunshea, F.R. Virtual reality environments on the sensory acceptability and emotional responses of no-and full-sugar chocolate. *LWT* **2020**, *137*, 110383. [CrossRef]
7. Wedel, M.; Bigné, E.; Zhang, J. Virtual and augmented reality: Advancing research in consumer marketing. *Int. J. Res. Mark.* **2020**, *37*, 443–465. [CrossRef]
8. Das, S.M.; Sridhara, V.; Khorashadi, B. Systems and Methods Involving Augmented Menu Using Mobile Device. U.S. Patent No. 9,179,278, 3 November 2015.
9. Crofton, E.C.; Botinestean, C.; Fenelon, M.; Gallagher, E. Potential applications for virtual and augmented reality technologies in sensory science. *Innov. Food Sci. Emerg. Technol.* **2019**, *56*, 102178. [CrossRef]
10. Kong, Y.; Sharma, C.; Kanala, M.; Thakur, M.; Li, L.; Xu, D.; Harrison, R.; Torrico, D.D. Virtual Reality and Immersive Environments on Sensory Perception of Chocolate Products: A Preliminary Study. *Foods* **2020**, *9*, 515. [CrossRef]
11. Sinesio, F.; Moneta, E.; Porcherot, C.; Abbà, S.; Dreyfuss, L.; Guillamet, K.; Bruyninckx, S.; Laporte, C.; Henneberg, S.; McEwan, J.A. Do immersive techniques help to capture consumer reality? *Food Qual. Prefer.* **2019**, *77*, 123–134. [CrossRef]
12. Zandstra, E.; Kaneko, D.; Dijksterhuis, G.; Vennik, E.; De Wijk, R. Implementing immersive technologies in consumer testing: Liking and Just-About-Right ratings in a laboratory, immersive simulated café and real café. *Food Qual. Prefer.* **2020**, *84*, 103934. [CrossRef]
13. Sinesio, F.; Saba, A.; Peparaio, M.; Civitelli, E.S.; Paoletti, F.; Moneta, E. Capturing consumer perception of vegetable freshness in a simulated real-life taste situation. *Food Res. Int.* **2018**, *105*, 764–771. [CrossRef] [PubMed]
14. Ammann, J.; Stucki, M.; Siegrist, M. True colours: Advantages and challenges of virtual reality in a sensory science experiment on the influence of colour on flavour identification. *Food Qual. Prefer.* **2020**, *86*, 103998. [CrossRef]
15. Gutiérrez, F.; Htun, N.N.; Charleer, S.; De Croon, R.; Verbert, K. Designing augmented reality applications for personal health decision-making. In Proceedings of the 52nd Hawaii International Conference on System Sciences, Maui, HI, USA, 8–11 January 2019.
16. Lawless, H.T.; Heymann, H. *Sensory Evaluation of Food: Principles and Practices*; Springer Science & Business Media: New York, NY, USA, 2010.
17. Li, B.; Hayes, J.E.; Ziegler, G.R. Just-about-right and ideal scaling provide similar insights into the influence of sensory attributes on liking. *Food Qual. Prefer.* **2014**, *37*, 71–78. [CrossRef]
18. Chetachukwu, A.S.; Thongraung, C.; Yupanqui, C.T. Development of reduced-fat coconut yoghurt: Physicochemical, rheological, microstructural and sensory properties. *Int. J. Dairy Technol.* **2019**, *72*, 524–535. [CrossRef]
19. Jaeger, S.R.; Lee, P.Y.; Xia, Y.; Chheang, S.L.; Roigard, C.M.; Ares, G. Using the emotion circumplex to uncover sensory drivers of emotional associations to products: Six case studies. *Food Qual. Prefer.* **2019**, *77*, 89–101. [CrossRef]
20. Janiaski, D.; Pimentel, T.; Cruz, A.; Prudencio, S. Strawberry-flavored yogurts and whey beverages: What is the sensory profile of the ideal product? *J. Dairy Sci.* **2016**, *99*, 5273–5283. [CrossRef]
21. Gutjar, S.; de Graaf, C.; Kooijman, V.; de Wijk, R.A.; Nys, A.; ter Horst, G.J.; Jager, G. The role of emotions in food choice and liking. *Food Res. Int.* **2015**, *76*, 216–223. [CrossRef]
22. Schouteten, J.J.; De Steur, H.; Sas, B.; De Bourdeaudhuij, I.; Gellynck, X. The effect of the research setting on the emotional and sensory profiling under blind, expected, and informed conditions: A study on premium and private label yogurt products. *J. Dairy Sci.* **2017**, *100*, 169–186. [CrossRef]
23. Nestrud, M.A.; Meiselman, H.L.; King, S.C.; Lesher, L.L.; Cardello, A.V. Development of EsSense25, a shorter version of the EsSense Profile®. *Food Qual. Prefer.* **2016**, *48*, 107–117. [CrossRef]
24. Sketchfab. Available online: https://sketchfab.com/3d-models/palms-f34a5b4ccc774d68b9fc51b3e59690db (accessed on 28 November 2020).

25. Sketchfab. Available online: https://sketchfab.com/3d-models/coconut-b68fd365187f45cf843e316d2ea23260 (accessed on 29 November 2020).
26. Pixabay. Available online: https://pixabay.com/photos/palm-trees-trees-tropical-palm-4968319/ (accessed on 29 November 2020).
27. YouTube. Available online: https://www.youtube.com/watch?v=QX4j_zHAlw8&t=146s (accessed on 29 November 2020).
28. Sketchfab. Available online: https://sketchfab.com/3d-models/cow-99d333e3b4e4470a8d7d38436489c001 (accessed on 28 November 2020).
29. Sketchfab. Available online: https://sketchfab.com/3d-models/milk-5e8d71045e2040ba8f6619d86d204cf5 (accessed on 28 November 2020).
30. Pixabay. Available online: https://pixabay.com/photos/cows-pasture-land-animals-meadow-3636937/ (accessed on 29 November 2020).
31. YouTube. Available online: https://www.youtube.com/watch?v=KjmuBo8xoCU (accessed on 29 November 2020).
32. Meullenet, J.F.; Xiong, R.; Findlay, C.J. *Multivariate and Probabilistic Analyses of Sensory Science Problems*; John Wiley & Sons: Hoboken, NJ, USA, 2008; Volume 25.
33. Pagès, J.; Berthelo, S.; Brossier, M.; Gourret, D. Statistical penalty analysis. *Food Qual. Prefer.* **2014**, *32*, 16–23. [CrossRef]
34. Velasco, C.; Salgado-Montejo, A.; Marmolejo-Ramos, F.; Spence, C. Predictive packaging design: Tasting shapes, typefaces, names, and sounds. *Food Qual. Prefer.* **2014**, *34*, 88–95. [CrossRef]
35. Kashima, H.; Hayashi, N. Basic taste stimuli elicit unique responses in facial skin blood flow. *PLoS ONE* **2011**, *6*, e28236. [CrossRef] [PubMed]
36. Robin, O.; Rousmans, S.; Dittmar, A.; Vernet-Maury, E. Gender influence on emotional responses to primary tastes. *Physiol. Behav.* **2003**, *78*, 385–393. [CrossRef]
37. Pomirleanu, N.; Gustafson, B.M.; Bi, S. Ooh, that's sour: An investigation of the role of sour taste and color saturation in consumer temptation avoidance. *Psychol. Mark.* **2020**, *37*, 1068–1081. [CrossRef]
38. Strohminger, N. Disgust talked about. *Philos. Compass* **2014**, *9*, 478–493. [CrossRef]
39. Bangcuyo, R.G.; Smith, K.J.; Zumach, J.L.; Pierce, A.M.; Guttman, G.A.; Simons, C.T. The use of immersive technologies to improve consumer testing: The role of ecological validity, context and engagement in evaluating coffee. *Food Qual. Prefer.* **2015**, *41*, 84–95. [CrossRef]
40. Campbell, M.C.; Manning, K.C.; Leonard, B.; Manning, H.M. Kids, cartoons, and cookies: Stereotype priming effects on children's food consumption. *J. Consum. Psychol.* **2016**, *26*, 257–264. [CrossRef]
41. Gupta, M.; Torrico, D.D.; Hepworth, G.; Gras, S.L.; Ong, L.; Cottrell, J.J.; Dunshea, F.R. Differences in Hedonic Responses, Facial Expressions and Self-Reported Emotions of Consumers Using Commercial Yogurts: A Cross-Cultural Study. *Foods* **2021**, *10*, 1237. [CrossRef] [PubMed]
42. Sharma, C.; Swaney-Stueve, M.; Severns, B.; Talavera, M. Using correspondence analysis to evaluate consumer terminology and understand the effects of smoking method and type of wood on the sensory perception of smoked meat. *J. Sens. Stud.* **2019**, *34*, e12535. [CrossRef]
43. Torrico, D.D.; Fuentes, S.; Viejo, C.G.; Ashman, H.; Dunshea, F.R. Cross-cultural effects of food product familiarity on sensory acceptability and non-invasive physiological responses of consumers. *Food Res. Int.* **2019**, *115*, 439–450. [CrossRef]

Article

'TeeBot': A High Throughput Robotic Fermentation and Sampling System

Nicholas van Holst Pellekaan [1,2], Michelle E. Walker [1], Tommaso L. Watson [1] and Vladimir Jiranek [1,2,*]

1 Department of Wine Science, The University of Adelaide, Glen Osmond, SA 5064, Australia; nick.vanholst@adelaide.edu.au (N.v.H.P.); michelle.walker@adelaide.edu.au (M.E.W.); tommaso.watson@adelaide.edu.au (T.L.W.)

2 Australian Research Council Training Centre for Innovative Wine Production, Adelaide, SA 5064, Australia

* Correspondence: vladimir.jiranek@adelaide.edu.au; Tel.: +61-8-313-5561

Abstract: When fermentation research requires the comparison of many strains or conditions, the major bottleneck is a technical one. Microplate approaches are not able to produce representative fermentative performance due to their inability to truly operate anaerobically, whilst more traditional methods do not facilitate sample density sufficient to assess enough candidates to be considered even medium throughput. Two robotic platforms have been developed that address these technological shortfalls. Both are built on commercially available liquid handling platforms fitted with custom labware. Results are presented detailing fermentation performance as compared to current best practice, i.e., shake flasks fitted with airlocks and sideports. The 'TeeBot' is capable sampling from 96 or 384 fermentations in 100 mL or 30 mL volumes, respectively, with airlock sealing and minimal headspace. Sampling and downstream analysis are facilitated by automated liquid handling, use of 96-well sample plate format and temporary cryo-storage (<0 °C).

Keywords: TeeBot; fermentation; high throughput; liquid handling robot; metabolite analysis

Citation: van Holst Pellekaan, N.; Walker, M.E.; Watson, T.L.; Jiranek, V. 'TeeBot': A High Throughput Robotic Fermentation and Sampling System. *Fermentation* **2021**, 7, 205. https://doi.org/10.3390/fermentation7040205

Academic Editors: Claudia Gonzalez Viejo and Sigfredo Fuentes

Received: 15 August 2021
Accepted: 18 September 2021
Published: 24 September 2021

Publisher's Note: MDPI stays neutral with regard to jurisdictional claims in published maps and institutional affiliations.

1. Introduction

The isolation or improvement of fermentatively useful strains requires that candidates be assessed in relation to each other and industry stalwarts. The options available to researchers working in this field are to date limited in terms of sample volume (Biolector-https://www.m2p-labs.com (accessed on 24 September 2021)), density, ease of use (100 mL shake flask with airlock), and cost (AMBR 15/250 https://www.sartorius.com/en/products/fermentation-bioreactors/ambr-multi-parallel-bioreactors/ambr-250-high-throughput (accessed on 24 September 2021)). 2mag AG (Muenchen, Germany) produce bioREACTOR 48, a block of 48 parallelized mini-reactors, which does offer some of the features of the system described here, but does not allow access to the sample volume for open platform measurement of analytes; however, it does offer gas control and measurement of pH and Dissolved Oxygen.

To address this technical deficit, two sampling robots along with custom labware have been developed. Both are based on the Tecan EVO200 deck (Tecan Australia Pty Ltd., Port Melbourne, VIC, Australia) with standard liquid handling arm (LiHa). The labware fit out of the instrument deck allows for strain comparison under self-anaerobic fermentation conditions, facilitates aseptic sampling into industry standardized, 96-well microplates (0.2–2 mL), with temporary storage (below zero) to inhibit continued metabolism, prior to retrieval for later analysis. The automated sampling is cost effective in terms of consumables (syringe needles), sterility and labour.

The two customized robots are novel, as to the best of the authors' knowledge there is no commercial system available with offline sampling in an industry standard assay format that offers the density of samples (96 or 384), allowing for replication, with sufficient sample volumes (100 mL or 30 mL, respectively) that more closely resemble current fermentation systems. Further the reusable bottles (100 mL) or disposable tubes allow the system to be run economically with per sample costs <3 AU$.

2. Materials and Methods

Design of Robotic Fermentation Platforms

Tecan Evo 200 liquid handling platforms were purchased from Tecan (Tecan AG, Switzerland), equipped with 'Liquid LiHa' arm, 5 mL diluters and either 2 (100 mL ferment volumes) or 4 (30 mL ferment volumes) standard (Teflon®-coated or stainless steel) washable tips (Tecan, as fitted to standard Liquid LiHa arms). All liquid handling was backed up with 20% ethanol (v/v) used as 'system liquid'.

Platform design is outlined as follows:

I. 96 ×100 mL ferments (Generation#1 instrument)

Four billet aluminum blocks were machined (DIEMOULD Engineering, Wingfield, SA, Australia) with each block holding 24 modified 100 mL Duran style bottles in a 6 × 4 array. The bottles were custom blown to increase actual volume to 140 mL, allowing for a standard ferment volume of 100 mL. Two discrete sets of coolant galleries were drilled both along and across the billets to allow for alternate crossflow connection to heater/chiller units facilitating temperature control. Airlocks (https://www.carbon3d.com/resources/case-study/tthandadelaide/ (accessed on 24 September 2021)) with replaceable silicon septa (Translucent 40 Duro 3 × 12 mm dia. Industrial Gaskets, Melrose Park, SA, Australia), were secured with GL45 aperture caps to the bottles with silicone GL45 gaskets. Retainer bars fitted to the top of the block stopped bottles being lifted by the sampling needle.

Aluminum blocks were attached to a custom-supplied, 24-position, 2mag AG (Muenchen, Germany) MIXdrive base, which allowed for mixing of bottle contents with magnetic octahedral stirrer bars (25 × 8 mm, PTFE; Cowie®, Middlesbrough, UK). A schematic of the block is shown in Figure 1 and "S1 Video Gen#1 operation.mov" video (Supplementary Materials).

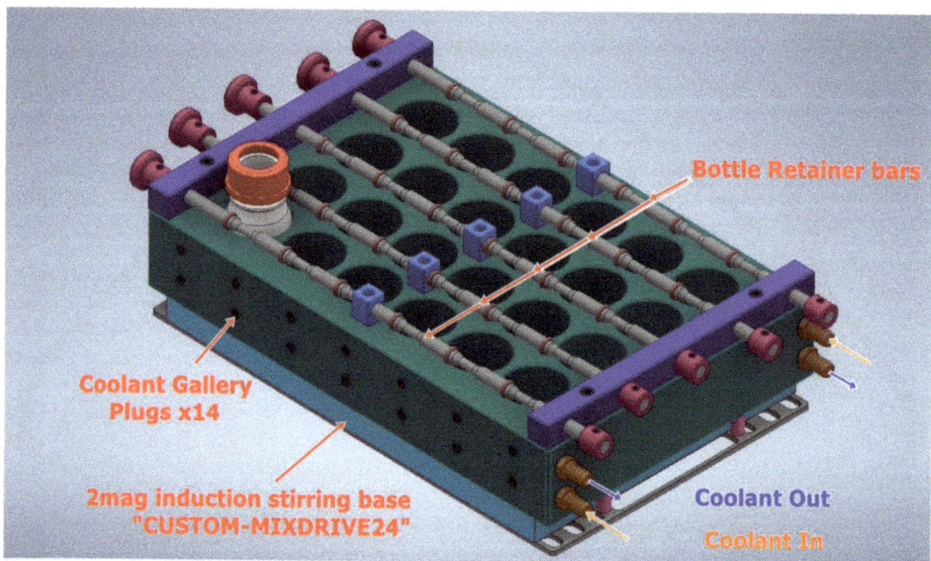

Figure 1. 24 × 100 mL temperature control and retention block (1 of 4) used to house 24 fermentation vessels.

II. 384 × 30 mL ferments (Generation#2)

Four frames were machined from sheet/billet aluminum (Witley Engineering, Gawler, SA, Australia) and assembled to hold an 8 × 12 array of (96) 50 mL screw cap polypropylene tubes (P50; Techno Plas, St Marys, SA, Australia). The 50 mL tubes have an ideal ferment volume <30 mL as attested to by the group's research [1]. A custom cut silicon gasket (Industrial Gaskets, Melrose Park, SA, Australia) provides airtight sealing between the tubes and each of 96 CLIP printed airlocks. Each airlock retains a silicon septum (3 × 12 mm) that is held in place by an aluminum retainer plate. This plate is subsequently torqued into place with a bolt/washer fastener from above. Assembly of the components is shown in Figure 2 and "S2 Video Gen#2 operation" video (Supplementary Materials).

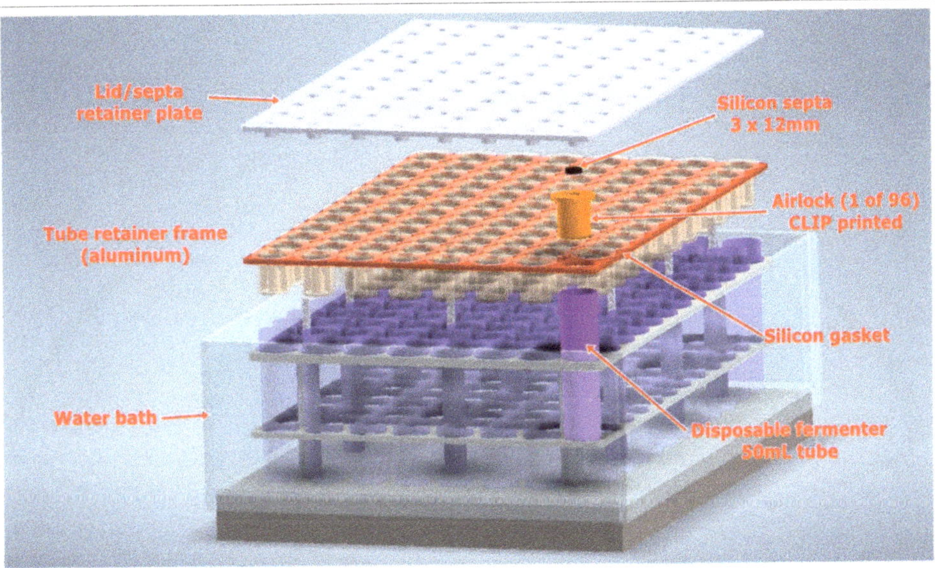

Figure 2. 96 × 30 mL rack and airlock assembly (1 of 4) used to house 96 fermentation vessels.

Temperature control for the 4 × 96 frames was maintained by immersion into indexed water baths (Witley Engineering, Gawler, SA, Australia) fitted with frame retainers to prevent lifting by the sampling needle. Frames were indexed/located into the bath on four lugs that key into the aluminum frame. Laminar flow in the bath was encouraged by a "parapet" spillway across the effluent end, with water returning to heater/chiller units housed beneath the instrument. Mixing was mediated by magnetic cross stirrer bars (8 × 20 mm, PFTE; Cowie®, Middlesbrough, UK), driven by a custom-supplied 96-position, 2mag AG MIXdrive base beneath the water bath, which is removable for ease of cleaning.

III. Airlock design

Several iterations of engineered airlocks were tested in the Generation#1 instrument leading to the adoption of CLIP printed airlocks (Printed using Carbon3d instruments by The Technology House, Streetsboro, Ohio USA) for both generations (Figure 3). These are dishwasher safe and autoclavable (15 min, 121 °C). Airlocks house a 3 × 12 mm silicon rubber septum either by a retainer plate with the GL45 collar (Generation#1) or a machined plate bolted to the frame. In either case, the septa and gaskets are compressed to prevent gas escape. Sterilization of the fermentation vessels was achieved by autoclaving, with 1–2 mL of sterile water added to each airlock prior to commencement of fermentation.

Figure 3. Comparison of fermentation vessels.

From Left rear: 250 mL shake flask, Generation #1 fermenter assembled (middle rear) and component parts (to front), Generation #2 fermenter (right rear) and component parts (to front) note gasket and septa retainer omitted as they are integral to the rack assembly.

IV. Temperature Control

Temperature regulation (for both instruments) was performed by two CORIO CD-200F (Julabo GmbH, Seelbach, Germany) recirculating heater/chiller units each. This allowed for two discrete fermentation temperatures to be tested per run, comprising 50% of the available fermentation vessels in each case (i.e., 48 or 192 vessels, respectively). Fermentations in the Generation#1 instrument are a 'closed loop' whilst for the Generation#2 instrument they are in an 'open bath' system. Uniformity was assessed using a HH147U 4 channel data logging temperature gauge fitted with K-type thermocouples. Readings were taken after 2 min to ensure stable readings.

V. Software

Sampling schemes, times and volumes, as well as sanitisation routines between sampling cycles were controlled using 'Tecan Evoware Std' software supplied with the instruments. Multiple sampling cycles could be scheduled to start at defined times in the future, limited only by the number of available sample deposition locations (microplates).

VI. Laboratory scale fermentations

Fermentation performance was conducted in 100 mL of 0.22 μM filter-sterilized Chemically Defined Grape Juice Medium (CDGJM) in a 'TeeBot' flask (Generation#1) or 250 mL shake flask. The CDGJM was as described in [2], except nitrogen was as ammonium chloride and amino acids. Fermentations were inoculated to a density of 5×10^6 cells mL^{-1} of *Saccharomyces cerevisiae* grown in CDGJM starter (50 g L^{-1} glucose, 50 g L^{-1} fructose, supplemented with 10 mg L^{-1} ergosterol and 540 mg L^{-1} Tween 80). Sugar consumption (glucose, fructose) was measured during the course of the fermentation by enzyme assay [2–15].

3. Results

Any engineered solution to the challenges posed by high throughput fermentation needs to prove empirically that it does not impose artefacts on the data produced. Furthermore, a range of potential issues need to be addressed: (i) maintenance of temperature uniformity, (ii) establishment of an anaerobic environment, (iii) preservation of sterility of individual samples per se during sampling, (iv) reproducibility of fermentations without artefact, comparable to or better than 'best practice' and (v) provision of sufficient sample density/throughput to address experimental requirements.

3.1. Maintenance of Temperature Uniformity

For the Generation#1 instrument, temperature uniformity was assessed using a HH147U (Omega instruments) 4 channel data logging thermocouple based digital thermometer. Flasks containing 100 mL of deionized (DI) water were allowed to equilibrate for 4–5 h at 30 °C with readings taken after 2 min stabilization time. Only the two outermost blocks were assessed, as they were the most likely to see variation through convective heat loss. Thermal exchange from the heater/chiller units was mediated by DI water, with temperature set to 30 °C (Figure 4). All four blocks were insulated with a 12 mm high density neoprene jacket to minimize heat exchange.

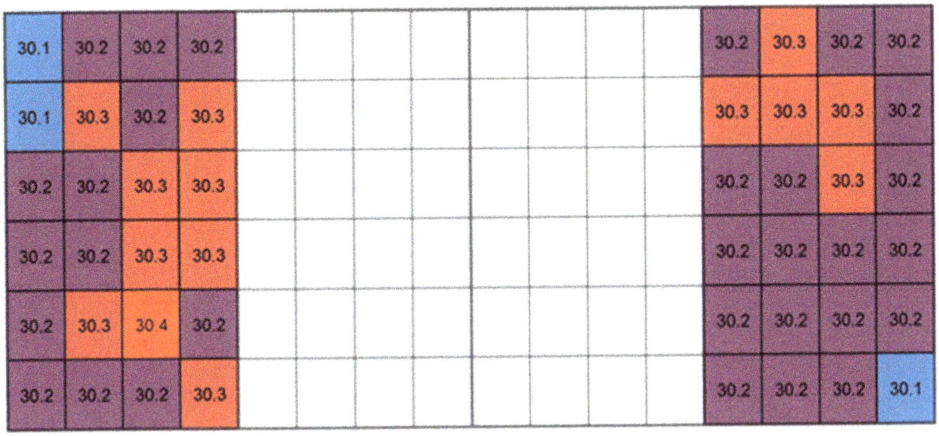

30.1	30.2	30.2	30.2									30.2	30.3	30.2	30.2
30.1	30.3	30.2	30.3									30.3	30.3	30.3	30.2
30.2	30.2	30.3	30.3									30.2	30.2	30.3	30.2
30.2	30.2	30.3	30.3									30.2	30.2	30.2	30.2
30.2	30.3	30.4	30.2									30.2	30.2	30.2	30.2
30.2	30.2	30.2	30.3									30.2	30.2	30.2	30.1

Figure 4. Heat-map of 'TeeBot' flask temperatures in Generation#1 instrument. An average temperature of 30.2 ± 0.1 °C was observed across wells held in a set temperature of 30 °C (n = 48).

Temperature uniformity for the Generation#2 system was not tested as the individual tubes were in a 'closed' system not accessible to the temperature probe. However, the immersion bath was found to be thermally stable using the digital thermometer used above. Flow rates of 5 L/min ensured complete exchange of liquid in the bath each minute (data not shown).

3.2. *Establishment of an Anaerobic Environment*

Resazurin (a redox indicator) was used to determine whether an anaerobic environment was established in the different vessels (flask, bottle, tube), following yeast inoculation in CDGJM (data not shown). The blue resazurin solution (0.1 g/100 mL water), when added to the culture at 1 mL L^{-1}, is irreversibly reduced to resorufin (pink), which is subsequently reduced to colourless dihydroresorufin in a reversible secondary reaction when anaerobic conditions are attained through yeast proliferation [16]. This occurs rapidly (within 4 to 6 h in flasks without N_2 sparging [15]. Furthermore, the presence of glucose and fructose (200 g L^{-1} total sugar), ensures that *Saccharomyces cerevisiae* undergo fermentation rather than respiration, as Crabtree positive yeast, producing ethanol [17] regardless of oxygen content [13].

3.3. *Preservation of Sterility of Individual Samples Per Se during Sampling*

The system cannot prove its utility unless it can be both sterilized at the outset of an experiment and subsequently not lead to serial contamination during sampling. The LiHa's two needles (Generation#1) or four needles (Generation#2) were run through a decontamination cycle in which methanol (100%) was used as a bactericidal/fungicidal wash between each sampling event. In addition, the tubing connected to the LiHa was flushed with 20% ethanol (*v*/*v*) between sampling from a 20 L plastic jerry can as a reservoir. Inclusion of regular medium-only (un-inoculated) controls has proven this method and the sampling regime to be aseptic even when using nutrient rich media such as CDGJM and YEPD (Figures 5 and 6) or filter-sterilized (0.2 μM) white grape juice [4,5].

3.4. *Reproducibility of Fermentations without Artefact, Comparable to or Better Than 'Best Practice'*

To be deemed a successful solution to the problem posed by high throughput fermentations, this system must be capable of producing fermentation data that is consistent and comparable to current best practices. Fermentation kinetics from sample to sample must be comparable (with small replica variance) and it must meet or exceed current lower throughput alternatives. The current 'best practice' fermentation analysis, which is lower throughput, uses ~100 mL of media in a 250 mL flask, orbitally agitated in a controlled temperature environment.

In an experiment comparing identical media, inoculation rates and yeast strains, similar fermentation kinetics should be expected. Differences between two distinct methods of incubation are expected as variations in headspace and absolute temperature are not able to be eliminated. A comparison of a single yeast (EC1118) in CDGJM (200 g L^{-1}, 450 mg L^{-1} FAN) was conducted either in a 'TeeBot' (30 replicates) or 100 mL shake flasks (40 replicates; Figure 5). Breaking of sample sterility was observed in two of the three flask replicates (no yeast; blue dashed line) as sugar consumption after ~50 h. Whilst there is an observed difference in fermentation kinetics, the relative variability is acceptable between the methodologies. The variance between the replicates with flasks was 2.3% (AUC; 4388.8 ± 100.3, n = 45) compared to 11% (AUC; 4930.6 ± 544.4, n = 30) using the 'TeeBot'. The larger variation with the latter is greatly compensated for by the advantages of the platform regarding sampling, which can be programmed remotely, with samples taken (0.2–1 mL), arrayed for high-throughput enzymatic analysis of metabolites and temporarily stored at 0 °C until they are removed for cryo-storage or direct analysis. Considerable intervention is required with flasks, which require manual sampling and clarification (usually of 1 mL samples) prior to cryo-storage or transfer into plates for cryo-storage and later analysis.

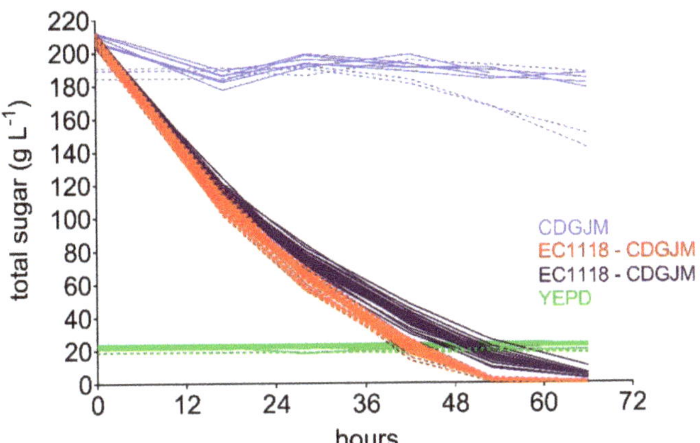

Figure 5. Comparison between shake flask and Generation #1 platform based fermentations using a single yeast (EC1118). The strain was fermented as technical replicates in 'Teebot' bottles (n = 30; purple solid line) and flasks (n = 40; red dashed line). Sterility controls (no yeast) were included as technical replicates for YEPD (3 flasks, green dashed line and 12 Teebot bottles, green solid line) and CDGJM (3 flasks, blue dashed line and 6 'Teebot' bottles, blue solid line).

Fermentation was completed more rapidly in shake flasks. This is most likely due to larger headspace and initial oxygen availability at the outset of fermentation, which would allow the yeast to build biomass more rapidly, enabling more rapid sugar consumption [18] although this was not confirmed here. In the Jiranek laboratory, ferment vessels are not routinely sparged with nitrogen (to displace the air) prior to fermentation. This study has shown that the ferments become anaerobic within 4 to 6 h of inoculation [15]. More important, is that the fermentation trend of a yeast in a given medium is reproducible and consistent within a particular apparatus setup.

3.5. Provision of Sample Density Sufficient to Address Experimental Needs

To address a need for high(er) throughput fermentation, several basics were indispensable in the design of the two fermentation platforms; consistency and transferability of data related to the fermentation performance under given condition(s). Sampling was semi-automated, with removal of deposition plates limited to when they needed to be replaced (up to 2 days depending upon sampling regime and platform). Transferability of data was evaluated as a single experiment where the fermentation performance of four commercial wine yeast was compared in CDGJM (Figure 6).

The sugar curves produced using the Generation#1 and #2 platforms were very similar (Figure 6C), although increased variation was noted for Generation#2 (Figure 6B), which is most likely a result of the number of replicates and limitations of the sugar analysis itself. Importantly, the Area Under the Curve values [6] for each strain were similar (Table 1), indicative of fermentation phenotype being reproducible when scaled up as common with screening large numbers.

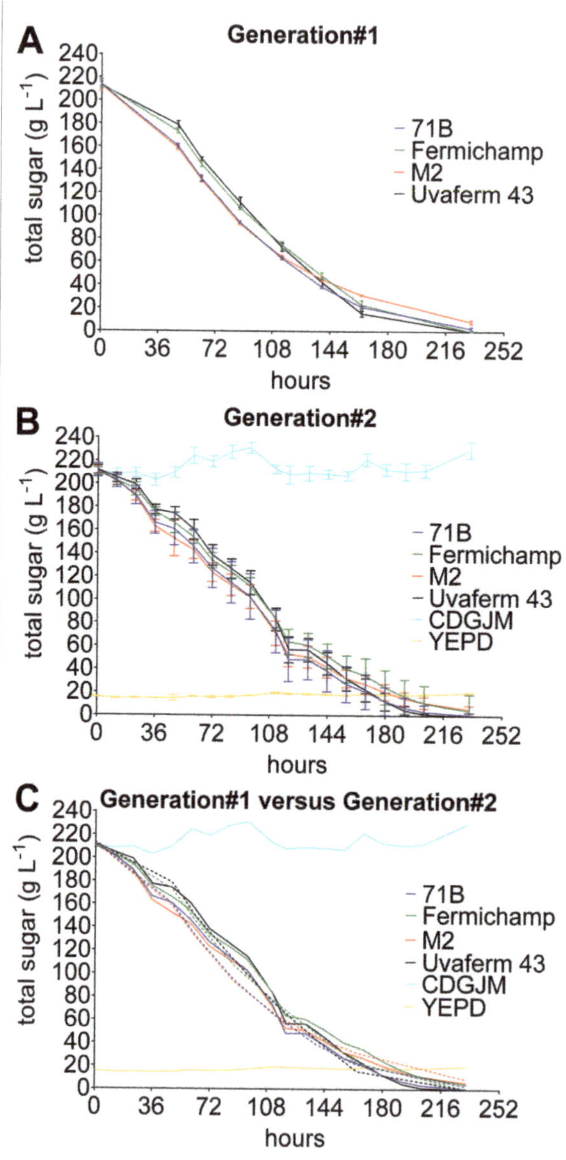

Figure 6. Comparison of sugar consumption by four commercial yeasts fermented as 100 mL cultures (Generation#1) versus 30 mL cultures (Generation#2) in CDGJM. Four commercial wine yeast were inoculated from YEPD starter cultures at 1% in CDGJM (115 g L^{-1} glucose, 115 g L^{-1} fructose; 350 mg L^{-1} FAN). Graphs depict residual sugar (g L^{-1}) over time (hours). (**A**) 3 replicates per strain (Generation#1). (**B**) 19 replicates per strain (Generation#2). (**C**) Overlay of the sugar curves produced from the two 'Teebot' platforms. Sterility controls (9 replicates) were included in the Generation#2 experiment (YEPD, 20 g L^{-1} sugar and CDGJM, 230 g L^{-1} sugar).

Table 1. Comparison of fermentation performance of four wine yeast fermented in CDGJM (230 g L^{-1} sugar, 350 mg L^{-1} FAN) in 100 mL (Generation#1) and 30 mL (Generation#2) volumes. The mean Area Under the Curve (AUC) value was calculated for each strain, together with the standard deviation from the mean.

Strain	Generation#1	Generation#2
	AUC ± Stdev (*n* = 3)	AUC ± Stdev (*n* = 19)
71B	18,887 ± 101	19,261 ± 524
Fermichamp	20,383 ± 196	21,269 ± 448
M2	19,608 ± 155	19,641 ± 311
Uvaferm43	20,181 ± 206	20,693 ± 285

The platform format chosen for a particular experiment will be dependent not only on sample number but also downstream analysis requirements. For example, Generation#1 and #2 are suitable for several analyses requiring <1 mL samples such as for organic acids, sugars and nitrogen (Megazyme assay kits), which are useful for monitoring alcoholic fermentation (sugar and nitrogen) and secondary malolactic fermentation (malic acid, lactic acid). End-of-fermentation sample volume requirements range from ≤1 mL for HPLC analysis (sugars, organic acids, acetaldehyde, glycerol and ethanol; [9–14] and ≥10 mL for GCMS analysis of secondary metabolites or volatiles (Generation#1; [5]; [8] and #2; [4]. Whilst determination of sulfur dioxide by aspiration/titration [11] and ethanol by Alcolyzer (Anton Paar GmbH, Graz, Austria) requires ~40–50 mL (Generation#1 only). These fermentation platforms have been successfully used to complement large-scale screenings using microplate (0.2 mL) scale fermentations, where only sugar (and nitrogen) consumption can be monitored [7]. Complete Generation#1 and Generation#2 platforms are shown in Figure 7.

Generation#1 platform – 96 fermentations

Generation#2 platform – 384 fermentations

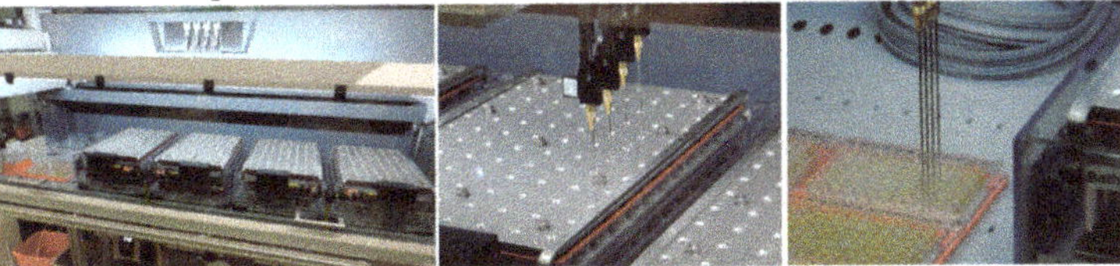

Figure 7. Generation#1 and #2 'Teebot' fermentation and sampling systems.

4. Discussion

With the successful application of both generations of automated fermentation platforms and the publication of several reports [2,5–10] this study achieved all the criteria defined earlier in the design and implementation of a high throughput fermentation platform. There are no other systems available that match the 'Teebots' for performance in their ability to screen large numbers of candidate fermentative organisms. In terms of flexibility in throughput, scale of fermentation is important as the larger volumes avoid the shortcomings of fermentation in micro-titer plates, namely inoculation rate, evaporation, and replication. Care is needed to prevent evaporation, with the micro-titer plates requiring adequate sealing and humidity, as the overall volume size is only 0.2–1 mL. Achieving consistent inoculation can be problematic, with inoculation volumes typically being only 5–10 μL per 200 μL of media, which equates to ~5 × 10^6 yeast cells mL^{-1} recommended for industrial (https://www.awri.com.au/industry_support/winemaking_resources/wine_fermentation/yeast-rehydration/ (accessed on 24 September 2021)) and laboratory scale fermentation [10]. Monitoring of fermentation requires several replicates of a single plate, which is labour intensive, especially with large numbers, such that replication is often forgone until later in the screening process [10,14]. These compromises can be avoided through the use of the 'TeeBot' platforms, which allow for larger scale volumes, more suited for inoculation by cell density (cells per mL) and greater analysis of metabolites and fermentation kinetics as well as automated inoculation "S3 Video Gen#2 inoculation" (Supplementary Materials). Most importantly, the tedious but important task of evaluating new wine yeast and bacteria can be done with minimal intervention by the researcher.

5. Emerging Trends and Future Prospects

The process of bioprospecting for novel strains for application in food production is an ever-increasing sphere of research. The ability to collect strains from the environment is considerably constrained by the ability of researchers to characterise the strains and assess their suitability for application. In the field of rapid characterisation of large pools of strains either obtained after evolutionary experiments, mutagenesis, breeding or bioprospecting, there is a need to be able to assess fermentative performance of sufficiently replicated samples in volumes larger than those represented in a microplate. Cost is always a concern, and with the reusable (100 mL) or disposable (30 mL) ferment vessels neither are prohibitively expensive once the capital outlay has been achieved. As such, systems such as the 'TeeBot' robotic systems described here will become standard equipment in a fermentation-related research laboratory.

Supplementary Materials: The following are available online at https://www.mdpi.com/article/10.3390/fermentation7040205/s1, "S1 Video Gen#1 operation.mov": Sampling for generation#1, "S2 Video Gen#2 operation.m4v": Sampling for generation#2. "S3 Video Gen#2 inoculation.m4v": Inoculation for generation#2.

Author Contributions: The following contributions to the project were made by the co-authors: conceptualization, V.J. and T.L.W.; instrument development, T.L.W.; validation, T.L.W.; writing—original draft preparation, N.v.H.P. and M.E.W.; writing—review and editing, M.E.W., V.J. and N.v.H.P.; funding acquisition and supervision, V.J. All authors have read and agreed to the published version of the manuscript.

Funding: This project was supported by funding from Wine Australia [UA1101, UA1302, UA 1803-2.1] and The Australian Research Council Training Centre for Innovative Wine Production (www.ARCwinecentre.org.au; project number IC170100008), which is funded by the Australian Government with additional support from Wine Australia and industry partners. Wine Australia invests in and manages research, development and extension on behalf of Australia's winegrowers and winemakers and the Australian Government.

Institutional Review Board Statement: Not applicable.

Informed Consent Statement: Not applicable.

Data Availability Statement: Data is published within paper or available on request to the corresponding author.

Acknowledgments: The authors acknowledge the contribution from the Wine Microbiology laboratory of Vladimir Jiranek, in providing data from experiments using the two robotic platforms.

Conflicts of Interest: The authors declare there are no conflicts of interest regarding the study.

Abbreviations

CLIP—Continuous Liquid Interface Production; DO—dissolved oxygen; FAN—free amino nitrogen

References

1. Bartle, L.; Peltier, E.; Sundstrom, J.F.; Sumby, K.; Mitchell, J.G.; Jiranek, V.; Marullo, P. QTL mapping: An innovative method for investigating the genetic determinism of yeast-bacteria interactions in wine. *Appl. Microbiol. Biotechnol.* **2021**, *105*, 5053–5066. [CrossRef] [PubMed]
2. McBryde, C.; Gardner, J.M.; de Barros Lopes, M.; Jiranek, V. Generation of novel wine yeast strains by adaptive evolution. *Am. J. Enol. Vitic.* **2006**, *57*, 423.
3. *Boehringer-Mannheim, Methods of Biochemical Analysis and Food Analysis Using Test-Combinations, D-Glucose/D-Fructose: UV Method*; Boehringer-Mannheim GmbH Biochemicals: Mannheim, Germany, 1989; pp. 50–55.
4. Gardner, J.M.; Walker, M.E.; Boss, P.K.; Jiranek, V. The effect of grape juice dilution on oenological fermentation. *bioRxiv* **2020**. [CrossRef]
5. Hranilovic, A.; Gambetta, J.M.; Schmidtke, L.; Boss, P.K.; Grbin, P.R.; Masneuf-Pomarede, I.; Bely, M.; Albertin, W.; Jiranek, V. Oenological traits of *Lachancea thermotolerans* show signs of domestication and allopatric differentiation. *Sci. Rep.* **2018**, *8*, 14812. [CrossRef] [PubMed]
6. Liccioli, T.; Chambers, P.J.; Jiranek, V. A novel methodology independent of fermentation rate for assessment of the fructophilic character of wine yeast strains. *J. Ind. Microbiol. Biotechnol.* **2011**, *38*, 833–843. [CrossRef] [PubMed]
7. Liccioli, T.; Tran, T.M.T.; Cozzolino, D.; Jiranek, V.; Chambers, P.J.; Schmidt, S.A. Microvinification—How small can we go? *Appl. Microbiol. Biotechnol.* **2010**, *89*, 1621–1628. [CrossRef] [PubMed]
8. Morgan, S.C.; Haggerty, J.J.; Jiranek, V.; Durall, D.M. Competition between *Saccharomyces cerevisiae* and *Saccharomyces uvarum* in controlled Chardonnay wine fermentations. *Am. J. Enol. Vitic.* **2020**, *71*, 198–207. [CrossRef]
9. Morgan, S.C.; Haggerty, J.J.; Johnston, B.; Jiranek, V.; Durall, D.M. Response to sulfur dioxide addition by two commercial *Saccharomyces cerevisiae* strains. *Fermentation* **2019**, *5*, 69. [CrossRef]
10. Peter, J.J.; Watson, T.L.; E Walker, M.; Gardner, J.; A Lang, T.; Borneman, A.; Forgan, A.; Tran, T.; Jiranek, V. Use of a wine yeast deletion collection reveals genes that influence fermentation performance under low-nitrogen conditions. *FEMS Yeast Res.* **2018**, *18*, foy009. [CrossRef] [PubMed]
11. Rankine, B.; Pocock, K. Alkalimetric determination of sulphur dioxide in wine. *Aust. Wine Brew. Spirit Review* **1970**, *88*, 40–44.
12. Sumby, K.; Niimi, J.; Betteridge, A.; Jiranek, V. Ethanol-tolerant lactic acid bacteria strains as a basis for efficient malolactic fermentation in wine: Evaluation of experimentally evolved lactic acid bacteria and winery isolates. *Aust. J. Grape Wine Res.* **2019**, *25*, 404–413. [CrossRef]
13. Tai, S.L.; Boer, V.M.; Daran-Lapujade, P.; Walsh, M.C.; de Winde, J.H.; Daran, J.M.; Pronk, J.T. Two-dimensional transcriptome analysis in chemostat cultures. Combinatorial effects of oxygen availability and macronutrient limitation in *Saccharomyces cerevisiae*. *J. Biol. Chem.* **2005**, *280*, 437–447. [CrossRef] [PubMed]
14. Walker, M.E.; Gardner, J.M.; Vystavelova, A.; McBryde, C.; Lopes, M.D.B.; Jiranek, V. Application of the reuseable, KanMX selectable marker to industrial yeast: Construction and evaluation of heterothallic wine strains of *Saccharomyces cerevisiae*, possessing minimal foreign DNA sequences. *FEMS Yeast Res.* **2003**, *4*, 339–347. [CrossRef]
15. Walker, M.E.; Nguyen, T.D.; Liccioli, T.; Schmid, F.; Kalatzis, N.; Sundstrom, J.F.; Gardner, J.M.; Jiranek, V. Genome-wide identification of the Fermentome; genes required for successful and timely completion of wine-like fermentation by *Saccharomyces cerevisiae*. *BMC Genom.* **2014**, *15*, 552. [CrossRef] [PubMed]
16. Bitacura, J.G. The use of baker's yeast in the Resazurin Reduction Test: A simple, low-cost method for determining cell viability in proliferation and cytotoxicity assays. *J. Microbiol. Biol. Educ.* **2018**, *19*, 19–87. [CrossRef] [PubMed]
17. Wardrop, F.; Liti, G.; Cardinali, G.; Walker, G. Physiological responses of Crabtree positive and Crabtree negative yeasts to glucose upshifts in a chemostat. *Ann. Microbiol.* **2004**, *54*, 103–114.
18. Aceituno, F.F.; Orellana, M.; Torres, J.; Mendoza, S.; Slater, A.W.; Melo, F.; Agosin, E. Oxygen response of the wine yeast *Saccharomyces cerevisiae* EC1118 grown under carbon-sufficient, nitrogen-limited enological conditions. *Appl. Environ. Microbiol.* **2012**, *78*, 8340–8352. [CrossRef] [PubMed]

Article

Dynamic Optimisation of Beer Fermentation under Parametric Uncertainty

Satyajeet Bhonsale, Wannes Mores and Jan Van Impe *

BioTeC+, Department of Chemical Engineering, KU Leuven Technology Campus Ghent, 9000 Ghent, Belgium; satyajeetsheetal.bhonsale@kuleuven.be (S.B.); wannes.mores@kuleuven.be (W.M.)
* Correspondence: jan.vanimpe@kuleuven.be

Abstract: Fermentation is one of the most important stages in the entire brewing process. In fermentation, the sugars are converted by the brewing yeast into alcohol, carbon dioxide, and a variety of by-products which affect the flavour of the beer. Fermentation temperature profile plays an essential role in the progression of fermentation and heavily influences the flavour. In this paper, the fermentation temperature profile is optimised. As every process model contains experimentally determined parameters, uncertainty on these parameters is unavoidable. This paper presents approaches to consider the effect of uncertain parameters in optimisation. Three methods for uncertainty propagation (linearisation, sigma points, and polynomial chaos expansion) are used to determine the influence of parametric uncertainty on the process model. Using these methods, an optimisation formulation considering parametric uncertainty is presented. It is shown that for the non-linear beer fermentation model, the linearisation approach performed worst amongst the three methods, while second-order polynomial chaos worked the best. Using the techniques described below, a fermentation process can be optimised for ensuring high alcohol content or low fermentation time while ensuring the quality constraints. As we explicitly consider uncertainty in the process, the solution, even though conservative, will be more robust to parametric uncertainties in the model.

Keywords: beer fermentation; stochastic dynamic optimisation; uncertainty

Citation: Bhonsale, S.; Mores, W.; Van Impe, J. Dynamic Optimisation of Beer Fermentation under Parametric Uncertainty. *Fermentation* **2021**, *7*, 285. https://doi.org/10.3390/fermentation7040285

Academic Editors: Claudia Gonzalez Viejo and Sigfredo Fuentes

Received: 4 November 2021
Accepted: 25 November 2021
Published: 28 November 2021

1. Introduction

Consumption of alcoholic beverages like beer, wine, and spirits has always played an important role in food security and health [1]. With its history extending to more than 7 millennia [2], beer is one of the most popular beverages in the world. The global beer consumption in 2019 was estimated at around 189.05 million kilolitres [3]. The beer sector is a major contributor to the European Union's economy with more than 10,300 active breweries providing over 130,000 jobs. In 2018, the beer sector contributed over €55 billion to the EU's economic growth [4]. With 340 breweries in Belgium producing over 1500 different beers [5], the beer culture of Belgium has been inscribed on UNESCO's Intangible Cultural Heritage of Humanity list [6]. According to Belgian law [7], beer is "*the beverage obtained after alcoholic fermentation of a wort prepared primarily from starch and sugary raw materials, of which at least 60% barley or wheat malt, as well as hops (possibly in processed form) and brewing water*". Thus, the main ingredients which form beer are only barley malt (i.e., a starch source), hops, yeast, and water.

Nevertheless, beer production is a complex process with a multitude of processing steps involved. The barley is converted into malt during the malting process. Malting converts the hard barley grains to friable malt by producing and activating various enzymes. The malt is then milled, and mixed with water to convert the starch and proteins into fermentable sugars through a process known as mashing. The mashed product, known as wort, is then boiled with hops (although hops can be added at different stages of the beer production). The residual hops and other products coagulated from the boiling process (called the trub) are removed from the wort. The clarified wort is cooled and transferred to a fermentation tank. There it is pitched with yeast and fermented to produce ethanol, carbon dioxide, and secondary metabolites.

Although the entire process involves only four ingredients, a plethora of chemical compounds are involved in the entire brewing process. The interaction of these chemical species gives each beer its characteristic flavour. According to Amerine et al. [8], flavour is described as "*the sum of perceptions resulting from stimulation of the sense ends that are grouped together at the entrance of the alimentary and respiratory tracts*". Flavour comprises (i) odour: perception of volatile compounds in the nasal cavity, (ii) aroma: perception of volatiles which pass via the nasopharyngeal passage to the olfactory epithelium, (iii) taste: perception of soluble substances on the taste buds on the surface of the tongue, and (iv) mouthfeel: physical perception of beer in the mouth [9]. Although only trace amounts of volatile esters are present in beer, they have a large influence on its flavour profile. Some common esters in beer are ethyl acetate, ethyl caprylate, ethyl octanoate, isoamyl acetate, etc. [10–13]. Apart from the esters, other volatile compounds include higher alcohols (e.g., amyl alcohol), carbonyl compounds (e.g., 2,3-butanedione, commonly called diacetyl), ketones, aldehydes, etc. [11]. Although all these compounds are produced throughout the brewing process, a majority are produced as metabolic intermediates or by-products during the fermentation step [12]. This makes fermentation a key process in the brewing chain.

Fermentation is an exothermic process in which the yeast converts the sugars in the wort to ethanol, carbon dioxide, and many other flavour-inducing compounds. The cooled wort is transferred to a cylindro-conical fermentation tank and the yeast is pitched. Along with the pitching rate and dissolved oxygen, fermentation is influenced by the temperature [14]. Temperature has a strong effect on yeast metabolism. Most brewing yeasts have optimum growth temperatures between 30 and 34 °C. However, fermentation is carried out at much lower temperatures. Typically, lagers are fermented around 10 °C, while ales are fermented at 22 °C. At elevated temperatures, the fermentation is vigorous and leads to excessive loss of volatiles and formation of undesired by-products [15]. However at higher temperatures, the time required for fermentations is reduced. Thus, the brewmaster must control the fermentation temperature such that fermentation is accelerated while still maintaining the beer quality profile. Traditionally, brewers have relied on their experience and traditional recipes for temperature control. In this paper, a computer-aided optimal control profile for the fermentation temperature is proposed.

Such an optimisation of the temperature profile is not novel. Several authors have proposed optimisation strategies to obtain a dynamic temperature profile by considering a variety of objectives. An objective describes the goal of the optimisation. In context of beer fermentation, some of the possible objectives are maximisation of ethanol concentration, minimisation of by-product formation, minimisation of batch time, or even a combination of these. Carrillo-Ureta et al. [16] used genetic algorithms to determine the temperature profiles. Their objective was the combination of five different objectives: maximise ethanol concentration, minimise two by-products, minimise batch time, and minimise jumps in the temperature profile. Xiao et al. [17] used a stochastic ant colony algorithm with similar objectives. In Rodman and Gerogiorgis [18], a weighted objective of maximum ethanol and minimum batch time was optimised using a combination of simulated annealing and high fidelity simulations. Bosse and Griewank [19] made use of forward–backward sweeping methods based on Pontryagin's maximum principle to solve a modified version of the problem considered in Carrillo-Ureta et al. [16]. Apart from the single objective studies, several multiobjective studies have also been reported. Andrés-Toro et al. [20] also used genetic algorithms to solve the multiobjective optimisation problem by considering a variety of objectives. Rodman and Gerogiorgis [14] used a simplified weighted-sum-type approach for multiobjective optimisation with two objectives: maximum ethanol and minimum batch time. In Rodman et al. [21], a stochastic "*Strawberry*" algorithm is used to determine the Pareto front between the contradicting objectives.

A common requirement for all the optimisation studies mentioned above is the process model. A process model is a mathematical abstraction of the reality. Describing the plethora of chemical species produced during fermentation [12,22] would lead to an extremely complex mathematical model. It is thus common to use reduced-order models which only consider key chemical compounds. A variety of process models have been proposed to describe the fermentation process [23–26]. The model developed in de Andrés-Toro et al. [23] is based on extensive experimentation on industrial-scale fermentation and has been widely used in optimisation studies. All models contain parameters which have to estimate from experimental data.

As all experiments inherently have noise (either measurement noise, or in the case of yeast, biological variability), the parameters estimated from the data are uncertain. Although it is possible to reduce the uncertainty (i.e., improve the parameter accuracy) by performing more and better experiments, and by using better sensors, it is impossible to completely eliminate the uncertainty. Thus, any optimisation study based on a mathematical model must take this parametric uncertainty into consideration. Use of inaccurate parameters can lead to constraint violations, which in reality might affect the safety of the process or the quality of the product. In this paper, an optimisation strategy to determine the fermentation temperature profile by including the parametric uncertainty in the process model is presented. Following all the optimisation studies mentioned earlier, the model developed by de Andrés-Toro et al. [23] is used.

In the following sections, the dynamic optimisation formulation under parametric uncertainty is presented. The concept and techniques of uncertainty propagation are introduced. This is followed by the description of the de Andres-Toro model for beer fermentation. The in-house tool used for the dynamic optimisation is briefly presented. The temperature profile obtained after the optimisation is then discussed in the results section. Finally, the conclusions section summarises the main findings of this paper.

2. Materials and Methods

In this section the fermentation model used for optimisation is described. Then, the robustified dynamic optimisation formulation is presented. The three techniques for uncertainty propagation are discussed. These are the linearisation, sigma point, and polynomial chaos expansion methods. Next, the approach to evaluate the solutions based on Monte Carlo simulation is presented. Finally, the software used is briefly discussed.

2.1. Fermentation Model

As mentioned, the fermentation model developed by de Andrés-Toro et al. [23] is used in this study. It is assumed that the yeast pitched to the wort is immediately suspended and consists of active, latent, and dead yeast cells. The latent yeast cells become active over time and eventually die. Only the active cells contribute to the fermentation. The fermentation is differentiated into two phases: a lag phase in which the latent cells convert to active cells and the active phase in which fermentation occurs. The active phase begins when 80% of the latent cells are activated. The total concentration of yeast cells suspended in the wort ($C_{X,sus}$) at any time is given by

$$C_{X,sus}(t) = C_{X,act}(t) + C_{X,lat}(t) + C_{X,dead}(t) \tag{1}$$

where the subscripts act, lat, and dead correspond to the concentration of active, latent, and dead yeast cells.

In the lag phase, the dead cells settle in the wort and the latent cells are activated. The rate of settling (μ_{SD}) depends on the wort density and concentration of carbon dioxide. The wort density is related to the initial concentration (C_{S0}), while the carbon dioxide concentration is related to ethanol concentration (C_{eth}). The rate of yeast activation (μ_L) is related to the temperature via an Arrhenius-type equation. The lag phase can be modelled as,

$$\frac{dC_{X,sus}(t)}{dt} = -\mu_{SD}\, C_{X,dead}(t), \qquad t < t_{lag} \tag{2}$$

$$\frac{dC_{X,act}(t)}{dt} = \mu_L\, C_{X,lat}(t), \qquad t < t_{lag} \tag{3}$$

$$\frac{dC_{X,dead}(t)}{dt} = \mu_{DT}\, C_{X,act}(t), \qquad t < t_{lag} \tag{4}$$

Once the active phase begins, the active yeast cells start reproducing and dying (with a rate, μ_{DT}), while the latent cells continue to get activated. The dead yeast cells continue to settle to the bottom of the fermentation tank. The biomass dynamics in the active phase are described as

$$\frac{dC_{X,act}(t)}{dt} = \mu_X\, C_{X,act}(t) - \mu_{DT}\, C_{X,act}(t) + \mu_L\, C_{X,lat}(t), \qquad t > t_{lag} \tag{5}$$

$$\frac{dC_{X,dead}(t)}{dt} = \mu_{DT}\, C_{X,act}(t) - \mu_{SD}\, C_{X,dead}(t), \qquad t > t_{lag} \tag{6}$$

The sugar consumption is proportional to the active yeast cell concentration and the uptake rate μ_S follows Michaelis–Menten kinetics.

$$\frac{dC_S(t)}{dt} = -\mu_S\, C_{X,act}(t) \tag{7}$$

Similarly, the rate of ethanol formation μ_{eth} also follows Michaelis–Menten kinetics. However, as experimental data showed a decrease in the rate with time, an inhibition factor (f) is included. The evolution of ethanol concentration is expressed as

$$\frac{dC_{eth}(t)}{dt} = f\, \mu_{eth}\, C_{X,act}(t) \qquad t > t_{lag} \tag{8}$$

Two by-products are considered: diacetyl (C_{DY}) and ethyl acetate (C_{EA}). Diacetyl or 2,3-butanedione is a carbonyl compound which is produced primarily in the early phases of fermentation. At a later stage, diacetyl is converted into other products. The rate expressions for the formation and then conversion of diacetyl used in the model are obtained experimentally. Although an important compound for beer quality, it adversely affects the flavour in high concentrations. At concentrations over 0.1 ppm, it brings a buttery flavour and rancid mouthfeel to the beer [11]. Ethyl acetate is an ester with the highest concentration in beer. It is normally used as an indicator for all the esters present and has a buttery-solvent-like aroma [14]. The by-product evolution is described by

$$\frac{dC_{EA}(t)}{dt} = Y_{EA}\, \mu_X\, C_{X,act} \qquad t > t_{lag} \tag{9}$$

$$\frac{dC_{DY}(t)}{dt} = \mu_{DY}\, C_S(t) C_{X,act}(t) - \mu_{AB}\, C_{DY}(t)\, C_{eth}(t) \qquad t > t_{lag} \tag{10}$$

All the rate expressions are described in Table 1. The rate constants depend on the temperature Arrhenius-type relation described by

$$\mu_i = \exp\left(A_i + \frac{B_i}{T(t)}\right) \tag{11}$$

The parameter values used are reported in Table 2.

Table 1. Rate expressions used in the dynamic model.

Component	Rate Expression
Settling of dead cells	$\mu_{SD} = \frac{0.5\ \mu_{SD,0}\ C_{S,0}}{0.5C_{S,0} + C_{eth}}$
Biomass growth	$\mu_X = \frac{\mu_{X,0}\ C_S}{k_s + C_S}$
Sugar consumption	$\mu_S = \frac{\mu_{S,0}\ C_S}{k_s + C_S}$
Ethanol formation	$\mu_X = \frac{\mu_{eth,0}\ C_S}{k_s + C_S}$
Ethanol inhibition	$f = 1 - \frac{C_{eth}}{0.5\ C_{S,0}}$
Diacetyl formation	$\mu_{DY} = 1.2762 \times 10^{-4}$
Diacetyl conversion	$\mu_{AB} = 1.13864 \times 10^{-3}$

Table 2. Parameters in the Arrhenius-type rate constants.

Description	A_i	B_i
Maximum settling rate of dead cells, $\mu_{SD,0}$	33.82	−10,033.28
Maximum growth rate, $\mu_{X,0}$	108.31	−31,934.09
Maximum sugar consumption rate, $\mu_{S,0}$	41.92	11,654.64
Maximum ethanol production rate, $\mu_{eth,0}$	3.27	−12,667.24
Yeast death rate, μ_{DT}	130.16	−38,313.00
Yeast activation rate, $\mu_{SD,0}$	30.72	−9501.54
Affinity constant, k_s	−119.63	34,203.95
Ethyl acetate production, Y_{EA}	89.92	−26.589

2.2. *Dynamic Optimisation under Parametric Uncertainty*

Any process model can be represented in a general form as $\dot{\mathbf{x}} = f(\mathbf{x}, \mathbf{u}, \boldsymbol{\theta}, t)$ with $\mathbf{x} \in \mathbb{R}^n$ as the process state vector, $\mathbf{u} \in \mathbb{R}^m$ as the control vector, and $\boldsymbol{\theta} \in \mathbb{R}^p$ as the parameter vector. The general optimisation problem, which minimises an objective J in the interval $t \in [0, t_f]$, is written as

$$\begin{aligned} &\min_{\mathbf{x},\mathbf{u},t_f} J \\ &\text{subject to,} \\ &\quad \dot{\mathbf{x}} = f(\mathbf{x}, \mathbf{u}, \boldsymbol{\theta}, t) \\ &\quad \mathbf{c}(\mathbf{x}, \mathbf{u}, \boldsymbol{\theta}, t) \leq \mathbf{0} \end{aligned} \tag{12}$$

As mentioned in the introduction, the parameters $\boldsymbol{\theta}$ are estimated from experimental data and are inherently uncertain. The uncertainty in the parameters is typically described by a (possibly unknown) probability distribution. The aim is now to ensure that critical constraints ($c(\mathbf{x}, \mathbf{u}, \boldsymbol{\theta}, t)$) are not violated. Two approaches are possible. In the robust approach, the worst case scenario is considered and the problem is reformulated in terms of the worst case [27]. However, it is possible that the worst case scenario occurs at a very low probability and optimising for this scenario gives a very conservative solution [28]. In the stochastic approach, the constraints are reformulated as chance constraints which express a limit on the probability of constraint violation [29]. These constraints are expressed as

$$\Pr\Big(\mathbf{c}(\mathbf{x}, \mathbf{u}, \boldsymbol{\theta}, t) \geq 0\Big) \leq \beta \tag{13}$$

Computationally tractable formulation of such probabilistic chances constraints is difficult [30]. Thus, it is common to approximate them using deterministic constraints of the form

$$\mathbb{E}\big[\mathbf{c}\big] + \alpha_{\mathbf{c}}\sqrt{\mathbb{V}\big[\mathbf{c}\big]} \leq \mathbf{0} \tag{14}$$

where $\mathbb{E}[\cdot]$ and $\mathbb{V}[\cdot]$ are the expectation and variance operators, respectively. $\alpha_\mathbf{c}$ is termed the *backoff parameter*. The choice of the backoff parameter depends on the allowed probability of constraint violation β. The backoff parameter can be chosen via the Cantelli–Chebyshev inequality [31]. Although this parameter is valid for any probability distribution, it leads to an excessively large backoff parameter, which can then lead to infeasibility in the optimisation problem.

Another approach to compute the backoff parameter utilises the quantiles of the model output distribution [32]. However, this requires an assumption on the underlying distribution of the model output under uncertainty. It is common to assume a normal distribution [29,32,33]. Under the assumption of normality, a 5% allowance for constraint violation, for example, leads to a backoff parameter value of 1.65.

Similar to the constraints, the objective can also be reformulated by allowing for a backoff parameter. The reformulated optimisation problem can now be expressed as

$$\begin{aligned} &\min_{\mathbf{x},\mathbf{u},t_f} \mathbb{E}[J] + \alpha_J\sqrt{\mathbb{V}[J]} \\ &\text{subject to,} \\ &\quad \dot{\mathbf{x}} = f(\mathbf{x},\mathbf{u},\boldsymbol{\theta},t) \\ &\quad \mathbb{E}[\mathbf{c}] + \alpha_\mathbf{c}\sqrt{\mathbb{V}[\mathbf{c}]} \leq \mathbf{0} \end{aligned} \tag{15}$$

Note that it is not necessary to reformulate all constraints as chance constraints. Normal constraints and approximated chance constraints can both be present in the formulation.

2.3. Uncertainty Propagation Techniques

The formulation presented in Equation (15) requires the expectation and variance of the constraints and the objective to be computed efficiently. The uncertainty in the parameters propagates through the nonlinear model onto the model output. When the probability density function of the uncertain parameters $\rho_\theta(\theta)$ is known, it is possible to compute the expectation and variance. For a hypothetical nonlinear function $y = f(\theta)$, the expectation and variance can be computed as

$$\mathbb{E}[y] = \int_{\mathbb{R}^\theta} f(\theta)\rho_\theta(\theta)d\theta \tag{16}$$

$$\mathbb{V}[y] = \int_{\mathbb{R}^\theta} \mathbb{E}\left[(y - \mathbb{E}[y])(y - \mathbb{E}[y])^\top\right]\rho_\theta(\theta)d\theta \tag{17}$$

However, evaluating these integrals is numerically challenging. Thus, various approaches to estimate the expectation and the variance are used.

Linearisation is a popular approach based on using first-order approximation of the process model.

$$y = f(x) \approx f(\theta_0) + S_{\theta_0}d\theta \tag{18}$$

where S_{θ_0} is the sensitivity matrix ($S = \partial f/\partial\theta$) evaluated at the nominal parameter value x_0. Thus, when the parameters in the nonlinear model follow a normal distribution, the variance–covariance matrix of the model response can be approximated by $V_y = S_{\theta_0}V_\theta S_{\theta_0}^\top$. Linearisation performs well when the uncertainty is small compared to the model curvature. When the model is highly nonlinear, the approach fails to provide a correct estimate of the variance [34,35].

The other two techniques utilised in this paper are the sigma point method [36] and polynomial chaos expansion method [30,37]. The sigma point method relies on a smart selection of sampling points (called the sigma points), which are then propagated through the nonlinear model to obtain a response set. The expectation and the variance are then estimated from this response set. The sigma points are computed based on the parameter variance $\mathbb{V}[\theta]$.

$$\theta_\sigma = \theta \pm \sigma \quad \text{, where, } \; \sigma \leftarrow \sqrt{(n+\kappa)\mathbb{V}[\theta]} \tag{19}$$

For p uncertain parameters, the set of sigma points contains $2p$ elements. This set is then transformed by the nonlinear map (i.e., the nonlinear process model in this case) to obtain a set of model responses,

$$\mathbf{y}_0 = f(\mathbf{x}(\theta_0)) \quad \text{, and, } \; \mathbf{y}_\sigma = f(\mathbf{x}(\theta_\sigma)) \tag{20}$$

The expectation and the variance–covariance matrix are computed using this response set.

$$\bar{\mathbf{y}}_{\text{sig}} = \frac{1}{n+\kappa}\left[\kappa\mathbf{y}_0 + \frac{1}{2}\sum^{2n}\mathbf{y}_\sigma\right] \tag{21}$$

$$\mathbf{V}_y = \frac{1}{n+\kappa}\left[\kappa(\mathbf{y}_0 - \bar{\mathbf{y}}_{\text{sig}})(\mathbf{y}_0 - \bar{\mathbf{y}}_{\text{sig}})^\top + \frac{1}{2}\sum^{2n}(\mathbf{y}_\sigma - \bar{\mathbf{y}}_{\text{sig}})(\mathbf{y}_\sigma - \bar{\mathbf{y}}_{\text{sig}})^\top\right] \tag{22}$$

The polynomial chaos expansion aims to approximate the model response by a truncated sum of orthogonal polynomials. The model response is approximated as

$$y = f(\theta) \approx \sum_{i=0}^{M} a_i \Psi_i(\theta) \tag{23}$$

where $\Psi_i(\theta)$ are multivariate orthogonal polynomials. The total number of terms required in the approximation (M) depends on the number of uncertain parameters (p) and the order of the polynomials (o) used.

$$M + 1 = \frac{(p+o)!}{p!\,o!} \tag{24}$$

If the parameters are assumed to be independent, the multivariate polynomials can be constructed from univariate orthogonal polynomials. The Wiener–Askey scheme [37] lists some commonly used probability distributions that are associated with certain univariate orthogonal polynomials. Other approaches based on statistical moments [38,39] or the Gram–Schmidt orthogonalisation [40,41] are available to generate the monovariate polynomials when the parameters are correlated or do not follow a distribution in the Wiener–Askey scheme.

Once the polynomials are generated, the coefficients in the expansion, a_is, need to be computed. Approaches to compute these coefficients are classified into intrusive and non-intrusive approaches. Intrusive approaches use Galerkin projections on the process model to generate the coefficients [42,43]. Non-intrusive methods, however, treat the model as a black-box and rely on sampling [44]. In this paper, the non-intrusive approach based on least-squares regression is used. Both Nimmegeers et al. [29] (supplementary file) and Bhonsale et al. [38] elaborate on this approach.

Due to the certain attributes of orthogonal polynomials [42], the expectation and variance can be estimated directly from the coefficients in the model response expansion.

$$\mathbb{E}[y] = a_0 \tag{25}$$

$$\mathrm{V}_y = \sum_{i=1}^{M} a_i^2 \langle \Psi_i^2 \rangle \tag{26}$$

Regardless of the method used for uncertainty propagation, the optimisation problem described in Equation (15) has to be augmented with extra states so that the expectation and variance can be estimated. The augmented states depend on the technique used. For linearisation, the augmented state is the sensitivity matrix, while for sigma points and polynomial chaos the augmented states are the original states evaluated at the parameter sampling point.

2.4. Implementation

In all cases, the dynamic optimisation problem has to be solved numerically. In this paper the direct collocation approach [45] is utilised. In this method, both the state and control vectors are fully discretised in time leading to a nonlinear program (NLP). For the states, between every discretised time interval four collocation points are used. These collocation points obey the model equation through equality constraints. A cubic Lagrange polynomial with collocation points situated at the Radau roots is used at each interval. The control variable is discretised as piecewise constant on each time interval. The NLP is solved using the interior point method as implemented in IPOPT [46].

All the dynamic optimisation problems are implemented in a Python-based tool developed within the BioTeC+ team called POMODORO [47]. POMODORO utilises CasADi [48] to obtain the gradient and Hessian information required for the optimisation. CasADi can provide exact Hessian information with an automatic differentiation method. While the default NLP solver in POMODORO is IPOPT, other (commercial) solvers can also be used via their CasADi interface. POMODORO is available freely for academic use via the website https://cit.kuleuven.be/biotec/software/pomodoro (accessed on 12 November 2021).

3. Results and Discussion

The fermentation model described in Section 2.1 is optimised for two objectives: (i) maximising ethanol production, and (ii) minimising batch time. A constraint on minimum ethanol concentration is imposed. The minimum concentration at the end of the batch is set to 25 g/L. Similarly, an end time constraint on the suspended active cells is also imposed. The maximum concentration of active cells should be less than 0.5 g/L. The two major quality constraints imposed are on the concentrations of diacetyl and ethyl acetate. The maximum concentration of these by-products should be less than 0.2 ppm and 2 ppm, respectively [11,21]. Without accounting for the uncertainty on any of the parameters, the optimised temperature trajectories are depicted in Figure 1a,b.

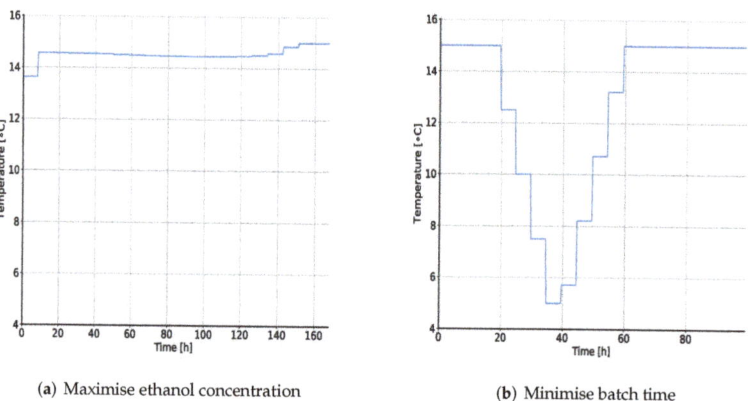

(**a**) Maximise ethanol concentration (**b**) Minimise batch time

Figure 1. Optimised temperature profiles for the nominal case with two different objectives: maximising ethanol concentration and minimising batch time.

When maximising the ethanol concentration at the end of the batch, the wort is immediately heated to a relatively high temperature to accelerate the conversion of latent cells to active cells and to enhance the growth of these active cells. The temperature profile stays relatively constant to boost the production of ethanol. This also enhances byproduct formation. The temperature is optimised such that the concentration of ethyl acetate does not exceed 2 ppm. As there is no limit on the batch time, the fermentation continues until all the active yeast cells are consumed. A maximum ethanol concentration of 60.24 g/L is reached in 168 h.

When minimising the batch time, the wort is first heated at the maximum possible temperature (15 °C) to accelerate the the conversion to and the growth of active yeast cells. However, this also accelerates the production of ethyl acetate. Thus after some time, the wort is cooled too rapidly. The cooling occurs in steps because the consecutive temperature jumps are constrained to 2.5 °C/h. This slows down the byproduct formation; however, the ethanol production continues. The evolution of the concentration profiles is depicted in Figure 2b. Towards the end of the batch, the temperature increases again rapidly so as to ensure the active cells are killed and the concentration of yeast cells is below the threshold of 0.5 g/L. The batch stops as soon as this threshold limit is reached. The ethanol concentration achieved is 48.11 g/L in 100 h.

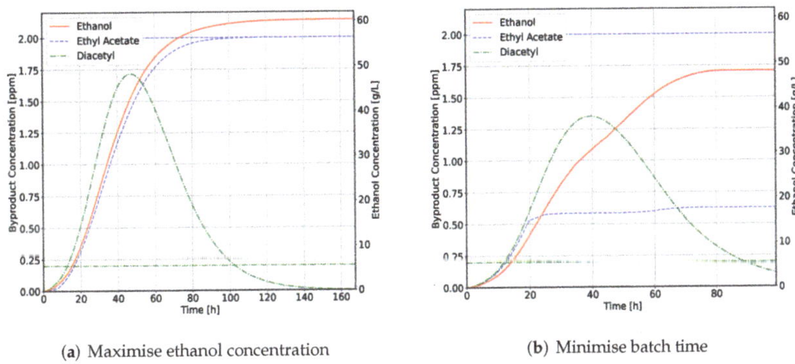

(**a**) Maximise ethanol concentration (**b**) Minimise batch time

Figure 2. Concentration profiles of ethanol and the two byproducts for the nominal case with two different objectives: maximising ethanol concentration and minimising batch time. The byproducts are represented on the left y-axis and ethanol on the right y-axis. The horizontal lines depict the constraints on the byproducts.

3.1. *Influence of Uncertainty*

The fermentation model used contains 18 parameters, each of which are determined experimentally. However, considering the uncertainty on each parameter would lead to a very large optimisation problem which would be numerically infeasible. Thus, based on a sensitivity analysis, four parameters which have the most influence on the states of interest (quality parameters) are considered. These are as follows: the maximum growth rate ($\mu_{X,0}$), yeast death rate (μ_{DT}), maximum sugar consumption rate ($\mu_{S,0}$), and ethyl acetate production (Y_{EA}). The only quality parameters considered are the byproducts and the ethanol concentration. Under in silico uncertainty, Figure 3 shows the trajectories of 500 Monte Carlo simulations with the optimal temperature profile obtained for maximising ethanol concentration. The diacetyl concentrations are always below the threshold, while the ethyl acetate concentrations show a large variation and large constraint violation. In total, 50.4% of Monte Carlo simulations violated the ethyl acetate constraint. This exemplifies the need to incorporate the uncertainty into any optimisation study.

3.2. *Optimisation under Uncertainty*

The uncertainty is incorporated in the optimisation study using the three approaches described earlier. For polynomial chaos expansion, two polynomial orders are used: first order and second order. As the parameter uncertainty is assumed to be Gaussian, Hermite polynomials are used. A desired confidence interval for the constraints is set according to the quantiles of a normal distribution. The backoff parameters used in this study are reported in Table 3. The optimised control profiles are then tested via a Monte Carlo simulation with randomly generated parameter values.

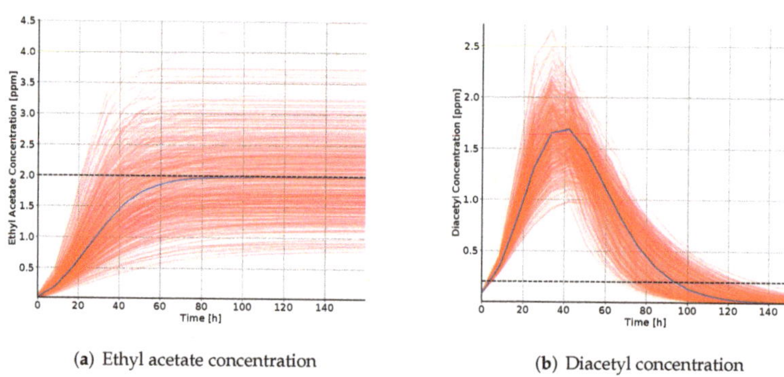

(**a**) Ethyl acetate concentration

(**b**) Diacetyl concentration

Figure 3. Concentration profiles of the two byproducts for the nominal case with maximising ethanol concentration as the objective, and minimising batch time. The blue trajectory represents the simulation with the nominal parameter value and the red trajectories depict the Monte Carlo simulations. The horizontal dashed lines depict the constraints on the byproducts.

Table 3. Backoff parameter (α) values corresponding to the confidence intervals.

Confidence Level	90%	95%	97.50%
Quantile	0.1	0.05	0.025
α	1.28	1.65	1.96

3.2.1. Maximising Ethanol

Figure 4 depicts the results obtained by all the three methods for 95% confidence on the constraint violation. Although concentration of ethanol obtained at the end of the fermentation is similar (slightly lower) to the nominal case, the concentration trajectories of ethyl acetate and diacetyl show differences in the optimised results. Amongst the methods, PCE2 gives the most conservative solution while PCE1 and LIN operate much closer to the concentration bound. This variation is attributed to the approximation of the variance of the concentration values.

It is impossible to judge *a priori* which approach provides the best approximation. Thus, the results obtained are assessed using a Monte Carlo simulation. The result with lowest percentage of constraint violation is preferred. The constraint violation as a percentage of the total simulation is reported in Table 4. All methods lead to lower constraint violations than the nominal case. However, second-order polynomial chaos seems to perform the best. The probability densities of the two constraints via the trajectories optimised using all methods are depicted in Figure 5. The majority of constraint violations are caused due to overproduction of ethyl acetate. This is expected from the process dynamics. As concentration evolution of both ethanol and ethyl acetate is described by similar mathematical expressions, maximising ethanol boosts the production of ethyl acetate. However, the conversion/inhibition factors and the production rates differ and are influenced by the temperature differently.

3.2.2. Minimising Batch Time

Figure 6 depicts the results obtained by all the three methods for 95% confidence on the constraint violation. It is seen that all the methods result in longer fermentation batches than the nominal case. As is the case with maximising ethanol, PCE2 and SP give the most conservative solutions while PCE1 and LIN result in fermentation times similar to the nominal case.

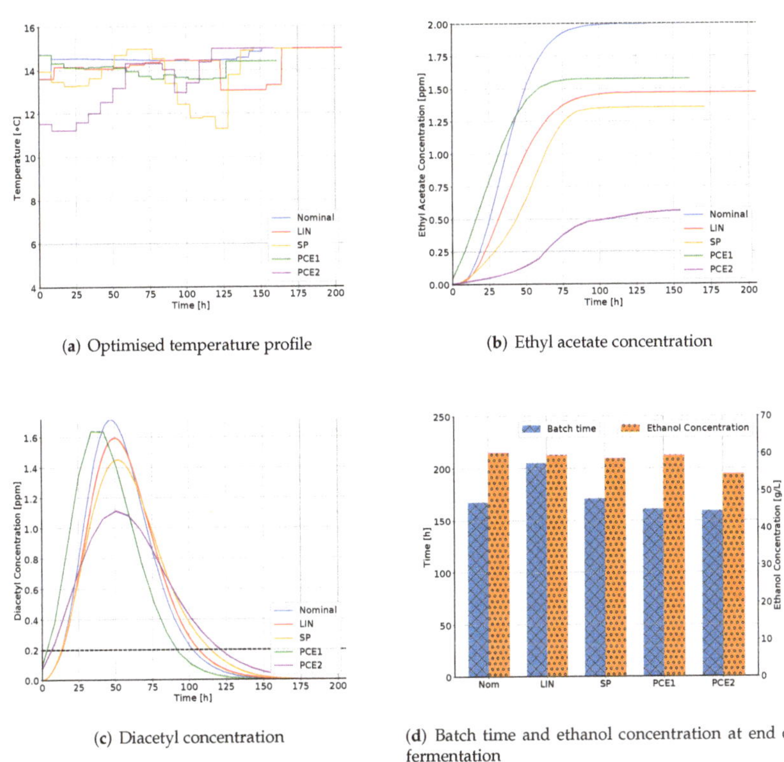

(**a**) Optimised temperature profile (**b**) Ethyl acetate concentration

(**c**) Diacetyl concentration (**d**) Batch time and ethanol concentration at end of fermentation

Figure 4. Optimisation results for maximising ethanol with all three methods and $\alpha = 1.65$. The concentration trajectories depicted were obtained by applying the optimised temperature trajectory to the model with nominal parameters.

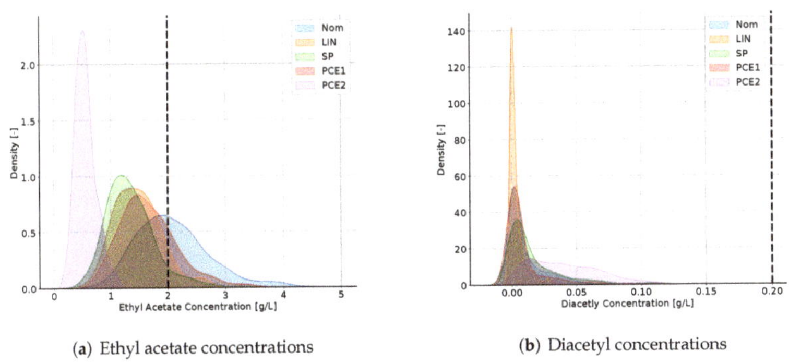

(**a**) Ethyl acetate concentrations (**b**) Diacetyl concentrations

Figure 5. Density plots for concentrations of ethyl acetate and diacetyl based on Monte Carlo simulations for optimal trajectory obtained with maximising ethanol. The vertical dashed line depicts the concentration threshold.

Table 4. Constraint violations obtained from Monte Carlo simulations for the maximising ethanol objective via all the methods and backoff parameter values.

	Linearisation	Sigma Points	PCE 1	PCE 2	Nominal
$\alpha = 1.28$	19.6%	18.5%	35.3%	9.5%	
$\alpha = 1.65$	13.6%	13.0%	27.6%	6.7%	50.4%
$\alpha = 1.96$	11.1%	9.6%	21.6%	5.4%	

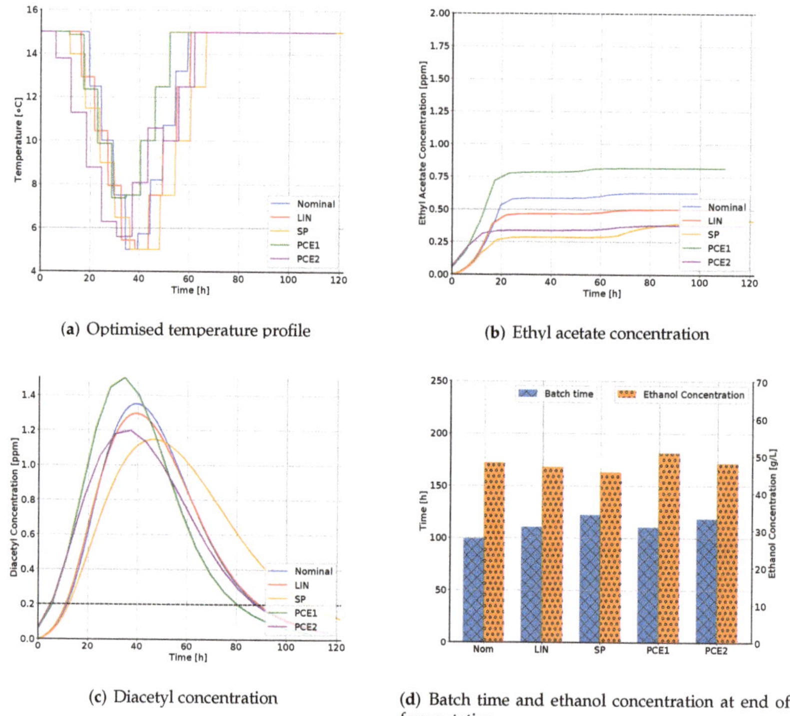

Figure 6. Optimisation results for minimising time with all three methods and $\alpha = 1.65$. The concentration trajectories depicted were obtained by applying the optimised temperature trajectory to the model with nominal parameters.

The constraint violation percentage is reported in Table 5. All the robustification approaches lead to a significant reduction in constraint violation. Again, the sigma points approach and PCE2 perform the best in terms of constraint violation. From the concentration density plots (Figure 7) for ethyl acetate and diacetyl, it is observed that when the objective is to minimise batch time, the majority of constraint violations are caused due to excess diacetyl concentration. For the case of maximising ethanol, the constraint violations are caused due to excess ethyl acetate. This can be explained by the dynamics of the process. If the fermentation is allowed to proceed uninterrupted, diacetyl decomposes into a variety of other products. As minimising the batch time essentially interrupts the fermentation, it influences the diacetyl decomposition.

Table 5. Constraint violations obtained from Monte Carlo simulations for the minimising time objective via all the methods and backoff parameter values.

	Linearisation	Sigma Points	PCE 1	PCE 2	Nominal
$\alpha = 1.28$	35.5%	17.2%	37.7%	17.2%	
$\alpha = 1.65$	29.3%	8.4%	20.1%	8.2%	70.4%
$\alpha = 1.96$	24.5%	4.6%	14.4%	4.4%	

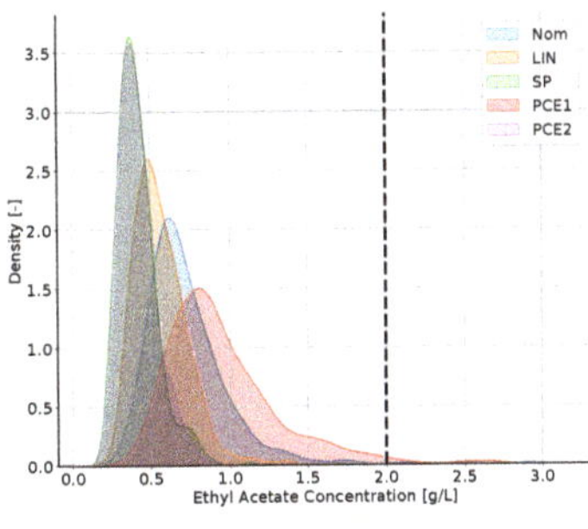

(**a**) Ethyl Acetate concentrations

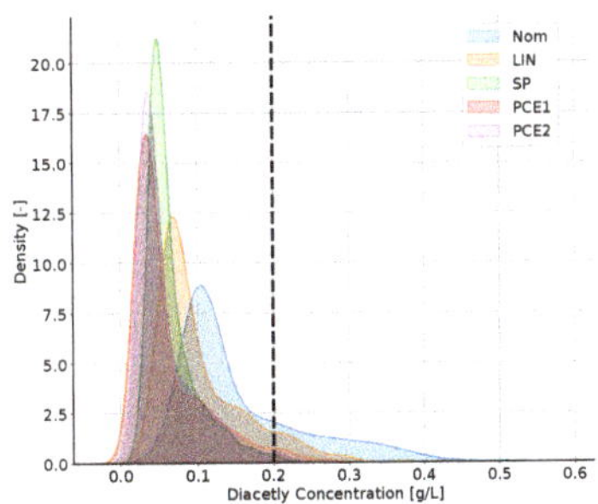

(**b**) Diacetyl concentrations

Figure 7. Density plots for concentrations of ethyl acetate and diacetyl based on Monte Carlo simulations for optimal trajectory obtained with minimising time. The vertical dashed line depicts the concentration threshold.

3.2.3. Comparison of Robustification Approaches

For both objectives, the sigma points approach and PCE2 approach perform best in terms of constraint violation. The two first-order methods, linearisation and PCE1, both lead to relatively high constraint violation percentages. As both these methods rely on local linear approximations of a nonlinear model, these methods can be expected to perform poorly when the underlying process model is very nonlinear. However, due to the simplicity of implementation, linearisation is often used despite the poor performance. In the case of maximising ethanol, linearisation performs surprisingly well. This can be explained by the optimised profile. The optimised profile consistently hovers around 14 °C and does not induce any abrupt changes to the process dynamics, leading to a smooth evolution of concentration profiles. In such cases, linearisation is known to perform well. However, when an optimised profile excites the nonlinearity in the model by a big jump (as is the case in minimising time), linearisation over(or under)estimates the variances drastically. This is corroborated by a comparison of the estimated variances of the concentrations with the variance obtained from Monte Carlo simulations reported in Table 6. For maximising ethanol, the variance is estimated with good accuracy, while for minimising batch time, the accuracy reduces. This is consistent with the results of Bhonsale et al. [35], which demonstrate the failure of linearisation when a step change in input is applied to the process exciting its nonlinear dynamics.

A polynomial chaos expansion is fundamentally a reduced-order model of the original process model. In PCE1, the expansion is truncated after the first-order terms making the model linear. The weak performance of PCE1 can be attributed to the local nonlinearity of the fermentation model, which makes the variance approximation poor. Adding a second-order term in PCE2 significantly improves the variance approximation and subsequently the performance of the optimisation under uncertainty.

Table 6. Comparison of variances approximated using linearisation with variances approximated by Monte Carlo simulations for $\alpha = 1.65$.

Variance	Maximise Ethanol		Minimise Batch Time	
	Linearisation	MC	Linearisation	MC
$\mathbb{V}[C_{EA}]$	1.56×10^{-1}	1.79×10^{-1}	1.20×10^{-2}	1.70×10^{-2}
$\mathbb{V}[C_{DY}]$	5.16×10^{-4}	5.38×10^{-4}	2.60×10^{-3}	1.57×10^{-3}
$\mathbb{V}[C_{eth}]$	1.19×10^{1}	1.42×10^{1}	1.54×10^{1}	2.29×10^{1}

3.2.4. Effect of Backoff Parameter

As mentioned earlier, the backoff parameter is based on the quantiles of the output distribution. As expected, increasing the value of the back-off parameter (i.e., constraint violations tolerated) leads to a reduction in the percentage of constraint violations for all the methods. This comes at the cost of a more conservation solution. For all methods, an increase in backoff parameter leads to a reduction in the maximum ethanol concentration obtained and longer batch times when the batch time is minimised. It should be noted that in all cases, the tolerances set for the constraint violation are not met. For example, a backoff parameter of 1.65 corresponds to 5% tolerance for constraint violation. However, PCE2, which performs the best for both objectives, still leads to 6.7% and 8.2% violations for maximising ethanol and minimising batch time objectives, respectively. This is because the backoff parameter is based on the quantiles of a normal distribution. It is evident from Figures 5 and 7 that the model output does not follow a Gaussian distribution. Hence, the quantiles used do not exactly correspond to the tolerances. As only two moments, the expectation and the variance, can be obtained from either of the three methods, information on the distribution is impossible to obtain without Monte Carlo simulations. Nevertheless, the reduction in constraint violations with increasing backoff parameters under the assumption of Gaussianity is still significant.

4. Conclusions

Dynamic optimisation has tremendous potential to improve fermentation operation. However, models used in optimisation are inherently uncertain due to their estimation from experimental (noisy) data. It has been shown that if the parametric uncertainty is not included in the optimisation, the state constraints can be violated. Hence, to avoid bad fermentation batches it is necessary to include the uncertainty information available in the optimisation framework.

In this study, a beer fermentation process was optimised while accounting for parametric uncertainty using three techniques. The results obtained with the three techniques were compared for different backoff parameter values. Two objectives, maximising the ethanol concentration and minimising the fermentation time, were considered. While all the methods led to a significant improvement in constraint violation when compared to the nominal case, second-order PCE performed the best. The main advantage of the PCE approach is that no assumption on the distribution of uncertain parameters needs to be made. For independent parameters with certain distributions, the polynomials involved in the PCE can be obtained through the Wiener–Askey scheme [37]. If the parameters are not independent, or do not follow a distribution from the Wiener–Askey scheme, the polynomials can be obtained by utilising the statistical moments and Gram–Schmidt orthogonalisation. However, PCE is computationally expensive as it leads to a significantly bigger optimisation problem. The sigma point approach leads to a smaller optimisation problem while still performing well. However, the direct application of the sigma point approach is limited to parameters with symmetric unimodal distributions. For asymmetric distributions, transformation functions need to be utilised to implement the sigma point approach [49]. Linearisation performs the worst amongst the three approaches. This is expected due to the nonlinearity of the process. However, when the objective is to maximise ethanol linearisation, it performs similarly to the sigma point approach. The study has also emphasised the need for appropriate selection of the backoff parameter.

Author Contributions: Conceptualisation, S.B.; methodology, S.B. and J.V.I.; software, S.B. and W.M.; validation, S.B., W.M., and J.V.I.; formal analysis, W.M. and S.B.; investigation, W.M.; resources, J.V.I.; writing—original draft preparation, S.B.; writing—review and editing, S.B., W.M., and J.V.I.; visualisation, S.B. and W.M.; supervision, S.B. and J.V.I.; project administration, J.V.I.; funding acquisition, S.B. and J.V.I. All authors have read and agreed to the published version of the manuscript.

Funding: This work was supported by KU Leuven Center-of-Excellence Optimization in Engineering (OPTEC), projects G086318N and G0B4121N of the Fund for Scientific Research Flanders (FWO), the European Commission within the framework of the Erasmus+ FOOD4S Programme (Erasmus Mundus Joint Master Degree in Food Systems Engineering, Technology and Business 619864-EPP-1-2020-1-BE-EPPKA1-JMD-MOB) and by the European Union's Horizon 2020 Research and Innovation programme under the Marie Skłodowska-Curie Grant Agreement N956126 (E-MUSE Complex microbial Ecosystems MUltiScale modElling) and 813329 (PROTECT Predictive mOdelling Tools to evaluate the Effects of Climate change on food safeTy and spoilage).

Institutional Review Board Statement: Not applicable.

Informed Consent Statement: Not applicable.

Conflicts of Interest: The authors declare no conflict of interest.

Abbreviations

The following abbreviations are used in this manuscript:

LIN	Linearisation
MC	Monte Carlo
NLP	Non-linear programme
Nom	Nominal
PCE	Polynomial chaos expansion
SP	Sigma point

References

1. Anderson, K.; Meloni, G.; Swinnen, J. Global Alcohol Markets: Evolving Consumption Patterns, Regulations, and Industrial Organizations. *Annu. Rev. Resour. Econ.* **2018**, *10*, 105–132. [CrossRef]
2. Nelson, M. *The Barbarian's Beverage: A History of Beer in Ancient Europe*; Routledge: London, UK, 2005.
3. Kirin Beer University. Global Beer Consumption by Country in 2019. 2020. Available online: https://www.kirinholdings.com/en/newsroom/release/2020/1229_01.pdf (accessed on 12 November 2021).
4. The Brewers of Europe. The Contribution Made by Beer to the European Economy. 2020. Available online: https://brewersofeurope.org/uploads/mycms-files/documents/publications/2020/contribution-made-by-beer-to-EU-economy-2020.pdf (accessed on 12 November 2021).
5. Belgian Brewers. Annual Report. 2019. Available online: http://www.belgianbrewers.be/en/economy/article/annual-report (accessed on 12 November 2021).
6. UNESCO. Decision of the Intergovernmental Committee: 11.COM 10.B.5. 2016. Available online: https://ich.unesco.org/en/decisions/11.COM/10.B.5 (accessed on 12 November 2021).
7. Ministerie Van Economische Zaken. Volksgezondheid en Leefmilieu. In *Economische Zaken: Koninklijk Besluit Betreffende Bier*. 1993. Available online: http://www.ejustice.just.fgov.be/cgi_loi/change_lg.pl?language=nl&la=N&cn=1993033131&table_name=wet (accessed on 12 November 2021).
8. Amerine, M.; Pangborn, R.; Roessler, E. *Principles of Sensory Evaluation of Food*; Academic Press: New York, NY, USA, 1965.
9. Trueba, P.B. Beer Flavour Instability: Unravelling Formation and/or Release of Staling Aldehydes. Ph.D. Thesis, Arenberg Doctoral School, KU Leuven, Leuven, Belgium, 2020.
10. Kobayashi, M.; Shimizu, H.; Shioya, S. Beer Volatile Compounds and Their Application to Low-Malt Beer Fermentation. *J. Biosci. Bioeng.* **2008**, *106*, 317–323. [CrossRef] [PubMed]
11. Humia, B.V.; Santos, K.S.; Barbosa, A.M.; Sawata, M.; da Costa Mendonça, M.; Padilha, F.F. Beer Molecules and Its Sensory and Biological Properties: A Review. *Molecules* **2019**, *24*, 1568. [CrossRef] [PubMed]
12. Pires, E.J.; Teixeira, J.A.; Brányik, T.; Vicente, A.A. Yeast: The soul of beer's aroma—A review of flavour-active esters and higher alcohols produced by the brewing yeast. *Appl. Microbiol. Biotechnol.* **2014**, *98*, 1937–1949. [CrossRef]

13. Verstrepen, K.J.; Derdelinckx, G.; Dufour, J.P.; Winderickx, J.; Thevelein, J.M.; Pretorius, I.S.; Delvaux, F.R. Flavor-active esters: Adding fruitiness to beer. *J. Biosci. Bioeng.* **2003**, *96*, 110–118. [CrossRef]
14. Rodman, A.D.; Gerogiorgis, D.I. Multi-objective process optimisation of beer fermentation via dynamic simulation. *Food Bioprod. Process.* **2016**, *100*, 255–274. [CrossRef]
15. Briggs, D.E. *Brewing: Science and Practice*; CRC Press Woodhead Pub: Cambridge, UK, 2004.
16. Carrillo-Ureta, G.; Roberts, P.; Becerra, V. Genetic algorithms for optimal control of beer fermentation. In Proceedings of the 2001 IEEE International Symposium on Intelligent Control (ISIC '01) (Cat. No.01CH37206), Mexico City, Mexico, 5–7 September 2001. [CrossRef]
17. Xiao, J.; Zhou, Z.; Zhang, G. Ant colony system algorithm for the optimization of beer fermentation control. *J. Zhejiang Uni. Sci. A* **2004**, *5*, 1597–1603. [CrossRef] [PubMed]
18. Rodman, A.D.; Gerogiorgis, D.I. Dynamic optimization of beer fermentation: Sensitivity analysis of attainable performance vs. product flavour constraints. *Comput. Chem. Eng.* **2017**, *106*, 582–595. [CrossRef]
19. Bosse, T.; Griewank, A. Optimal control of beer fermentation processes with Lipschitz-constraint on the control. *J. Inst. Brew.* **2014**, *120*, 444–458. [CrossRef]
20. Andrés-Toro, B.; Girón-Sierra, J.M.; Fernández-Blanco, P.; López-Orozco, J.; Besada-Portas, E. Multiobjective optimization and multivariable control of the beer fermentation process with the use of evolutionary algorithms. *J. Zhejiang Univ. Sci.* **2004**, *5*, 378–389. [CrossRef]
21. Rodman, A.D.; Fraga, E.S.; Gerogiorgis, D. On the application of a nature-inspired stochastic evolutionary algorithm to constrained multi-objective beer fermentation optimisation. *Comput. Chem. Eng.* **2018**, *108*, 448–459. [CrossRef]
22. Vanderhaegen, B.; Neven, H.; Verachtert, H.; Derdelinckx, G. The chemistry of beer aging—A critical review. *Food Chem.* **2006**, *95*, 357–381. [CrossRef]
23. De Andrés-Toro, B.; Girón-Sierra, J.; López-Orozco, J.; Fernández-Conde, C.; Peinado, J.; Garcia-Ochoa, F. A kinetic model for beer production under industrial operational conditions. *Math. Comput. Simul.* **1998**, *48*, 65–74. [CrossRef]
24. Gee, D.A.; Ramirez, W.F. Optimal temperature control for batch beer fermentation. *Biotechnol. Bioeng.* **1988**, *31*, 224–234. [CrossRef] [PubMed]
25. Ramirez, W.F.; Maciejowski, J. Optimal Beer Fermentation. *J. Inst. Brew.* **2007**, *113*, 325–333. [CrossRef]
26. Trelea, I.C.; Titica, M.; Landaud, S.; Latrille, E.; Corrieu, G.; Cheruy, A. Predictive modelling of brewing fermentation: From knowledge-based to black-box models. *Math. Comput. Simul.* **2001**, *56*, 405–424. [CrossRef]
27. Houska, B.; Logist, F.; Van Impe, J.; Diehl, M. Robust optimization of nonlinear dynamic systems with application to a jacketed tubular reactor. *J. Process Control* **2012**, *22*, 1152–1160. [CrossRef]
28. Srinivasan, B.; Bonvin, D.; Visser, E.; Palanki, S. Dynamic optimization of batch processes. *Comput. Chem. Eng.* **2003**, *27*, 27–44. [CrossRef]
29. Nimmegeers, P.; Telen, D.; Logist, F.; Van Impe, J. Dynamic optimization of biological networks under parametric uncertainty. *BMC Syst. Biol.* **2016**, *10*. [CrossRef]
30. Mesbah, A.; Streif, S.; Findeisen, R.; Braatz, R.D. Stochastic nonlinear model predictive control with probabilistic constraints. In Proceedings of the 2014 American Control Conference, Portland, OR, USA, 4–6 June 2014. [CrossRef]
31. Mesbah, A.; Streif, S. A Probabilistic Approach to Robust Optimal Experiment Design with Chance Constraints. *IFAC-PapersOnLine* **2015**, *48*, 100–105. [CrossRef]
32. Telen, D.; Vallerio, M.; Cabianca, L.; Houska, B.; Van Impe, J.; Logist, F. Approximate robust optimization of nonlinear systems under parametric uncertainty and process noise. *J. Process Control* **2015**, *33*, 140–154. [CrossRef]
33. Vallerio, M.; Telen, D.; Cabianca, L.; Manenti, F.; Van Impe, J.; Logist, F. Robust multi-objective dynamic optimization of chemical processes using the Sigma Point method. *Chem. Eng. Sci.* **2016**, *140*, 201–216. [CrossRef]
34. Akkermans, S.; Nimmegeers, P.; Van Impe, J.F. A tutorial on uncertainty propagation techniques for predictive microbiology models: A critical analysis of state-of-the-art techniques. *Int. J. Food Microbiol.* **2018**, *282*, 1–8. [CrossRef] [PubMed]
35. Bhonsale, S.; Telen, D.; Stokbroekx, B.; Van Impe, J. An Analysis of Uncertainty Propagation Methods Applied to Breakage Population Balance. *Processes* **2018**, *6*, 255. [CrossRef]
36. Julier, S.; Uhlmann, J.K. *A General Method for Approximating Nonlinear Transformations of Probability Distributions*; Technical Report; Robotics Research Group, Department of Engineering Science, University of Oxford: Oxford, UK, 1996.
37. Xiu, D.; Karniadakis, G. The Wiener–Askey polynomial chaos for stochastic differential equations. *SIAM J. Sci. Comput.* **2002**, *24*, 619–644. [CrossRef]
38. Bhonsale, S.; Muñoz López, C.A.; Van Impe, J. Global Sensitivity Analysis of a Spray Drying Process. *Processes* **2019**, *7*, 562. [CrossRef]
39. Yang, S.; Xiong, F.; Wang, F. Polynomial Chaos Expansion for Probabilistic Uncertainty Propagation. In *Uncertainty Quantification and Model Calibration*; InTech Open: London, UK, 2017. [CrossRef]
40. Navarro, M.; Witteveen, J.; Blom, J. Polynomial Chaos Expansion for general multivariate distributions with correlated variables. *arXiv* **2014**, arXiv:1406.5483
41. Paulson, J.A.; Buehler, E.A.; Mesbah, A. Arbitrary Polynomial Chaos for Uncertainty Propagation of Correlated Random Variables in Dynamic Systems. *IFAC-PapersOnLine* **2017**, *50*, 3548–3553. [CrossRef]
42. Ghanem, R.; Spanos, P. *Stochastic Finite Elements—A Spectral Approach*; Springer: Berlin/Heidelberg, Germany, 1991.

43. Debusschere, B.; Najm, H.; Pébay, P.; Knio, O.; Ghanem, R.; Maitre, O.L. Numerical challenges in the use of polynomial chaos representations for stochastic processes. *SIAM J. Sci. Comput.* **2004**, *26*, 698–719. [CrossRef]
44. Tatang, M.; Pan, W.; Prinn, R.; McRae, G. An efficient method for parametric uncertainty analysis of numerical geophysical models. *J. Geophys. Res. Atmos.* **1997**, *102*, 21925–21932. [CrossRef]
45. Biegler, L.T. An overview of simultaneous strategies for dynamic optimization. *Chem. Eng. Process. Process Intensif.* **2007**, *46*, 1043–1053. [CrossRef]
46. Wächter, A.; Biegler, L.T. On the implementation of an interior-point filter line-search algorithm for large-scale nonlinear programming. *Math. Program.* **2005**, *106*, 25–57. [CrossRef]
47. Bhonsale, S.; Telen, D.; Vercammen, D.; Vallerio, M.; Hufkens, J.; Nimmegeers, P.; Logist, F.; Van Impe, J. Pomodoro: A Novel Toolkit for Dynamic (MultiObjective) Optimization, and Model Based Control and Estimation. *IFAC-PapersOnLine* **2018**, *51*, 719–724. [CrossRef]
48. Andersson, J.A.E.; Gillis, J.; Horn, G.; Rawlings, J.B.; Diehl, M. CasADi—A software framework for nonlinear optimization and optimal control. *Mathe. Program. Comput.* **2019**, *11*, 1–36. [CrossRef]
49. Neumann, T.; Dutschk, B.; Schenkendorf, R. Analyzing uncertainties in model response using the point estimate method: Applications from railway asset management. *Proc. Inst. Mech. Eng. Part O J. Risk Reliab.* **2019**, *233*, 761–774. [CrossRef]

Article

Effect of Temperature and Time on Oxygen Consumption by Olive Fruit: Empirical Study and Simulation in a Non-Ventilated Container

Eddy Plasquy [1,*], María C. Florido [2], Rafael Rubén Sola-Guirado [3], José María García Martos [1] and Juan Francisco García Martín [4,*]

[1] Department of Biochemistry and Technology of Plant Foods, Instituto de la Grasa, Spanish National Research Council, Ctra. Sevilla Utrera, km 1, Edifice 46, 41013 Seville, Spain; jmgarcia@cica.es
[2] Department of Crystallography, Mineralogy and Agricultural Chemistry, Escuela Técnica Superior de Ingeniería Agronómica, Universidad de Sevilla, Ctra. Sevilla Utrera, km 1, 41013 Seville, Spain; florido@us.es
[3] Department of Mechanic, Campus Rabanales, University of Córdoba, Ed. Leonardo Da Vinci, Ctra. Nacional IV, km 396, 14014 Córdoba, Spain; ir2sogur@uco.es
[4] Departamento de Ingeniería Química, Facultad de Química, Universidad de Sevilla, 41012 Seville, Spain
* Correspondence: eddy.plasquy@telenet.be (E.P.); jfgarmar@us.es (J.F.G.M.)

Abstract: Fermentation processes within olive fruit jeopardize the quality of the extracted oil. Aeration, temperature, and time play a crucial role in attaining the critical threshold at which an aerobic respiration shifts towards anaerobic. In this work, the O_2 consumption and CO_2 production of olive fruit kept in a closed container at different temperatures (5–45 °C) were measured over 7 h. The data allowed us to describe the relationship between the temperature and the respiration rate as an Arrhenius function and simulate the oxygen consumption in the inner part of a container full of fruit with low aeration, considering the generated respiration heat over time. The simulation revealed that olives risk shifting to anaerobic respiration after 3 h at 25 °C and less than 2 h at 35 °C when kept in a non-ventilated environment. The results underline the irreversible damage that high day temperatures can produce during the time before fruit processing, especially during transport. Lowering, as soon as possible, the field temperature thus comes to the fore as a necessary strategy to guarantee the quality of the olives before their processing, like most of the fruit that is harvested at excessive temperatures.

Keywords: Fermentation; *Olea europaea*; respiration rate; storage conditions; transport

Citation: Plasquy, E.; Florido, M.C.; Sola-Guirado, R.R.; García Martos, J.M.; García Martín, J.F. Effect of Temperature and Time on Oxygen Consumption by Olive Fruit: Empirical Study and Simulation in a Non-Ventilated Container. *Fermentation* **2021**, 7, 200. https://doi.org/10.3390/fermentation7040200

Academic Editors: Claudia Gonzalez Viejo and Sigfredo Fuentes

Received: 1 September 2021
Accepted: 20 September 2021
Published: 23 September 2021

1. Introduction

Currently, the implementation of super high-density hedgerow orchards is accelerating at a fast rate. For example, it increased roughly 116% in Spain in 2016 when compared to 2015 [1]. While still being a fraction of the 1.5 Mha cultivated in Andalusia (the world's leading olive-producing region), the shift towards fully mechanized harvesting and pruning is setting [2]. The 'Arbequina' cv. is currently by far the most established (90%) variety in super high-density hedgerow orchards, along with 'Arbosana', 'Koroneiki', and the recently introduced 'Sikitita' [3,4]. These cultivars are characterized by an early ripening that starts early November. However, the season often starts several weeks earlier because extra virgin olive oil (EVOO) with the highest quality is achieved from green olives. Consequently, harvesting is carried out in a period when the day temperatures are well above 30 °C. It can be expected that olive production will cope with major and profound challenges according to the predictions of climate change in the Mediterranean Basin [5–8]. The flowering and ripening of the fruit are expected to be brought forward, so it is necessary to start the harvest earlier to obtain an optimal quality of oil [9–12].

Harvesting under these conditions implies that the temperature of the fruit can quickly attain excessive levels, especially when the fruit are collected in containers or trailers that are fully exposed to the sun. Temperature is considered to be the most important factor affecting postharvest shelf-life, along with atmospheric composition and physical stress [13]. Within the physiological temperature range of most crops (0–30 °C), higher temperatures cause an exponential rise in respiration. Aerobic respiration involves three interconnected metabolic pathways: glycolysis, the Krebs or tricarboxylic acid (TCA) cycle, and oxidative phosphorylation. During glycolysis, pyruvate is produced out of glucose, which is then further broken down in the TCA cycle to produce CO_2, flavin adenine dinucleotide ($FADH_2$), and nicotinamide adenine dinucleotide (NADH). During the electron transfer, the energy contained in NADH and $FADH_2$ is released, producing two and three adenosine triphosphate (APT) molecules, respectively. In the absence of O_2, NADH and $FADH_2$ are accumulated, and the concentration of the oxidized forms (NAD+ and FAD) decreases. The TCA cycle is interrupted, and glycolysis becomes the only way to obtain ATP. The regeneration of NAD+ in the absence of oxygen takes place through fermentative metabolism by the reductive decarboxylation of pyruvate to ethanol. The shift from predominantly aerobic to predominantly anaerobic respiration is known as the anaerobic compensation point or the fermentative threshold and varies the gas diffusion and respiration rates within the different tissues [13]. Values between 1.5 and 3.5% of O_2 have been reported for apples and mushrooms, respectively, as the critical limit at which an increased CO_2 production marks the offset of an anaerobic respiration [14].

The fermentative processes in olives provokes the development of negative sensory attributes, such as winey, vinegary or muddy sediment, in the resulting olive oils. This means that these oils cannot be commercially considered as EVOO. The presence of alcohols such as methanol or ethanol facilitates the esterification of free fatty acids and the formation of fatty acid alkyl esters (FAAE) in olive oil, the presence of which shows that the olives have undergone a fermentation process [15]. Other causes of oil deterioration such as oxidation or freezing do not induce the increase of these compounds [16]. Methanol can also be formed by the metabolization of pectin from the cell wall during olive ripening, while ethanol can have this origin only in some very specific varieties, so its formation is more related to fermentation [17]. As a consequence, the presence of fatty acid ethyl esters (FAEE) was considered in 2013 as a marker of oil deterioration and is included in the European Union Regulation of olive oil quality.

The hydrolytic and oxidative deterioration of olive fruit is strongly influenced by inappropriate practices during its harvest, storage and transport to industry. The content of ethanol was significantly higher in ground-picked olives than in fruit harvested directly from the tree, although the concentration of FAEE in the extracted oil stayed below the official EVOO limit [18]. Jabeur et al. showed that a relevant factor to explain high levels of FAEE in stored olives is the type of container used for post-harvest storage, reporting higher levels in closed plastic bags than in perforated plastic boxes [19]. Although the fruit in closed plastic bags was brought under conditions of increased asphyxiation, no record was kept of their actual O_2 consumption or CO_2 production during the storage at room temperature up to 25 days. As the fruit was kept at room temperature, no effect of the temperature on the respiration rate was studied, either.

The available O_2 and the storage temperature are both equally important factors in the formation of ethanol in the fruit [20]. Beltrán et al. measured a six-fold increase in ethanol content for both washed and unwashed olive fruit after 12 h of storage in silos, although without reporting the storage temperature [21]. The production of ethanol is directly related to anaerobic respiration, while the temperature has a direct impact on the rate of respiration [21]. In conditions where ventilation is hindered, high temperatures will speed up respiration by generating more heat and accelerating the consumption of oxygen. Ultimately, when all the available oxygen is consumed, anaerobic respiration will take place. It can thus be expected that, in a closed system, the higher the respiration rate, the earlier the processes of fermentation will start to produce ethanol. Thus far, the respiration rate of olive fruit was studied jointly with its ethylene production to document the effects of cold storage on olive fruit [22–26]. Only one study reported the respiration rate and the physicochemical changes that occur in olive fruit picked for table olives and exposed to air at temperatures between 10 and 40 °C [27]. The respiration quotient or the ratio of CO_2 produced to O_2 consumed varied between 2.5 and 1.3 for the Gordal cultivar, and between 2.2 and 1.2 for the 'Manzanilla' cv. However, the experiment took place in an open system, so the evolution of the oxygen consumption was not taken into consideration.

The rates of gas diffusion and respiration may vary significantly when the fruit are piled up, as when olives are collected and transported in large containers. Under such conditions, some olives in the containers risk becoming anaerobic while others remain aerobic. The reduction of O_2 in the storage atmosphere, which determines the ethanol formation threshold of olive fruit in an unventilated environment, has never been determined thus far. This study aims to determine experimentally the time at which the olive fruit attains this formation threshold when kept in a closed system at distinct temperatures.

2. Materials and Methods

2.1. Olive Samples

Roughly 5 kg of olive fruits ('Picual' variety) were hand-picked at a farm in Bollullos par del Condado (Spain) in November 2019, and immediately brought to the facilities of the Instituto de la Grasa (Spanish National Research Council, Seville, Spain) in a perforated plastic box. The olives were healthy, without any visible infection or bruising, and green (ripening index 2.3). On arrival, the fruit was stored in a refrigerator set at 5 °C (±1 °C).

2.2. Experimental Set-Up

The experiment consisted in measuring the CO_2 and O_2 concentrations in three hermetically sealed glass jars (approx. 125 mL) containing a known amount of olives (approx. 45 g) hourly for 7 consecutive hours, while keeping the temperature constant at 5, 10, 15, 20, 23, 30, 35, 40, or 45 °C. The CO_2 and O_2 contents of the headspace of these jars were determined with a G100 portable gas analyzer (Geotechnical Instrument Ltd., Leamington Spa, UK). The tests took place at the facilities and laboratories of the Instituto de la Grasa.

The measuring device consisted of a support, three glass jars with screw caps, and a gas analyzer (Figure 1). The support was made of a wooden board (30 × 30 × 1 cm; four legs of 5 cm), with three holes (7 cm) to hold the glass jars. Through two holes in the screw cap, flexible silicone tubes of 15 cm (diameter of 1 cm) were fixed at 1 cm inside and sealed with silicone. The full tightness of the recipient (jar with the screw cap with tubes) was tested by immersion in a bowl of water. Each tube was connected to two plastic valves, placed one behind the other. Both valves were fixed at the wooden platform to create a stable structure. A silicone tube of 5 cm was connected to the second valve to provide a connection with the gas analyzer.

The temperatures of 5, 10, 15, and 20 °C were obtained in a cooling chamber, adjusted to these temperatures (±1 °C) 24 h before the trials. Stable room temperature allowed for measurements at 23 °C. A FD720 incubator (BINDER GmbH, Tuttlingen, Germany), with a volume of 740 dm^3 was used to attain the temperatures within the range of 30 °C and 45 °C.

The exact volume of each jar and the connected tubes (including the two valves) was determined by weighting the needed water to fill the closed system (precision 0.1 g). On the day of the trial, fruit was taken out of the refrigerator to fill the three glass jars as much as possible without damaging the fruit. The olives of each jar were weighted separately on a laboratory balance (precision 0.1 g). Thereafter, the three jars (A, B, C) filled with the olives were placed without the cap in the thermostatic environment (cooling room or incubator) to attain the desired temperature. After one hour, the screw caps were closed and a first measurement (t_0) was performed with the gas analyzer. In the following 7 h, the O_2 and CO_2 concentration (% *v/v*) was measured. This period was taken as the maximum timespan between the harvesting and the discharge of the fruit in the mill. The samples in the incubator were taken outside for the time of the measurement (approx. 360 s). The ones kept in the cooling room were measured there.

The temperature of each set of three jars (T_i) was set to 5, 10, 15, 20, 23, 30, 35, or 40 °C. The measurements (t_j) took place at 0, 1, 2, 3 4, 5, 6, and 7 h.

The volume of the air (V_{air}) in the closed system with the olive fruit was deduced from the measured volumes of the closed jars (V_{jarr}) minus the volume of olives (V_{olive}) obtained from the weight of the olives, knowing the density of the olive fruit. The volume fractions of the O_2 (V_{O_2}) and CO_2 (V_{CO_2}) in each jar could be calculated by multiplying the V_{air} by the volumetric concentrations of each gas ($y_{O_2}(T_i, t_j)$ and $y_{CO_2}(T_i, t_j)$) in % (*v/v*), which were measured with the gas analyzer at T_i and t_j.

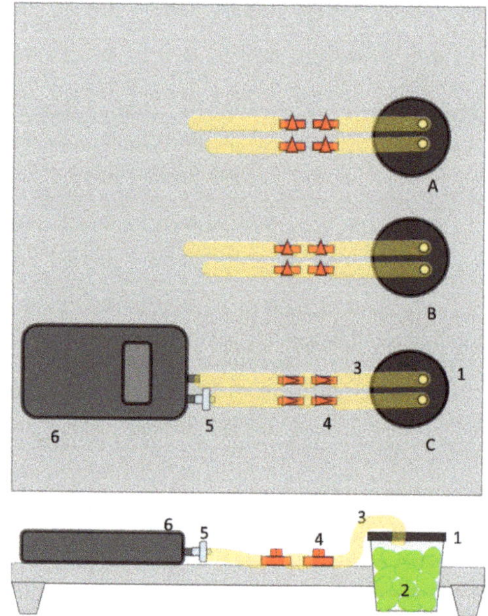

Figure 1. Representation of the measuring device to register the consumed O_2 and produced CO_2 in a closed system by portable gas analyzer. Elements of the measuring device: 1. glass jar (125 mL) with 2 holes (1 cm Ø); 2. olive fruit (±40 g); 3. silicone tube (1 cm Ø); 4. plastic valve; 5. humidity filter; 6. G100 portable gas analyzer. A, B, C are the three samples of the triplicate design.

2.3. *Theory and Calculus*

The amount of each gas ($W(T_i, t_j)$) expressed in mg per g of fruit (mg/g_{olive}) at temperature T and time t (Equations (1) and (2)) could be calculated given the molecular masses of 32 g/mole for O_2 and 44 g/mole for CO_2, and the molar volume of a gas (Equation (3)).

$$W_{O_2}(T_i, t_j)(mg/g_{olive}) = 32 \cdot V_{O_2}(T_i, t_j) \cdot 10^3 \cdot W_{olive}^{-1} \cdot V_m(T_i)^{-1} \quad (1)$$

$$W_{CO_2}(T_i, t_j)(mg/g_{olive}) = 44 \cdot V_{CO_2}(T_i, t_j) \cdot 10^3 \cdot W_{olive}^{-1} \cdot V_m(T_i)^{-1} \quad (2)$$

where

$$V_m(T_i) = \frac{RT_i}{P} \quad (3)$$

with V_m (T_i) = molar volume of a gas (dm^3) at temperature T_i.

R: gas constant (0.082 dm^3 atm K^{-1} mol^{-1}) of the ideal gas law.

T_i: temperature (K).

The respiration rate is the velocity by which CO_2 is produced (RR_{CO_2}) and O_2 is consumed (RR_{O_2}) at a given temperature T_i, both in $cm^3\ kg^{-1}h^{-1}$, and can be calculated given the values of $y_{O_2}(T_i,\ t_j)$ and two different t_j:

$$RR_{CO_2}(T_i) = \left[y_{CO_2}(T_i, t) - y_{CO_2}(T_i,\ t_j)\right] \cdot V_f / \left[\left(t(T_i) - t_j(T_i)\right) \cdot W_{olive} \cdot 100\right] \tag{4}$$

$$RR_{O_2}(T_i) = -\left[y_{O_2}(T_i, t) - y_{O_2}(T_i,\ t_j)\right] \cdot V_f / \left[\left(t(T_i) - t_j(T_i)\right) \cdot W_{olive} \cdot 100\right] \tag{5}$$

That can be expressed in terms of moles ($mM\ kg^{-1}h^{-1}$) as follows:

$$RR_{CO_2}(T_i) = RR_{CO_2}(T_i)\left(cm^3\ kg^{-1}h^{-1}\right) \cdot 44 \cdot V_m(T_i)^{-1} \tag{6}$$

$$RR_{O_2}(T_i) = RR_{O_2}(T_i)\left(cm^3\ kg^{-1}h^{-1}\right) \cdot 32 \cdot V_m(T_i)^{-1} \tag{7}$$

The respiration rate can be characterized by using either the CO_2 production rate or the O_2 consumption rate. In the closed system under study, oxygen acted as the limiting reagent and was therefore given preference over CO_2 production. The Arrhenius model allows for the determination of the respiration rate at a given temperature from data obtained at other temperatures, assuming that the reaction is of zero-order and that the respiration rate is therefore constant, equal to the specific rate constant [14].

$$RR_{O_2} = A_{O_2} \cdot e^{-Ea_{O_2} \cdot \frac{1}{RT}} \tag{8}$$

with A_{O_2} ($cm^3\ kg^{-1}\ h^{-1}$): pre-exponential factor for O_2 consumption.

Ea_{O_2} ($J\ mol^{-1}$): activation energy for the O_2 consumption.

R: gas constant ($8.314\ J\ mol^{-1}\ K^{-1}$) of the ideal gas law.

A_{O_2} ($cm^3\ kg^{-1}\ h^{-1}$) and Ea_{O_2} ($J\ mol^{-1}$) can be calculated when knowing the RR_{O_2} at two different T_i solving the two Equations (9) and (10), and by having Equations (11) and (16).

$$RR_{O_2}(T_1) = A_{O_2} \cdot e^{-Ea_{O_2} \cdot \frac{1}{RT_1}} \tag{9}$$

$$RR_{O_2}(T_2) = A_{O_2} \cdot e^{-Ea_{O_2} \cdot \frac{1}{RT_2}} \tag{10}$$

$$A_{O_2} = RR_{O_2}(T_1) \cdot e^{Ea_{O_2} \cdot \frac{1}{RT_1}} \tag{11}$$

$$RR_{O_2}(T_2) = RR_{O_2}(T_1) \cdot e^{Ea_{O_2} \cdot \frac{1}{RT_1}} \cdot e^{-Ea_{O_2} \cdot \frac{1}{RT_2}} \tag{12}$$

$$\frac{RR_{O_2}(T_2)}{RR_{O_2}(T_1)} = e^{Ea_{O_2} \cdot \frac{1}{RT_1} - Ea_{O_2} \cdot \frac{1}{RT_2}} \tag{13}$$

$$\ln\left(\frac{RR_{O_2}(T_2)}{RR_{O_2}(T_1)}\right) = Ea_{O_2} \cdot \frac{1}{RT_1} - Ea_{O_2} \cdot \frac{1}{RT_2} \tag{14}$$

$$\ln\left(\frac{RR_{O_2}(T_2)}{RR_{O_2}(T_1)}\right) = Ea_{O_2}\left(\frac{1}{RT_1} - \frac{1}{RT_2}\right) \tag{15}$$

$$Ea_{O_2} = \frac{\ln\left(\frac{RR_{O_2}(T_2)}{RR_{O_2}(T_1)}\right)}{\left(\frac{1}{RT_1} - \frac{1}{RT_2}\right)} \tag{16}$$

2.4. Simulation Model

After verifying that the Arrhenius model could be used to estimate the consumption rate at the different temperatures, a simulation study was performed to calculate the time needed to consume up to 3% of the O_2 present in a closed system filled with fruit at a given temperature. This limit was set arbitrarily, and was considered the thresholds as reported by Fonseca et al. [14].

The closed system was a 1-m^3 cube situated in the lower part of a container used to transport olives with dimensions set at 6 m long, 2.5 m wide, and 2.5 m high (Figure 2). The cube was situated at 0.75 m from the lateral walls to guarantee that direct heat exchange with the outside was negligible. A distance of 1.5 m between the upper layers of the fruit and the upper site, together with the absence of a lateral or vertical airflow, prevented the entrance of O_2 into the cube of fruit under study.

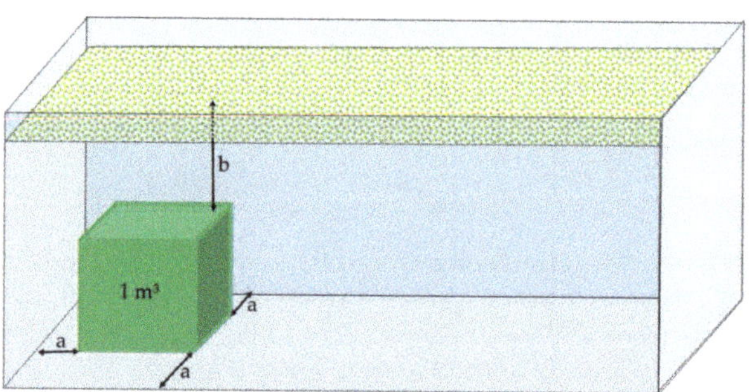

Figure 2. Situation of a closed system of 1 m^3 volume within a container, with a distance of 0.75 m (a) from the lateral walls and 1.5 m between upper layers of the fruit and the upper site (b).

The biometrical characteristics of olives ('Picual' cv.) used in another experiment were used to calculate the air volume (V_{air}) in 1 m^3 [28,29]. The mean weight was 4.4 g and the volume, calculated after measuring the width and height and assuming the volume to be a prolate spheroid, was 4.14 cm^3. By weighting the number of olives that entered in a known volume, it was determined that 500 g of olives (of 4.4 g per olive) entered in a volume of 1 dm^3, and 500 kg, i.e., 113,636 olives, in 1 m^3. The volume occupied by the fruit thus accounted for 0.470 m^3. The total volume of O_2, expressed in dm^3, in 1 m^3 filled with 500 kg of olive fruit, amounted to (1 m^3–0.470 m^3) · 0.209 O_2 dm^3 O_2/ dm^3 air · 1000 dm^3/m^3, i.e., 110.7 dm^3.

The amount of O_2 consumed for a given time interval ($t_{j+1} - t_j$) can be calculated using the respiration rate at each temperature, as obtained by the Arrhenius equation.

$$V_{O_2}(T_i, t_{j+1}) = V_{O_2}(T_i, t_j) - RR_{O_2}(t_j) \cdot \left(\frac{500 \text{ kg}}{1000 \text{ kg}}\right) \cdot (t_{j+1} - t_j) \tag{17}$$

During aerobic respiration, energy is produced at a rate of 2667 kJ for each mole of glucose that reacts with 6 moles of O_2 [13]:

$$C_6H_{12}O_6(s) + 6O_2(g) \rightarrow 6CO_2(g) + 6H_2O\ (l) + 2667\ \text{kJ} \tag{18}$$

where 5% of that energy is consumed during the breaking down of the glucose molecule, 41% is used to produce 38 ATP molecules, and 57% is lost as heat [13]. Therefore, for each mole of O_2 consumed, 253 kJ are added to the system. Under adiabatic conditions, the heat generated during a time t needs to be taken into consideration as it added up to the initial temperature t under which the fruit was stored. As the production of this heat follows the consumption of O_2, the Arrhenius model can be used to calculate the rate of heat production, $l(T)$, expressed in kJ kg^{-1} h^{-1}, at a given temperature $T(t)$ (Equation (19)). As it refers to the same reaction, the activation energy (Ea) is the same, although the pre-exponential factor B (J kg^{-1} h^{-1}) is different.

$$l(T) = B \cdot e^{-Ea\frac{1}{RT}} \tag{19}$$

At time t, the temperature of the system (m) and C_p increase by

$$T(t-t_0)=\frac{l(T_0,\ t_0)}{m{\cdot}C_p}{\cdot}t \tag{20}$$

The temperature *T(t)* at time t thus becomes:

$$T(t)=T(t_0)+T(t-t_0) \tag{21}$$

Considering:

$$RR_{O_2}(t)=RR_{O_2}(T_0) \text{ and } RR_{O_2}(t_0)=RR_{O_2}(0) \tag{22}$$

It can be deduced that:

$$RR_{O_2}(t)-RR_{O_2}(0)=RR_{O_2}(T_0)(t-t_0) \tag{23}$$

$$RR_{O_2}(t)=RR_{O_2}(0)+Ea^{\frac{-Ea}{R\cdot T_0}}(t-t_0) \tag{24}$$

Therefore:

$$RR_{O_2}(t)=RR_{O_2}(0)+Ea_{O_2}{}^{\frac{-Ea_{O_2}}{R(T_0+\frac{I}{mC_p}t)}}(t-t_0) \tag{25}$$

At each time interval (*t*–*t*$_0$), the rate of consumption of O_2 was calculated, considering the initial temperature and the accumulated heat produced during that interval. According to the Arrhenius equation, the reaction rate increases with increases in temperature, thus the increased respiration rate can then be accordingly adjusted. This corrected respiration rate allowed a precise estimation of the time it takes for the described closed system, at an initial temperature T to attain the critical threshold of 3% of O_2.

3. Results

3.1. Concentrations of O_2 and CO_2 over Time

The calculated values of W_{O_2} and W_{CO_2} over 7 h allowed us to appreciate the differences among the samples kept at different temperatures (Table 1). At 5 °C, the consumption of the O_2 hardly changed over the studied time. As the temperature increased, the amount of O_2 decreases rapidly and we recorded an increase of CO_2. At 20 °C, the amount of O_2 and CO_2 balanced after 7 h, while at 30 °C the balance occurred after 3 h, and at 40 °C it balanced in less than 2 h. At 45 °C, all the O_2 in the jars was consumed in less than 3 h.

Table 1. Amount of mg O_2 consumed and mg CO_2 produced per gram of olive fruit kept in a closed system at a temperature between 5 and 45 °C, over a time span of 7 h. Each value indicates the mean value of 3 replicates ± standard deviation.

Time (h)	5 °C	10 °C	15 °C	20 °C	O_2 (mg/g$_{olive}$) 23 °C	30 °C	35 °C	40 °C	45 °C
0	0.55 ± 0.02	0.46 ± 0.00	0.50 ± 0.01	0.53 ± 0.02	0.53 ± 0.01	0.53 ± 0.02	0.48 ± 0.01	0.48 ± 0.01	0.40 ± 0.02
1	0.53 ± 0.02	0.42 ± 0.01	0.47 ± 0.01	0.49 ± 0.02	0.49 ± 0.04	0.42 ± 0.01	0.39 ± 0.01	0.36 ± 0.01	0.27 ± 0.02
2	0.51 ± 0.02	0.41 ± 0.00	0.44 ± 0.01	0.42 ± 0.02	0.45 ± 0.04	0.35 ± 0.01	0.27 ± 0.02	0.26 ± 0.03	0.19 ± 0.02
3	0.50 ± 0.02	0.39 ± 0.01	0.42 ± 0.01	0.37 ± 0.02	0.40 ± 0.03	0.30 ± 0.00	0.19 ± 0.03	0.16 ± 0.01	0.00 ± 0.00
4	0.50 ± 0.02	0.40 ± 0.01	0.39 ± 0.01	0.33 ± 0.01	0.36 ± 0.04	0.22 ± 0.00	0.15 ± 0.03	0.07 ± 0.02	0.00 ± 0.00
5	0.49 ± 0.02	0.37 ± 0.01	0.37 ± 0.02	0.28 ± 0.02	0.32 ± 0.04	0.16 ± 0.00	0.13 ± 0.03	0.04 ± 0.01	0.00 ± 0.00
6	0.47 ± 0.03	0.37 ± 0.01	0.35 ± 0.02	0.26 ± 0.02	0.28 ± 0.03	0.13 ± 0.00	0.12 ± 0.03	0.03 ± 0.01	0.00 ± 0.00
7	0.48 ± 0.02	0.35 ± 0.01	0.33 ± 0.02	0.24 ± 0.02	0.25 ± 0.03	0.11 ± 0.00	0.10 ± 0.03	0.04 ± 0.01	0.00 ± 0.00
Time (h)	**5 °C**	**10 °C**	**15 °C**	**20 °C**	**CO_2 (mg/g$_{olive}$) 23 °C**	**30 °C**	**35 °C**	**40 °C**	**45 °C**
0	0.01 ± 0.00	0.02 ± 0.00	0.01 ± 0.00	0.00 ± 0.00	0.01 ± 0.00	0.00 ± 0.00	0.02 ± 0.00	0.01 ± 0.00	0.01 ± 0.00
1	0.04 ± 0.00	0.06 ± 0.00	0.04 ± 0.00	0.04 ± 0.00	0.06 ± 0.00	0.15 ± 0.00	0.17 ± 0.00	0.20 ± 0.00	0.22 ± 0.00
2	0.04 ± 0.00	0.07 ± 0.00	0.07 ± 0.00	0.09 ± 0.00	0.11 ± 0.00	0.23 ± 0.00	0.35 ± 0.00	0.34 ± 0.00	0.53 ± 0.00
3	0.05 ± 0.00	0.09 ± 0.00	0.10 ± 0.00	0.13 ± 0.00	0.16 ± 0.00	0.29 ± 0.00	0.49 ± 0.00	0.48 ± 0.00	0.63 ± 0.00
4	0.05 ± 0.00	0.10 ± 0.00	0.13 ± 0.00	0.17 ± 0.00	0.21 ± 0.00	0.39 ± 0.00	0.60 ± 0.00	0.64 ± 0.00	0.63 ± 0.00
5	0.06 ± 0.00	0.11 ± 0.00	0.15 ± 0.00	0.20 ± 0.00	0.25 ± 0.00	0.46 ± 0.00	0.65 ± 0.00	0.73 ± 0.00	0.63 ± 0.00
6	0.06 ± 0.00	0.11 ± 0.00	0.17 ± 0.00	0.23 ± 0.00	0.28 ± 0.00	0.49 ± 0.00	0.68 ± 0.00	0.73 ± 0.00	0.63 ± 0.00
7	0.06 ± 0.00	0.12 ± 0.00	0.19 ± 0.00	0.24 ± 0.00	0.31 ± 0.00	0.52 ± 0.00	0.70 ± 0.00	0.74 ± 0.00	0.63 ± 0.00

The ratio of CO_2 produced to O_2 consumed (RQ) fell within the range of 0.6 (5 and 10 °C) and 1.4 (30 °C) during the full time of the experiment for the samples that were kept at temperatures below 35 °C. From then on, the RQ started to increase before the end of the trials, with values above 3 after 6 h at 35 °C, 3.6 after 5 h at 40 °C, and 3.7 after 2 h at 45 °C. The limitations of the gas analyzer used impeded the registration of CO_2 concentrations above 20%, so no accurate values could be obtained once all the O_2 was consumed.

3.2. *Influence of Temperature on the Respiration Rate*

The influence of the temperature on the respiration rate became even more clear when the consumption of O_2 (Equation (4)) was plotted against the temperature (Figure 3). The respiration rate doubled between 10 and 20 °C and between 20 and 30 °C. The values of the O_2 increased exponentially ($y = 12.801e^{0.0495x}$; $R^2 = 0.97$) between 5 and 40 °C. However, above 40 °C the respiration rate did not increase further and it maintained an almost equal level. The CO_2 production (Equation (5)) increased similarly to the O_2 up to 30 °C, after which the values continued to increase at a much higher velocity than the O_2.

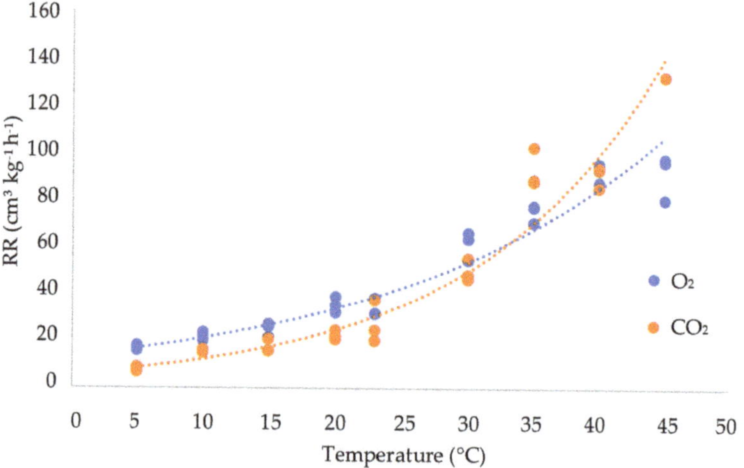

Figure 3. Respiration rate expressed as the O_2 consumed and CO_2 produced by olive fruit in a closed system kept at temperatures between 5 and 45 °C.

The data allowed the calculation of the pre-exponential factor (A) and the activation energy (*Ea*) of the Ahrrenius equation at each temperature, once the respiration rate, expressed in mM $kg^{-1}h^{-1}$ of O_2 consumed, was assessed following Equations (11) and (16). Both coefficients were determined with distinct intervals of (t–t_0), after which the calculated values of RR_{O_2} were compared with the rates obtained in the experiments. The data registered after 1 and 2 h were the most satisfactory when compared with the ones between 2 and 3, 3 and 4, and 1 and 4 h. A correlation coefficient of 0.99 was obtained when the data at 45 °C were not considered (data not shown). Therefore, the following Arrhenius function (Equation (26)) was withheld:

$$RR_{O_2}(T) = 1.9\ 10^7 \cdot e^{-31.4 \cdot \frac{1}{RT}} \quad (26)$$

Up to 40 °C, the Arrhenius function closely matched the observed data (Figure 4). At 45 °C, the oxygen consumption was nil after 3 h because all the oxygen in the jars had been already consumed within those three hours (Table 1).

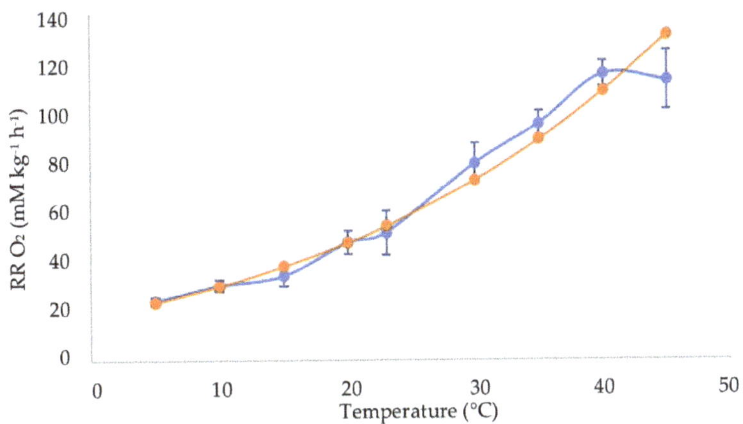

Figure 4. Experimental mean respiration rates (blue points), expressed as O_2 consumed, and theoretical calculated values according to the Arrhenius equation (orange points) at temperatures between 5 and 45 °C.

3.3. Heat Generated by Fruit Respiration

To take into account the additional heat generated by the reaction itself, the pre-exponential factor B of the Arrhenius function, as defined in Equation (19), was calculated under the assumption that the reaction produced 253 kJ for each mole of O_2 consumed, resulting in:

$$l(T) = 5.98\ 10^5 \cdot e^{-30.7 \cdot \frac{1}{RT}} \tag{27}$$

To estimate this additional effect, a simulation was performed following Equations (21)–(25), assuming a closed system of 1 m^3 containing 500 kg of olives, and a free volume of O_2 of 110.7 dm^3 over 7 h. The obtained curves at each T_0 are all defined by an adjusted quadratic function, presenting increasing acceleration with a higher initial temperature (Figure 5). It can be observed that the respiration rate at an initial temperature of 30 °C rose after 4 h to a level that equaled the one that exists at a T_0 of 35 °C. At an initial temperature of 35 °C, the respiration rate at 40 °C was attained after 3 h.

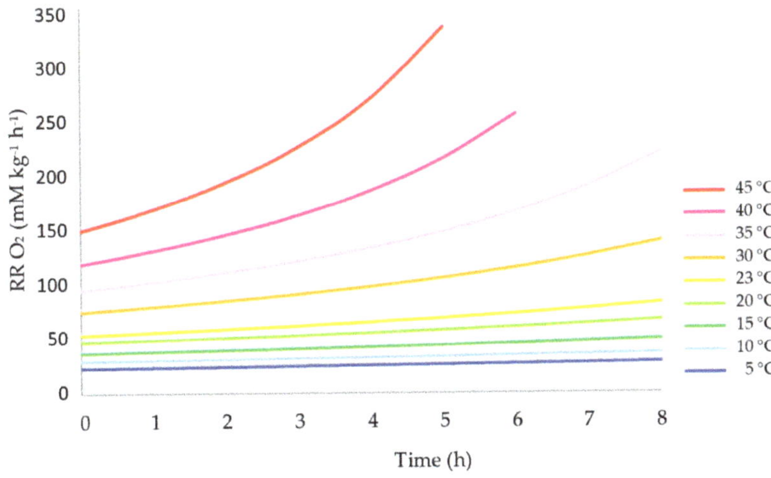

Figure 5. Simulation of the increase of the respiration rate of olive fruits kept in a closed system, expressed as consumed O_2 (mM kg^{-1} h^{-1}), at different temperatures over 8 h.

This theoretical estimation did not consider the decreasing amount of O_2 necessary to keep the aerobic reaction going. As such, the calculated respiration rates at high initial temperatures after three or more hours are not valid, as no oxygen will be present in the closed system.

3.4. Consumption of O_2 during Storage under Adiabatic Conditions

The adjusted respiration rates facilitated a simulation of the consumption of oxygen within the described closed system (Figure 6). To underscore the effect of the heat generated by the fruit itself, a dotted line is added to represent the decrease of O_2 without considering this additional heat. The limit where aerobic respiration becomes critical is set at 3% and marked as a horizontal dashed line. According to this simulation, the olives kept at a temperature above 35 °C, will attain that limit in less than 2 h. Fruit kept between 20 and 30 °C risk fermentation in less than 4 h, while fruit stored between 5 and 15 °C can be stored between 7.5 and 5 h without inducing anaerobic respiration. With 3% of O_2 left, the added heat reduced the needed time between 7.3% (5 °C) and 8.5% (45 °C) (data not shown). However, as it takes longer to attain this threshold at lower temperatures, the reduction of the time is equally greater: 35 min at 5 °C, 16 min at 23 °C, and 7 min at 45 °C.

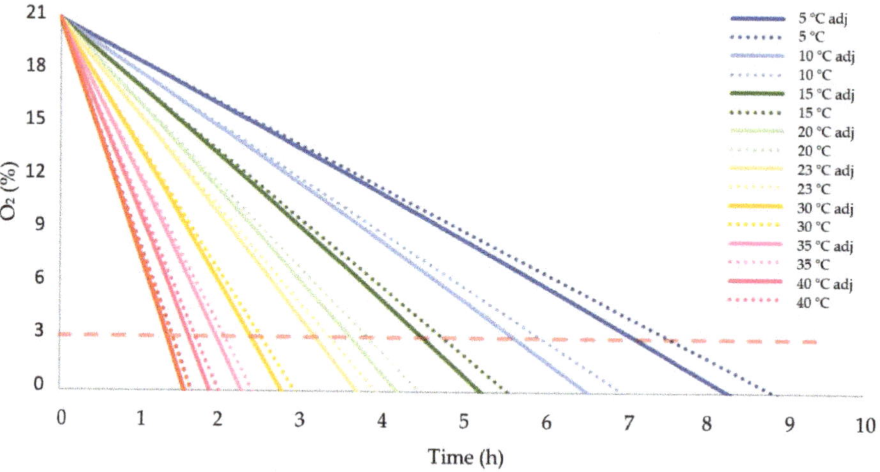

Figure 6. Consumption of available O_2 within the described closed system at different temperatures, taking into account the additional heat produced by the fruit itself (adj).

Plotting the different temperatures against the time needed to consume up to 3% of O_2 revealed an exponential curve ($y = 67.962\ e^{-0.341x}$) with an R^2 of 0.99 (Figure 7). The graphic indicates that when the fruit is kept at 25 °C under the described conditions, it will take 3 h to attain this threshold of 3%, while it will less than 2 h at 35 °C.

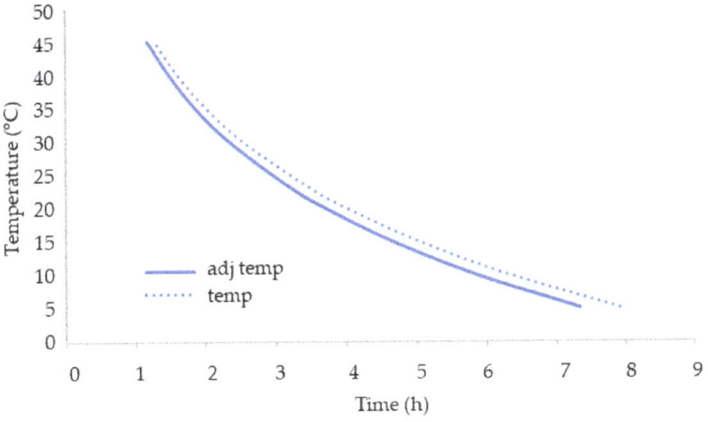

Figure 7. Time needed to attain the threshold of 3% O_2 in the described closed system kept at temperatures between 5 and 45 °C.

4. Discussion

The experiment provided concise data on the O_2 consumption and CO_2 production of olives in a closed system at different temperatures over a period of 7 h. The data allowed an estimation of the changes in respiration rate as well as the moment when a defined closed system would shift to anaerobic respiration. The limitations related to a closed or static system that were summed up by Fonseca et al. [14] were overcome or were negligible for the experiment. The gas volume could be accurately estimated, the experiment was designed to measure the effect of O_2 depletion and CO_2 production, the time between the sampling provided enough data to make generalized estimations, and no combination of gases was considered.

The obtained RQ values for the samples kept below 35 °C were within the normal values as reported in the literature, ranging from 0.7 to 1.3 [27,30]. The higher levels registered at an increasingly earlier time did indicate a shift towards anaerobic respiration. Values higher than 6 are reported for fruit and vegetables packed in low permeability films and causally linked to the presence of fermentative metabolism [14]. Although the limitations of the used gas analyzer did impede a precise registration of the produced CO_2, it is obvious that the registered values above 3 marked the change in metabolism.

The effect of time and temperature on respiration has been studied on various intact fruit as well as in fresh-cut produce [14,31–36]. In all of these studies, an Arrhenius-type equation was shown to be very reliable to describe the respiration rate as a function of the temperature. The calculated respiration rate of the olive fruit kept at different temperatures adjusted itself to an Arrhenius function up to 40 °C. The obtained preexponential factor and the respiration activation energy of olive fruit, being respectively 1.9 10^7 ($mg \cdot kg^{-1} \cdot h^{-1}$) and 31.4 ($kJ \cdot mol^{-1}$), lie within the normal range ($9.8 \cdot 10^5$–$1.2 \cdot 10^{17}$ $mg \cdot kg^{-1} \cdot h^{-1}$ and 29–93 $kJ \cdot mol^{-1}$), as reported in the literature for fruit and vegetables [37].

It was observed that biological reactions generally increase two or three-fold for every 10 °C raised in temperature for temperatures in the range of 10–30 °C [38]. At higher temperatures, enzymatic denaturation may occur, leading to a reduced respiration rate, while at lower temperatures, physiological injury can provoke an increase in respiration rate [39]. The obtained data did present comparable results in the aforementioned temperature range. However, the reduction of the respiration rate at the highest temperatures was not induced by an enzymatic denaturation but by the fact that oxygen acted as the limiting factor in the reaction, leading to a flattening of the respiration rate curve above 40 °C. At the lowest temperature, no physiological injuries were observed, nor were they expected as olive fruit can be stored at 5 °C without risk of physiological deterioration [40–42].

The fact that the respiration rate tends to decrease with storage time has been modelled in different ways. For example, Waghmare et al. used a first-order decay model [31], while Bhande et al. applied a model using principles of enzyme kinetics with uncompetitive inhibition [34]. However, all these models studied modified atmosphere packaging designs (MAP) and considered a far longer storage time. As those fruits and vegetables were kept for up to 5 days, developmental changes during storage needed to be considered. The experiment under discussion focused on the consumption of oxygen and did not consider these physiological changes. As the time frame was much shorter, the storage temperature was higher, and the fruit was not kept under MAP conditions, there was no need to model the effect of time on the respiration rate.

The cumulative effect of the heat produced during the respiration was given full consideration in the presented model and led to an adjusted Arrhenius equation in which the temperature was corrected with the heat generated by the respiration over a specified time interval. The model further assumed that the respiratory heat generation was not constant but a function of the heat at each moment, rendering a more precise estimation than suggested by Redding et al. [32]. Thus far, few studies have considered this variable, and it is often discarded it as being a neglectable factor in the overall heat transfer, suggesting that in a typical precooling process the heat added by respiration is approx. 0.5% [43–45]. However, the results indicate that, under specific conditions, this effect leads to a reduction of the estimated time by 7% before the critical minimum concentration of 3% is attained. The impact of the added heat is more pronounced at a lower temperature, simply because it takes longer to attain the threshold. As such, the results do confirm the suggestion made by Redding et al. [32] that the assumption of a negligible contribution of respiration rate is more likely to be untrue for lengthy precooling processes.

The model assumed that the 1 m^3 of fruit within the bulk transport could be considered a closed system, with neither fresh air entering nor a temperature change induced from outside. In a real bulk transport scenario, the top layers are not only exposed to fresh air, but also be submitted to higher temperatures if they are exposed to the sun. At the same time, the layers on the lateral sides of the container will be heated up as the walls can be warmed up during transport. As heat is brought into the system during the transport, the respiration rate will increase accordingly, while the entrance of fresh air will delay the moment when the critical threshold is reached. Specific measurements of the O_2 and CO_2 concentrations, as well as the temperature at different layers, are thus needed to further fine-tune the presented model. Alternatively, Computation Fluid Dynamics can be used to calculate the airflow and temperature changes within the defined container numerically, either with a porous-medium approach or a direct simulation [46]. Future research is also needed to confirm the presence of ethanol in fruit that is generated under the described critical conditions. It would also be interesting to calculate the behavior of damaged fruit, which can sometimes occur as a consequence of the detachment pattern generated by some harvester systems such as super-intensive harvesters [47]. As for now, the presented model cautions that to avoid fermentation processes in at least a substantial part of the fruit, it is necessary to consider the time of transport as an important risk factor.

5. Conclusions

The shift towards super high-density hedgerow orchards, together with the foreseeable impact of climatic changes, urges us to focus on the postharvest conditions of the olive fruit. The presented study clarified the direct relation between both variables and demonstrated that even at non-extreme ambient temperatures (between 30 and 40 °C), the threshold where anaerobic respiration becomes dominant can be attained within a couple of hours. To avoid detrimental fermentative processes, it is thus recommended to reduce the time between the harvest and the extraction as much as possible, as well as to contemplate the introduction of cooling systems to reduce the field heat prior to fruit processing.

Author Contributions: Conceptualization, E.P. and J.M.G.M.; methodology, E.P.; validation, R.R.S.-G.; formal analysis, E.P., M.d.C.F.F. and R.R.S.-G.; investigation, E.P.; resources, J.M.G.M.; data curation, E.P.; writing—original draft preparation, E.P., M.d.C.F.F., R.R.S.-G., J.M.G.M. and J.F.G.M.; writing—review and editing, E.P., M.d.C.F.F., J.M.G.M., R.R.S.-G. and J.F.G.M.; supervision, J.M.G.M., R.R.S.-G. and M.d.C.F.F.; project administration, J.M.G.M. All authors have read and agreed to the published version of the manuscript.

Funding: This research received no external funding.

Institutional Review Board Statement: Not applicable.

Informed Consent Statement: Not applicable.

Data Availability Statement: Not applicable.

Conflicts of Interest: The authors declare no conflict of interest.

References

1. Diez, C.M.; Moral, J.; Cabello, D.; Morello, P.; Rallo, L.; Barranco, D. Cultivar and tree density as key factors in the long-term performance of super high-density olive orchards. *Front. Plant Sci.* **2016**, *7*, 1–13. [CrossRef]
2. Servicio de Estudios y Estadísticas de la Junta de Andalucía. *Análisis de la Densidad en las Plantaciones de Olivar en Andalucía;* Junta de Andalucía: Seville, Spain, 2019.
3. Connor, D.J.; Gómez-del-Campo, M.; Rousseaux, M.C.; Searles, P.S. Structure, management and productivity of hedgerow olive orchards: A review. *Sci. Hortic.* **2014**, *169*, 71–93. [CrossRef]
4. Rallo, L.; Díez, C.M.; Morales-Sillero, A.; Miho, H.; Priego-Capote, F.; Rallo, P. Quality of olives: A focus on agricultural preharvest factors. *Sci. Hortic.* **2018**, *233*, 491–509. [CrossRef]
5. Fraga, H.; Pinto, J.G.; Santos, J.A. Climate change projections for chilling and heat forcing conditions in European vineyards and olive orchards: A multi-model assessment. *Clim. Change* **2019**, *152*, 179–193. [CrossRef]
6. Arenas-Castro, S.; Gonçalves, J.F.; Moreno, M.; Villar, R. Projected climate changes are expected to decrease the suitability and production of olive varieties in southern Spain. *Sci. Total Environ.* **2020**, *709*, 136161. [CrossRef]
7. Mihailescu, E.; Soares, M.B. The influence of climate on agricultural decisions for three European crops: A systematic review. *Front. Sustain. Food Syst.* **2020**, *4*, 64. [CrossRef]
8. Sousa, A.A.R.; Barandica, J.M.; Aguilera, P.A.; Rescia, A.J. Examining potential environmental consequences of climate change and other driving forces on the sustainability of Spanish olive groves under a socio-ecological approach. *Agriculture* **2020**, *10*, 509. [CrossRef]
9. Galán, C.; García-Mozo, H.; Vázquez, L.; Ruiz, L.; de La Guardia, C.D.; Trigo, M.M. Heat requirement for the onset of the *Olea europaea* L. pollen season in several sites in Andalusia and the effect of the expected future climate change. *Int. J. Biometeorol.* **2005**, *49*, 184–188. [CrossRef] [PubMed]
10. Avolio, E.; Orlandi, F.; Bellecci, C.; Fornaciari, M.; Federico, S. Assessment of the impact of climate change on the olive flowering in Calabria (southern Italy). *Theor. Appl. Climatol.* **2012**, *107*, 531–540. [CrossRef]
11. García-Mozo, H.; Oteros, J.; Galán, C. Phenological changes in olive (ola europaea l.) reproductive cycle in southern Spain due to climate change. *Ann. Agric. Environ. Med.* **2015**, *22*, 421–428. [CrossRef] [PubMed]
12. Gabaldón-Leal, C.; Ruiz-Ramos, M.; de la Rosa, R.; León, L.; Belaj, A.; Rodríguez, A.; Santos, C.; Lorite, I.J. Impact of changes in mean and extreme temperatures caused by climate change on olive flowering in southern Spain. *Int. J. Climatol.* **2017**, *37*, 940–957. [CrossRef]
13. Saltveit, M.E. Respiratory Metabolism. In *The Commercial Storage of Fruits, Vegetables, and Florist and Nursery Stocks. Agriculture Handbook 66*; Gross, K.C., Wang, C.Y., Saltveit, M., Eds.; United States Department of Agriculture: Washington, DC, USA, 2016; pp. 68–75.
14. Fonseca, S.C.; Oliveira, F.A.R.; Brecht, J.K. Modelling respiration rate of fresh fruits and vegetables for modified atmosphere packages: A review. *J. Food Eng.* **2002**, *52*, 99–119. [CrossRef]
15. Bendini, A.; Valli, E.; Cerretani, L.; Chiavaro, E.; Lercker, G. Study on the effects of heating of virgin olive oil blended with mildly deodorized olive oil: Focus on the hydrolytic and oxidative state. *J. Agric. Food Chem.* **2009**, *57*, 10055–10062. [CrossRef] [PubMed]
16. Gómez-Coca, R.B.; Cruz-Hidalgo, R.; Fernandes, G.D.; Pérez-Camino, M.D.C.; Moreda, W. Analysis of methanol and ethanol in virgin olive oil. *MethodsX* **2014**, *1*, e207–e211. [CrossRef]
17. García-Vico, L.; Belaj, A.; León, L.; de la Rosa, R.; Sanz, C.; Pérez, A.G. A survey of ethanol content in virgin olive oil. *Food Control* **2018**, *91*, 248–253. [CrossRef]
18. Beltran, G.; Sánchez, R.; Sánchez-Ortiz, A.; Aguilera, M.P.; Bejaoui, M.A.; Jimenez, A. How 'ground-picked' olive fruits affect virgin olive oil ethanol content, ethyl esters and quality. *J. Sci. Food Agric.* **2016**, *96*, 3801–3806. [CrossRef] [PubMed]
19. Jabeur, H.; Zribi, A.; Abdelhedi, R.; Bouaziz, M. Effect of olive storage conditions on Chemlali olive oil quality and the effective role of fatty acids alkyl esters in checking olive oils authenticity. *Food Chem.* **2015**, *169*, 289–296. [CrossRef] [PubMed]

20. Plasquy, E.; García Martos, J.M.; Florido, M.C.; Sola-guirado, R.R.; García Martín, J.F. Cold storage and temperature management of olive fruit: The impact on fruit physiology and olive oil quality—A review. *Processes* **2021**, *9*, 1543. [CrossRef]
21. Beltran, G.; Hueso, A.; Bejaoui, M.A.; Gila, A.M.; Costales, R.; Sánchez-Ortiz, A.; Aguilera, M.P.; Jimenez, A. How olive washing and storage affect fruit ethanol and virgin olive oil ethanol, ethyl esters and composition. *J. Sci. Food Agric.* **2021**, *101*, 3714–3722. [CrossRef]
22. García, J.M.; Streif, J. The effect of controlled atmosphere storage on fruit and oil quality of "Gordal" olives. *Gartenbauwissenschaft* **1991**, *56*, 233–238.
23. Agar, I.T.; Hess-Pierce, B.; Sourour, M.M.; Kader, A.A. Quality of Fruit and Oil of Black-Ripe Olives Is Influenced by Cultivar and Storage Period. *J. Agric. Food Chem.* **1998**, *46*, 3415–3421. [CrossRef]
24. Yousfi, K.; Cayuela, J.A.; García, J.M. Effect of Temperature, modified atmosphere and ethylene during olive storage on quality and bitterness level of the oil. *JAOCS J. Am. Oil Chem. Soc.* **2009**, *86*, 291–296. [CrossRef]
25. Morales-Sillero, A.; Pérez, A.G.; Casanova, L.; García, J.M. Cold storage of 'Manzanilla de Sevilla' and 'Manzanilla Cacereña' mill olives from super-high density orchards. *Food Chem.* **2017**, *237*, 1216–1225. [CrossRef] [PubMed]
26. Plasquy, E.; Blanco-Roldán, G.; Florido, M.C.; García, J.M. Effects of an integrated harvest system for small producers on the quality of the recollected olive fruit. *Grasas y Aceites* **2021**, *72*, e436.
27. García, P.; Brenes, M.; Romero, C.; Garrido, A. Respiration and physicochemical changes in harvested olive fruits. *J. Hortic. Sci.* **1995**, *70*, 925–933. [CrossRef]
28. Plasquy, E.; Garcia, J.M.; Florido, M.C.; Sola-guirado, R.R. Estimation of the cooling rate of six olive cultivars using thermal imaging. *Agriculture* **2021**, *11*, 164. [CrossRef]
29. Plasquy, E.; García Martos, J.M.; Florido Fernández, M.D.C.; Sola-Guirado, R.R.; García Martín, J.F. Adjustment of olive fruit temperature before grinding for olive oil extraction. experimental study and pilot plant trials. *Processes* **2021**, *9*, 586. [CrossRef]
30. Kader, A.A. Respiration and gas exchange of vegetables. In *Postharvest Physiology of Vegetables*; Weichmann, J., Ed.; Marcel Dekker: New York, NY, USA, 1987; pp. 25–43.
31. Waghmare, R.B.; Mahajan, P.V.; Annapure, U.S. Modelling the effect of time and temperature on respiration rate of selected fresh-cut produce. *Postharvest Biol. Technol.* **2013**, *80*, 25–30. [CrossRef]
32. Redding, G.P.; Yang, A.; Shim, Y.M.; Olatunji, J.; East, A. A review of the use and design of produce simulators for horticultural forced-air cooling studies. *J. Food Eng.* **2016**, *190*, 80–93. [CrossRef]
33. Bower, J.H.; Jobling, J.J.; Patterson, B.D.; Ryan, D.J. A method for measuring the respiration rate and respiratory quotient of detached plant tissues. *Postharvest Biol. Technol.* **1998**, *13*, 263–270. [CrossRef]
34. Bhande, S.D.; Ravindra, M.R.; Goswami, T.K. Respiration rate of banana fruit under aerobic conditions at different storage temperatures. *J. Food Eng.* **2008**, *87*, 116–123. [CrossRef]
35. Caleb, O.J.; Mahajan, P.V.; Opara, U.L.; Witthuhn, C.R. Modelling the respiration rates of pomegranate fruit and arils. *Postharvest Biol. Technol.* **2012**, *64*, 49–54. [CrossRef]
36. Iqbal, T.; Rodrigues, F.A.S.; Mahajan, P.V.; Kerry, J.P. Effect of Time, temperature, and slicing on respiration rate of mushrooms. *J. Food Sci.* **2009**, *74*, E298–E303. [CrossRef] [PubMed]
37. Exama, A.; Arul, J.; Lencki, R.W.; Lee, L.Z.; Toupin, C. Suitability of plastic films for modified atmosphere packaging of fruits and vegetables. *J. Food Sci.* **1993**, *58*, 1365–1370. [CrossRef]
38. Zagory, D.; Kader, A. Modified atmosphere packaging of fruits and vegetables. *Food Technol.* **1988**, *42*, 70–77.
39. Fidler, J.C.; North, C.J. The effect of conditions of storage on the respiration of apples. I. The effects of temperature and concentrations of carbon dioxide and oxygen on the production of carbon dioxide and uptake of oxygen. *J. Hortic. Sci.* **1967**, *42*, 189–206. [CrossRef]
40. García, J.M.; Yousfi, K. The postharvest of mill olives. *Grasas y Aceites* **2006**, *57*, 16–24. [CrossRef]
41. Clodoveo, M.L.; Delcuratolo, D.; Gomes, T.; Colelli, G. Effect of different temperatures and storage atmospheres on Coratina olive oil quality. *Food Chem.* **2007**, *102*, 571–576. [CrossRef]
42. Kalua, C.M.; Bedgood, D.R.; Bishop, A.G.; Prenzler, P.D. Changes in virgin olive oil quality during low-temperature fruit storage. *J. Agric. Food Chem.* **2008**, *56*, 2415–2422. [CrossRef]
43. Tanner, D.J.; Cleland, A.C.; Opara, L.U.; Robertson, T.R. A generalised mathematical modelling methodology for design of horticultural food packages exposed to refrigerated conditions: Part 1, formulation. *Int. J. Refrig.* **2002**, *25*, 33–42. [CrossRef]
44. Ferrua, M.J.; Singh, R.P. Modeling the forced-air cooling process of fresh strawberry packages, Part I: Numerical model. *Int. J. Refrig.* **2009**, *32*, 335–348. [CrossRef]
45. Dehghannya, J.; Ngadi, M.; Vigneault, C. Mathematical Modeling Procedures for Airflow, Heat and Mass Transfer during Forced Convection Cooling of Produce: A Review. *Food Eng. Rev.* **2010**, *2*, 227–243. [CrossRef]
46. Zhao, C.J.; Han, J.W.; Yang, X.T.; Qian, J.P.; Fan, B.L. A review of computational fluid dynamics for forced-air cooling process. *Appl. Energy* **2016**, *168*, 314–331. [CrossRef]
47. Sola-Guirado, R.R.; Aragon-Rodriguez, F.; Castro-Garcia, S.; Gil-Ribes, J. The vibration behaviour of hedgerow olive trees in response to mechanical harvesting with straddle harvester. *Biosyst. Eng.* **2019**, *184*, 81–89. [CrossRef]

MDPI
St. Alban-Anlage 66
4052 Basel
Switzerland
Tel. +41 61 683 77 34
Fax +41 61 302 89 18
www.mdpi.com

Fermentation Editorial Office
E-mail: fermentation@mdpi.com
www.mdpi.com/journal/fermentation

www.ingramcontent.com/pod-product-compliance
Lightning Source LLC
LaVergne TN
LVHW071245170726
843515LV00006BA/1337

* 9 7 8 3 0 3 6 5 3 6 5 5 2 *